Physics of the Lorentz Group (Second Edition)

Beyond high-energy physics and optics

Physics of the Lorentz Group (Second Edition)

Beyond high-energy physics and optics

Sibel Başkal
Department of Physics, Middle East Technical University, 06800 Ankara, Turkey

Young S Kim
Department of Physics, University of Maryland, College Park, MD 20742, USA

Marilyn E Noz
Department of Radiology, New York University School of Medicine, New York, NY 10016, USA

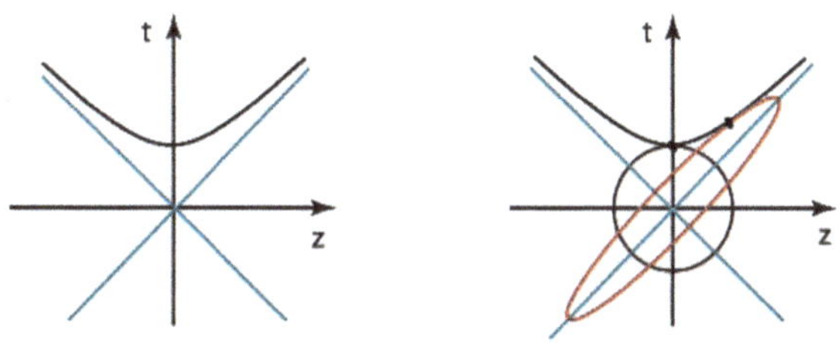

To Einstein's hyperbola of 1905,
we have added a circle and squeezed it.

IOP Publishing, Bristol, UK

ISBN 978-0-7503-3607-9 (ebook)
ISBN 978-0-7503-3605-5 (print)
ISBN 978-0-7503-3608-6 (myPrint)
ISBN 978-0-7503-3606-2 (mobi)

DOI 10.1088/978-0-7503-3607-9

Version: 20210501

IOP ebooks

British Library Cataloguing-in-Publication Data: A catalogue record for this book is available from the British Library.

Published by IOP Publishing, wholly owned by The Institute of Physics, London

IOP Publishing, Temple Circus, Temple Way, Bristol, BS1 6HG, UK

US Office: IOP Publishing, Inc., 190 North Independence Mall West, Suite 601, Philadelphia, PA 19106, USA

Contents

Preface ix

Author biographies xiii

1 Lorentz group and its representations **1-1**

1.1 Generators of the Lorentz group 1-2
1.2 Two-by-two representation of the Lorentz group 1-4
1.2.1 Properties of the Minkowski space 1-7
1.2.2 Properties of the energy–momentum four-vector 1-9
1.3 Conformal representation of the Lorentz group 1-9
1.4 Representations of the Poincaré group 1-10
1.5 Representations of the Lorentz group based on harmonic oscillators 1-12
1.5.1 Harmonic oscillators and Fock space 1-13
1.5.2 Bogoliubov transformation 1-14
1.5.3 Dirac's two-oscillator system 1-15
1.6 Wigner functions for the Lorentz group 1-16
References 1-17

2 Wigner's little groups for internal space–time symmetries **2-1**

2.1 Euler decomposition of Wigner's little group 2-2
2.2 $O(3)$-like little group for massive particles 2-2
2.3 $E(2)$-like little group for massless particles 2-3
2.3.1 Geometric aspects of the $E(2)$-like little group 2-6
2.3.2 Cylindrical interpretation of the $E(2)$-like little group 2-6
2.4 $O(2, 1)$-like little group for imaginary-mass particles 2-7
2.4.1 $O(2, 1)$-like little group in a fish tank—Cherenkov radiation 2-9
2.5 Further properties of Wigner's little groups 2-12
2.6 Little groups in the light-cone coordinate system 2-13
References 2-20

3 Group contractions **3-1**

3.1 Contraction with squeeze transformations 3-1
3.2 Contractions of the $O(3)$ rotation group 3-4
3.3 Contraction of the $O(2, 1)$ Lorentz group 3-7
3.4 Contraction of the Lorentz group 3-9
3.5 Tangential spheres 3-11
References 3-12

4	**Two-by-two representations of Wigner's little groups**	**4-1**
4.1	Transformation properties of the energy–momentum four-vector	4-2
4.2	Two-by-two representations of Wigner's little groups	4-3
4.3	Lorentz completion of the little groups	4-4
4.4	Bargmann and Wigner decompositions	4-6
4.5	Conjugate transformations	4-8
4.6	One little group with three branches	4-10
4.7	Classical damped harmonic oscillator	4-12
	References	4-14
5	**Relativistic spinors and polarization of photons and neutrinos**	**5-1**
5.1	Two-component spinors	5-2
5.2	Massive and massless particles	5-6
5.3	Dirac spinors and massless particles	5-7
5.4	Polarization of massless neutrinos	5-9
5.5	Scalars, vectors, tensors, and the polarization of photons	5-11
	5.5.1 Four-vectors	5-12
	5.5.2 Second-rank tensor	5-14
	5.5.3 Higher spins	5-17
	References	5-18
6	**Lorentz-covariant harmonic oscillators**	**6-1**
6.1	Dirac's plan to construct Lorentz-covariant quantum mechanics	6-2
6.2	Dirac's forms of relativistic dynamics	6-3
6.3	Running waves and standing waves	6-6
6.4	Little groups for relativistic extended particles	6-9
6.5	Further properties of covariant oscillator wave functions	6-10
6.6	Lorentz contraction of harmonic oscillators	6-12
6.7	Feynman's rest of the Universe	6-14
	References	6-17
7	**Quarks and partons in the Lorentz-covariant world**	**7-1**
7.1	Lorentz-covariant quark model	7-3
7.2	Feynman's parton picture	7-4
7.3	Proton structure function	7-6
7.4	Proton form factor and Lorentz coherence	7-10

7.5 Coherence in energy–momentum space 7-14
7.6 Hadronic temperature and boiling quarks 7-15
References 7-17

8 Wigner functions and their symmetries 8-1

8.1 Symmetries and the uncertainty principle in the Wigner phase space 8-3
8.2 Four-dimensional phase space 8-6
8.3 Canonical transformations 8-8
8.4 *SL*(4, *r*) symmetry 8-10
8.5 Dirac matrices for *O*(3, 3) 8-12
8.6 *O*(3, 3) symmetry 8-13
References 8-15

9 Coupled harmonic oscillators and squeezed states of light 9-1

9.1 Coupled oscillators 9-2
9.2 Lorentz-covariant oscillators 9-4
9.3 Squeezed states of light 9-7
9.4 Further notes on squeezed states 9-10
9.5 *O*(3, 2) symmetry from Dirac's coupled oscillators 9-14
9.6 Canonical and non-canonical transformations from the coupled oscillators 9-15
9.7 Entropy and the expanding Wigner phase space 9-17
References 9-18

10 Special relativity from quantum mechanics? 10-1

10.1 Definition of the problem 10-2
10.2 Symmetries of the single oscillator 10-4
10.3 Symmetries from two oscillators 10-8
10.4 Contraction of *O*(3, 2) to the inhomogeneous Lorentz group 10-12
References 10-16

11 Lorentz group in ray optics 11-1

11.1 The group of *ABCD* matrices applied to ray optics 11-2
11.2 Equi-diagonalization of the *ABCD* matrix 11-3
11.3 Decomposition of the *ABCD* matrix 11-5
11.4 Laser cavities 11-6

11.5	Composition of lens and translation matrices	11-8
11.6	Optical beam propagation through multilayers	11-11
11.7	Camera optics	11-15
	References	11-17
12	**Polarization optics**	**12-1**
12.1	Jones vectors	12-2
12.2	Squeeze transformation and phase shift	12-3
12.3	Rotation of the polarization axes	12-5
12.4	The $SL(2, c)$ group content of polarization optics	12-8
12.5	Optical activities	12-9
12.6	Correspondence to space–time symmetries	12-11
12.7	More optical filters from $E(2)$-like groups	12-13
	References	12-16
13	**Poincaré sphere**	**13-1**
13.1	Decoherence in polarization optics	13-2
13.2	Coherency matrix	13-3
13.3	Poincaré sphere	13-5
13.4	Two concentric Poincaré spheres	13-6
13.5	Symmetries derivable from the Poincaré sphere	13-8
13.6	$O(3, 2)$ symmetry for energy couplings	13-10
13.7	Entropy problem	13-11
	References	13-13
Appendix A	**Physics as art of synthesis**	**A-1**
A.1	Illustration of Hume, Kant, and Hegel	A-3
A.2	Kant and Einstein	A-5
A.3	Kantianism and Taoism	A-6
A.4	Einstein and Hegel	A-8
	References	A-11
Index		**I-1**

Preface

In the early years of the 20th century, Niels Bohr (1882–1962) was interested in the discrete energy levels of the hydrogen atom that led to the present form of quantum mechanics. Albert Einstein (1879–1955) was interested in how objects appear to moving observers that allowed him to formulate special relativity leading to his $E = mc^2$. The main purpose of this book is to examine how the hydrogen atom appears to moving observers. We are thus interested in further contents of Einstein's $E = mc^2$.

Paul A M Dirac (1902–1984) was interested in this problem and published many papers on this subject from 1927 to 1963. Eugene Paul Wigner (1902–1995) published two major papers providing the mathematical techniques for approaching this issue. His 1939 paper dealt with the internal space–time symmetries of particles where the mathematical tool in question was the Lorentz group. Later, with Erdal Inönü (1926–2007), Wigner published a paper on group contractions which allows us to understand abstract group theoretical concepts in terms of geometry. In fact the concept of group contraction can be as simple as considering a suitable tangential plane to a curved surface.

Both Dirac and Wigner had high expectations from mathematical approaches to physics. Let us hear what they said about this subject. Paul A M Dirac once said

> *It is quite clear that beauty does depend on one's culture and upbringing for certain kinds of beauty, pictures, literature, poetry and so on…But mathematical beauty is of a rather different kind. I should say perhaps it is of a completely different kind and transcends these personal factors. It is the same in all countries and at all periods of time*[1].

Eugene P Wigner once said

> *The miracle of the appropriateness of the language of mathematics for the formulation of the laws of physics is a wonderful gift which we neither understand nor deserve*[2].

Let us return to the physics of the moving hydrogen atom. During the time of Bohr and Einstein, there were no hydrogen atoms moving with speeds comparable with the

---

[1] *Basic Belief and fundamental Research.* Lecture by Paul A M Dirac, Center for Theoretical Studies, University of Miami, Coral Gables, Florida, 1972. Quote published among other places in Kragh, Helge (1990). Dirac: A Scientific Biography. Cambridge: Cambridge University Press. p. 288. ISBN 978-0-521-38089-8. OCLC 20013981.

[2] *The unreasonable effectiveness of mathematics in the natural sciences.* Richard Courant lecture in mathematical sciences delivered at New York University, May 11, 1959 by Eugene P Wigner. First published: Communications on Pure and Applied Mathematics 13(1):1–14, February 1960. https://doi.org/10.1002/cpa.3160130102.

velocity of light. Not even now. However, these days, high-energy accelerators routinely produce protons moving with speeds very close to that of light.

Those protons are not hydrogen atoms. Yet, the proton is, like the hydrogen atom, a quantum bound state of more fundamental particles called *quarks*, as Murray Gell-Mann (1929–2019) pointed out. Thus, the Bohr–Einstein issue becomes how the proton appears to the moving observer, or how the proton appears when it moves with a speed close to that of light, as illustrated in figure 0.1.

In 1969, Richard P Feynman (1918–1988) noted that the proton, when it moved with a speed close to that of light, appears like a collection of partons whose properties are totally different from the quarks inside the proton. Thus, the original Bohr–Einstein problem becomes whether the quarks and partons are two different aspects of the same entity. Here we address this issue, using the mathematical formalism developed from those of Dirac and Wigner.

This mathematical formalism for the resolution of the Bohr–Einstein issue is applicable to other branches of physics. There are many two-by-two matrices in optical sciences. While we are accustomed to the two-by-two matrices corresponding to rotations, there are many, in optics, which do not correspond to rotations. If not, those two-by-two matrices correspond to four-by-four matrices which perform Lorentz transformations applicable to the space–time coordinates (t, z, x, y). Thus, the Lorentz group is an indispensable mathematical instrument for optics, especially for the Poincaré sphere in polarization optics, and laser cavities, as well as in addressing decoherence problems, as illustrated in figure 0.2.

In quantum optics, the two-photon coherent state or squeezed state is one of the central issues. As early as in 1963, it was recognized that the algebra of two-photon states leads to the symmetry of the $O(3, 2)$ de Sitter group, which is the Lorentz group applicable to three space-like and two time-like coordinates.

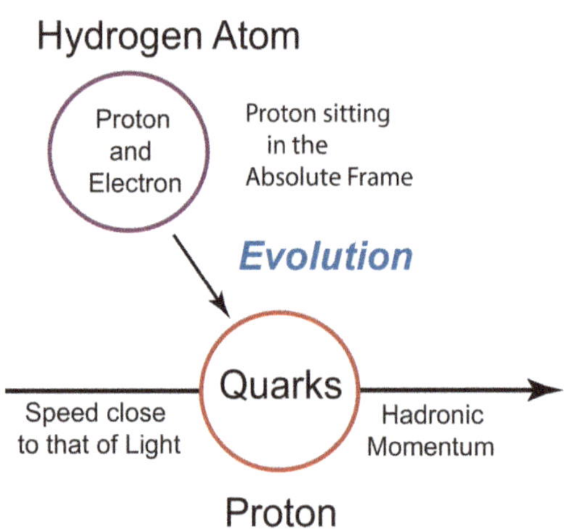

Figure 0.1. Evolution of the hydrogen atom. The proton and hydrogen atom share the same quantum mechanics of bound states. Unlike the hydrogen atom, the proton can be accelerated, and its speed can become extremely close to that of light.

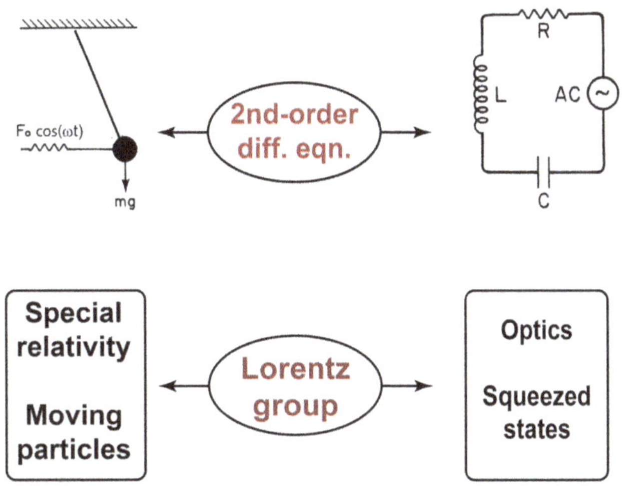

Figure 0.2. One mathematical language serving two branches of physics. A second-order differential equation can serve as the underlying mathematical language for both the damped harmonic oscillator and the resonance circuit. Likewise, the Lorentz group serves as the underlying language for both special relativity and modern optics.

In this five-dimensional space, there are three rotation generators, and three boost generators with respect to the first time variable. There are also three boost generators with respect to the second time variable. In addition, there is one rotation generator between the two time variables. There are four generators involving the second time variable. It is then possible to transform these generators into four translation generators in the four-dimensional space of (t, z, x, y). Thus the net result is the familiar Lorentz group applicable to the four-dimensional space of (t, z, x, y), plus translations along those four directions. The group containing all these transformations is called the inhomogeneous Lorentz group or the Poincaré group.

The language of photon creation and annihilation operators immediately originates from Heisenberg's uncertainty brackets. The best way to interpret this uncertainty relation is to use the Wigner phase space, as shown in figure 0.3. The amount of uncertainty is the area of the circle in the plane of x and p. As is well known, this uncertainty can be squeezed along the x- and p-axes. This squeeze operation can be rotated according to canonical transformations. Thus, the squeeze can be made along the direction which makes a 45° angle with the x and p-axes. As indicated in the same figure, these transformations correspond to the Lorentz group applicable to two space-like coordinates and one time-like coordinate.

It is important to note that the single pair of space and momentum uncertainty contains enough symmetry for the Lorentz group applicable to two space-like directions and one time-like direction. It is thus not surprising that two oscillators with their couplings produce the $O(3, 2)$ de Sitter group which can be contracted to the inhomogeneous Lorentz group from which $E = mc^2$ is shown to result. Indeed, Einstein's special relativity is derivable from the Heisenberg brackets.

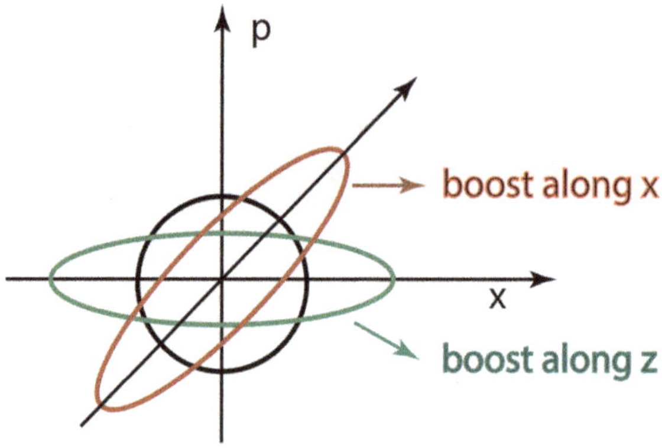

Figure 0.3. Uncertainty relation for the single pair of position and momentum variables. The circle can be rotated and squeezed. The squeeze operations correspond to the Lorentz boosts in two different directions. This constitutes the symmetry of the Lorentz group applicable to two space-like and one time-like directions.

This book is the second edition of the book entitled *Physics of the Lorentz group* published in 2015 jointly by IOP Publishing and Morgan & Claypool Publishers, San Rafael, California, USA.

Sibel Başkal
Young S Kim
Marilyn E Noz
December 2020

Author biographies

Sibel Başkal

Sibel Başkal is Professor Emerita of Physics at the Middle East Technical University. She is particularly interested in the manifestations of the Poincaré and little groups, and of group contractions in physical sciences. Her research interests extend to current problems in classical field theories, mostly on alternative approaches to Einstein's gravity. She has published more than 30 peer-reviewed papers and is the co-author of two books with Y S Kim and M E Noz. Professor Başkal can be reached at baskal@newton.physics.metu.edu.tr.

Young S Kim

Young S Kim came to the United States from South Korea in 1954 after high school graduation, to become a freshman at the Carnegie Institute of Technology (now called Carnegie Mellon University) in Pittsburgh. In 1958, he went to Princeton University for graduate study in physics and received his PhD degree in 1961. In 1962, he became an assistant professor at the University of Maryland at College Park near Washington, DC. After going through the academic ranks of associate and full professors, Dr Kim became a Professor Emeritus in 2007. This is still his position at the University of Maryland.

Dr Kim's thesis advisor at Princeton was Sam Treiman, but he had to go to Eugene Wigner whenever he had to face fundamental problems in physics. During this process, he became interested in Wigner's 1939 paper on internal space-time symmetries particles in Einstein's Lorentz-covariant world. Since 1973, his publications have been based primarily on constructing mathematical formulas for understanding Wigner's paper. In 1988, Dr Kim noted that the same set of mathematical devices are applicable to squeezed states in quantum optics. Since then, he has also been publishing papers on optical and information sciences. These days, Dr Kim publishes articles on the question of whether quantum mechanics and special relativity can be derived from the same basket of equations. Professor Kim can reached at yskim@umd.edu.

Marilyn E Noz

Marilyn E Noz is Professor Emerita in the Department of Radiology at NYU School of Medicine. Over the last more than 40 years, she has collaborated with Professor Kim on relativistic quantum mechanics using two-by-two matrices, harmonics oscillators, and the Lorentz group. She has contributed to over 50 peer-reviewed journal articles in elementary particle physics and optics. She has written three books with Professor Kim and two books with Professors Kim and Başkal. She continues to do research in elementary particle physics and quantum optics. Professor Noz can be reached at marilyn.noz@med.nyu.edu.

IOP Publishing

Physics of the Lorentz Group (Second Edition)

Beyond high-energy physics and optics

Sibel Başkal, Young S Kim, and Marilyn E Noz

Chapter 1

Lorentz group and its representations

The Lorentz group starts with four-by-four matrices performing Lorentz transformations on the four-dimensional Minkowski space (t, z, x, y). These Lorentz transformations leave the quantity $(t^2 - z^2 - x^2 - y^2)$ invariant. There are three rotations and three boosts making the Lorentz group a six-parameter group. It was Einstein who observed that this Lorentz group also applies to the four-dimensional energy–momentum space of (E, p_z, p_x, p_y). Thus he derived his Lorentz-covariant energy–momentum relation commonly known as $E = mc^2$. This transformation leaving the quantity $\left(E^2 - p_z^2 - p_x^2 - p_y^2\right)$ invariant, shows that the particle mass must also be a Lorentz invariant quantity.

The Lorentz group contains the rotation group as one of its subgroups. Both the rotation and Lorentz groups are continuous groups; therefore, they are *Lie groups* [12]. For these continuous groups, it is convenient to use their generators. For the three-dimensional rotation group, there are three generators, and they form a closed set of commutation relations. This closed commutator set is called the *Lie algebra* [12] of the rotation group. When the three boost generators are added to the rotation group, these six generators also form a closed set of commutation relations, which is the Lie algebra for the Lorentz group.

In section 1.1, we introduce generators in the form of four-by-four matrices applicable to the four-dimensional Minkowski space. These generators are also applicable to the four-dimensional energy–momentum space. In section 1.2, it is shown that similar to the rotation group generated by three two-by-two matrices, the Lorentz group is generated by six two-by-two matrices. Three of them are Hermitian and generate the rotation group known as $SU(2)$. The remaining three are anti-Hermitian and generate two-by-two matrices corresponding to Lorentz boosts in three different directions. The group generated by these Hermitian and anti-Hermitian generators is called $SL(2, c)$. Since it is unimodular (the determinant is equal to one), it has six independent parameters and shares the same Lie algebra as the Lorentz group. The four-component space–time and energy–momentum four-vectors are written in the form of two-by-two matrices. In section 1.2.1, we use examples to illustrate some properties of the Minkowski space–time four-vector in the two-by-two representation and, in section 1.2.2, we do the same for the energy–momentum four-vector.

Section 1.3 gives an introduction to the conformal representation of the Lorentz group. In section 1.4, we introduce the inhomogeneous Lorentz or Poincaré group and its representations. We show how it relates to the Lorentz group. In section 1.5, we study how we can transform functions. As an example, we define the annihilation and creation operators. We illustrate how the harmonic oscillator relates them to the Fock space where the number n represents the number of particles or photons in a given state. We then illustrate how the Bogoliubov transformation can be viewed as a Lorentz transformation. Additionally, we introduce Dirac's two-oscillator system in terms of these annihilation and creation operators.

In section 1.6, it is noted that the Wigner phase-space distribution function is an important tool for interpreting Heisenberg's uncertainty principle. It is pointed out that the Wigner function serves also as a tool for interpreting Einstein's Lorentz covariance.

1.1 Generators of the Lorentz group

Let us start with rotations applicable to the (t, z, x, y) coordinates. The four-by-four matrix for the rotation about the z-axis is

$$Z(\phi) = \begin{pmatrix} 1 & 0 & 0 & 0 \\ 0 & 1 & 0 & 0 \\ 0 & 0 & \cos\phi & -\sin\phi \\ 0 & 0 & \sin\phi & \cos\phi \end{pmatrix}. \tag{1.1}$$

$Z(\phi)$ can be written as

$$Z(\phi) = \exp(-i\phi J_3), \tag{1.2}$$

with the generator

$$J_3 = \begin{pmatrix} 0 & 0 & 0 & 0 \\ 0 & 0 & 0 & 0 \\ 0 & 0 & 0 & -i \\ 0 & 0 & i & 0 \end{pmatrix}. \tag{1.3}$$

The matrix J_3 is known as the generator of the rotation around the z-axis. It is not difficult to write the generators of rotations around the x- and y-axes, and they can be written as J_1 and J_2, respectively, with

$$J_1 = \begin{pmatrix} 0 & 0 & 0 & 0 \\ 0 & 0 & 0 & i \\ 0 & 0 & 0 & 0 \\ 0 & -i & 0 & 0 \end{pmatrix} \quad \text{and} \quad J_2 = \begin{pmatrix} 0 & 0 & 0 & 0 \\ 0 & 0 & -i & 0 \\ 0 & i & 0 & 0 \\ 0 & 0 & 0 & 0 \end{pmatrix}. \tag{1.4}$$

These three rotation generators satisfy the commutation relations

$$[J_i, J_j] = i\varepsilon_{ijk}J_k. \tag{1.5}$$

The matrix which performs the Lorentz boost along the z-direction is

$$B(\eta) = \begin{pmatrix} \cosh\eta & \sinh\eta & 0 & 0 \\ \sinh\eta & \cosh\eta & 0 & 0 \\ 0 & 0 & 1 & 0 \\ 0 & 0 & 0 & 1 \end{pmatrix}. \tag{1.6}$$

$B(\eta)$ can be written as

$$B(\eta) = \exp(-i\eta K_3), \tag{1.7}$$

with the generator

$$K_3 = \begin{pmatrix} 0 & i & 0 & 0 \\ i & 0 & 0 & 0 \\ 0 & 0 & 0 & 0 \\ 0 & 0 & 0 & 0 \end{pmatrix}. \tag{1.8}$$

It is then possible to write the matrices for the generators K_1 and K_2, as

$$K_1 = \begin{pmatrix} 0 & 0 & i & 0 \\ 0 & 0 & 0 & 0 \\ i & 0 & 0 & 0 \\ 0 & 0 & 0 & 0 \end{pmatrix} \quad \text{and} \quad K_2 = \begin{pmatrix} 0 & 0 & 0 & i \\ 0 & 0 & 0 & 0 \\ 0 & 0 & 0 & 0 \\ i & 0 & 0 & 0 \end{pmatrix}. \tag{1.9}$$

Then

$$[J_i, K_j] = i\varepsilon_{ijk}K_k \quad \text{and} \quad [K_i, K_j] = -i\varepsilon_{ijk}J_k. \tag{1.10}$$

There are six generators of the Lorentz group, and they satisfy the three sets of commutation relations given in equations (1.5) and (1.10). It is said that the Lie algebra of the Lorentz group consists of these sets of commutation relations.

These commutation relations are invariant under Hermitian conjugation. While the rotation generator is Hermitian, the boost generators are anti-Hermitian

$$J_i^\dagger = J_i \quad \text{while} \quad K_i^\dagger = -K_i. \tag{1.11}$$

Thus, it is possible to construct two four-by-four representations of the Lorentz group, one with K_i and the other with $-K_i$. For this purpose, we shall use the notation [4, 14]

$$\dot{K}_i = -K_i. \tag{1.12}$$

Since there are two representations, transformations with K_i are called the covariant transformations, while those with $\dot{K}_i$ are called contravariant transformations.

The generators of the Lorentz group are also applicable to an arbitrary function of the coordinate variables. Then the generators take the form

$$J_i = -i\varepsilon_{ijk}\left(x_j\frac{\partial}{x_k} - x_k\frac{\partial}{x_j}\right) \quad \text{and} \quad K_i = -i\left[t\left(\frac{\partial}{\partial x_i}\right) + x_i\left(\frac{\partial}{\partial t}\right)\right]. \tag{1.13}$$

The above can be combined into the covariant form

$$M_{\mu\nu} = i\left[x_\mu\left(\frac{\partial}{\partial x^\nu}\right) - x_\nu\left(\frac{\partial}{\partial x^\mu}\right)\right], \tag{1.14}$$

where $M_{\mu\nu} = -M_{\nu\mu}$. Here we use $x^\mu = (t, z, x, y)$ and $x_\mu = (t, -z, -x, -y)$. It is, therefore, possible to write

$$J_i = \varepsilon_{ijk}M_{kj} \quad \text{and} \quad K_i = M_{0i}. \tag{1.15}$$

We will use this formalism again in section 1.4. Throughout this book, we use Greek letters to designate all four indices, 0–3, and Latin letters to mean the indices 1, 2, 3 only.

1.2 Two-by-two representation of the Lorentz group

It is possible to construct the Lie algebra of the Lorentz group from the three Pauli matrices [2, 6, 14, 16]. Let us define

$$J_i = \frac{1}{2}\sigma_i \quad \text{and} \quad K_i = \frac{i}{2}\sigma_i, \tag{1.16}$$

where the Pauli spin matrices are defined as

$$\sigma_1 = \begin{pmatrix} 0 & 1 \\ 1 & 0 \end{pmatrix}, \quad \sigma_2 = \begin{pmatrix} 0 & -i \\ i & 0 \end{pmatrix}, \quad \text{and} \quad \sigma_3 = \begin{pmatrix} 1 & 0 \\ 0 & -1 \end{pmatrix}. \tag{1.17}$$

The two-by-two matrices given in equation (1.16) satisfy the Lie algebra of the Lorentz group given in equations (1.5) and (1.10). These Pauli matrices are also the well-known quantum mechanical operators associated with the spin of half-integer spin particles. Spin is a notable example of an internal space–time property of particles. Other forms of internal space–time properties and their symmetries will be discussed in detail in chapters 2 and 4.

These generators will lead to a two-by-two matrix of the form

$$G = \begin{pmatrix} \alpha & \beta \\ \gamma & \delta \end{pmatrix}, \tag{1.18}$$

with four complex matrix elements, thus eight real parameters. Since the determinant is fixed and is equal to one, there are six independent parameters. This six-parameter group is commonly called $SL(2, c)$. As the two-by-two representation of this six parameter group shares the same Lie algebra with the four-by-four representation for the Lorentz group, we say they are *locally isomorphic to each other* in the language of group theory. However, we shall simply say in this book that one is *like* the other. The matrices representing $SL(2, c)$ are the smallest matrices containing as many as six degrees of freedom.

For this reason, they occupy an important place in mathematics with applications in many branches of physics. Since the Lorentz group has six generators, this two-by-two matrix can serve as a representation of the Lorentz group. In the literature, it is said that $SL(2, c)$ is the covering group of the Lorentz group.

For each G matrix of $SL(2, c)$, there exists one four-by-four Lorentz transformation matrix. Thus, we can write the most general form of the four-by-four matrix in terms of the complex parameters α, β, γ, and δ [14]. Since it is unnecessary to use these complicated expressions here, we use only a subset of them. The correspondence between the two-by-two and four-by-four representations of the Lorentz group along with the generators that are most useful here are given in table 1.1. These representations can be used for coordinate or momentum transformations, and other four-vector quantities such as electromagnetic four-potentials.

Table 1.1. Transformation matrices of $SL(2, c)$ together with their generators. The two-by-two matrices of $SL(2, c)$ form a two-by-two representation of the (proper—all the determinants of the matrices are equal to one) Lorentz group. The last column has the corresponding four-by-four transformation matrices. These matrices apply to the Minkowski space–time four-vector (t, z, x, y) and to the energy–momentum four-vector (p_0, p_z, p_x, p_y).

Generators	Two-by-Two	Four-by-Four
$J_3 = \frac{1}{2}\begin{pmatrix} 1 & 0 \\ 0 & -1 \end{pmatrix}$	$\begin{pmatrix} \exp(i\phi/2) & 0 \\ 0 & \exp(-i\phi/2) \end{pmatrix}$	$\begin{pmatrix} 1 & 0 & 0 & 0 \\ 0 & 1 & 0 & 0 \\ 0 & 0 & \cos\phi & -\sin\phi \\ 0 & 0 & \sin\phi & \cos\phi \end{pmatrix}$
$K_3 = \frac{1}{2}\begin{pmatrix} i & 0 \\ 0 & -i \end{pmatrix}$	$\begin{pmatrix} \exp(\eta/2) & 0 \\ 0 & \exp(-\eta/2) \end{pmatrix}$	$\begin{pmatrix} \cosh\eta & \sinh\eta & 0 & 0 \\ \sinh\eta & \cosh\eta & 0 & 0 \\ 0 & 0 & 1 & 0 \\ 0 & 0 & 0 & 1 \end{pmatrix}$
$J_1 = \frac{1}{2}\begin{pmatrix} 0 & 1 \\ 1 & 0 \end{pmatrix}$	$\begin{pmatrix} \cos(\theta/2) & i\sin(\theta/2) \\ i\sin(\theta/2) & \cos(\theta/2) \end{pmatrix}$	$\begin{pmatrix} 1 & 0 & 0 & 0 \\ 0 & \cos\theta & 0 & \sin\theta \\ 0 & 0 & 1 & 0 \\ 0 & -\sin\theta & 0 & \cos\theta \end{pmatrix}$
$K_1 = \frac{1}{2}\begin{pmatrix} 0 & i \\ i & 0 \end{pmatrix}$	$\begin{pmatrix} \cosh(\lambda/2) & \sinh(\lambda/2) \\ \sinh(\lambda/2) & \cosh(\lambda/2) \end{pmatrix}$	$\begin{pmatrix} \cosh\lambda & 0 & \sinh\lambda & 0 \\ 0 & 1 & 0 & 0 \\ \sinh\lambda & 0 & \cosh\lambda & 0 \\ 0 & 0 & 0 & 1 \end{pmatrix}$
$J_2 = \frac{1}{2}\begin{pmatrix} 0 & -i \\ i & 0 \end{pmatrix}$	$\begin{pmatrix} \cos(\theta/2) & -\sin(\theta/2) \\ \sin(\theta/2) & \cos(\theta/2) \end{pmatrix}$	$\begin{pmatrix} 1 & 0 & 0 & 0 \\ 0 & \cos\theta & -\sin\theta & 0 \\ 0 & \sin\theta & \cos\theta & 0 \\ 0 & 0 & 0 & 1 \end{pmatrix}$
$K_2 = \frac{1}{2}\begin{pmatrix} 0 & 1 \\ -1 & 0 \end{pmatrix}$	$\begin{pmatrix} \cosh(\lambda/2) & -i\sinh(\lambda/2) \\ i\sinh(\lambda/2) & \cosh(\lambda/2) \end{pmatrix}$	$\begin{pmatrix} \cosh\lambda & 0 & 0 & \sinh\lambda \\ 0 & 1 & 0 & 0 \\ 0 & 0 & 1 & 0 \\ \sinh\lambda & 0 & 0 & \cosh\lambda \end{pmatrix}$

Since the dot or scalar product of two four-vectors, $p \cdot x$, is another important Lorentz-invariant quantity in physics, we can write it as:

$$p \cdot x = p_0 t - p_z z - p_x x - p_y y. \tag{1.19}$$

However, these four-vectors need to be written in the two-by-two formalism. We know that there are two-by-two matrices applicable to two-component column vectors called spinors. What we need to understand is how these four-vectors can be expressed in terms of $SL(2, c)$ spinors.

Let us start with the Minkowski space–time four-vector (t, z, x, y) which can be written as

$$X = \begin{pmatrix} t + z & x - iy \\ x + iy & t - z \end{pmatrix}, \tag{1.20}$$

and has the determinant

$$t^2 - z^2 - x^2 - y^2. \tag{1.21}$$

We can now consider the coordinate transformation

$$X' = G\, X\, G^{\dagger}. \tag{1.22}$$

The transformation of equation (1.22) can be explicitly written as

$$\begin{pmatrix} t' + z' & x' - iy' \\ x' + iy' & t' - z' \end{pmatrix} = \begin{pmatrix} \alpha & \beta \\ \gamma & \delta \end{pmatrix} \begin{pmatrix} t + z & x - iy \\ x + iy & t - z \end{pmatrix} \begin{pmatrix} \alpha^* & \gamma^* \\ \beta^* & \delta^* \end{pmatrix}. \tag{1.23}$$

We can now translate this formula into

$$\begin{pmatrix} t' + z' \\ t' - z' \\ x' - iy' \\ x' + iy' \end{pmatrix} = \begin{pmatrix} \alpha^*\alpha & \gamma^*\beta & \gamma^*\alpha & \alpha^*\beta \\ \beta^*\gamma & \delta^*\delta & \delta^*\gamma & \beta^*\delta \\ \beta^*\alpha & \delta^*\alpha & \beta^*\beta & \delta^*\beta \\ \alpha^*\gamma & \gamma^*\gamma & \alpha^*\delta & \gamma^*\delta \end{pmatrix} \begin{pmatrix} t + z \\ t - z \\ x - iy \\ x + iy \end{pmatrix}. \tag{1.24}$$

The above then leads to

$$\begin{pmatrix} t' \\ z' \\ x' \\ y' \end{pmatrix} = \frac{1}{2} \begin{pmatrix} 1 & 1 & 0 & 0 \\ 1 & -1 & 0 & 0 \\ 0 & 0 & 1 & 1 \\ 0 & 0 & i & -i \end{pmatrix} \begin{pmatrix} t' + z' \\ t' - z' \\ x' - iy' \\ x' + iy' \end{pmatrix}. \tag{1.25}$$

It is important to note that the transformation of equation (1.22) is not a similarity transformation. In the $SL(2, c)$ regime, not all the matrices are Hermitian [2].

Likewise, the two-by-two matrix for the energy–momentum four-vector of the particle takes the form

$$P = \begin{pmatrix} p_0 + p_z & p_x - ip_y \\ p_x + ip_y & p_0 - p_z \end{pmatrix} \tag{1.26}$$

with $p_0 = \sqrt{m^2 + p_z^2 + p_x^2 + p_y^2}$. The transformation of this matrix takes the same form as that of the space–time matrix given in equations (1.22) and (1.23). The determinant of this matrix is m^2 and remains invariant under Lorentz transformations. The explicit form of the transformation is

$$P' = G\ P\ G^\dagger = \begin{pmatrix} p_0' + p_z' & p_x' - ip_y' \\ p_x' + ip_y' & p_0' - p_z' \end{pmatrix}$$
$$= \begin{pmatrix} \alpha & \beta \\ \gamma & \delta \end{pmatrix} \begin{pmatrix} p_0 + p_z & p_x - ip_y \\ p_x + ip_y & p_0 - p_z \end{pmatrix} \begin{pmatrix} \alpha^* & \gamma^* \\ \beta^* & \delta^* \end{pmatrix}. \tag{1.27}$$

We can now write the matrix

$$P + X = \begin{pmatrix} p_0 + p_z + t + z & p_x - ip_y + x - iy \\ p_x + ip_y + x + iy & p_0 - p_z + t - z \end{pmatrix}. \tag{1.28}$$

The determinant of this expression gives [3, 17]

$$p \cdot x = \frac{1}{2}(\det(P + X) - \det(P) - \det(X)). \tag{1.29}$$

The above equation is equation (1.19) written in the two-by-two formalism. In the next section we obtain more results based on the two-by-two representation of the Lorentz group.

1.2.1 Properties of the Minkowski space

It seems reasonable that transformations on the four-dimensional space–time or energy–momentum four-vectors would be more straightforward using the four-by-four matrices defined in section 1.1. However, it is much easier to do physics with two-by-two matrices. Therefore we demonstrate some other properties of the two-by-two matrices derived in the previous section.

As a first example, we do some transformations on the Minkowski space–time four-vector written in the two-by-two formalism given in equation (1.20). Using table 1.1, we can extract the two-by-two boost matrix along the z-direction and call it $B(\eta)$:

$$B(\eta) = \begin{pmatrix} e^{\eta/2} & 0 \\ 0 & e^{-\eta/2} \end{pmatrix}. \tag{1.30}$$

By taking the matrix product $B(\eta)\ X\ B^\dagger(\eta)$, we obtain

$$B(\eta)XB^\dagger(\eta) = \begin{pmatrix} (t + z)e^{\eta} & x - iy \\ x + iy & (t - z)e^{-\eta} \end{pmatrix}. \tag{1.31}$$

From this, we obtain the transformation

$$t \to (\cosh\eta)t + (\sinh\eta)z, \quad \text{and} \quad z \to (\sinh\eta)t + (\cosh\eta)z, \tag{1.32}$$

which is a Lorentz boost along the z-direction.

We can also use the rotation matrix around the z-axis from table 1.1 which can be written as

$$Z(\phi) = \begin{pmatrix} e^{i\phi/2} & 0 \\ 0 & e^{-i\phi/2} \end{pmatrix}. \tag{1.33}$$

Then

$$Z(\phi)XZ^{\dagger}(\phi) = \begin{pmatrix} t+z & (x-iy)e^{i\phi} \\ (x+iy)e^{-i\phi} & t-z \end{pmatrix}. \tag{1.34}$$

The above is a rotation around the z-axis:

$$x \to (\cos\phi)x - (\sin\phi)y, \quad \text{and} \quad y \to (\sin\phi)x + (\cos\phi)y. \tag{1.35}$$

Finally, to demonstrate what happens when we choose to rotate about an axis perpendicular to the z-axis, we choose from table 1.1 the rotation around the y-axis. This rotation matrix has the form

$$R(\theta) = \begin{pmatrix} \cos(\theta/2) & -\sin(\theta/2) \\ \sin(\theta/2) & \cos(\theta/2) \end{pmatrix}. \tag{1.36}$$

When it is applied to the four-vector X in the same manner as $B(\eta)$ and $Z(\phi)$ were, the result is

$$R(\theta)XR^{\dagger}(\theta) = \begin{pmatrix} t+z' & x'-iy \\ x'+iy & t-z' \end{pmatrix}. \tag{1.37}$$

Here we have used

$$\begin{pmatrix} z' \\ x' \end{pmatrix} = \begin{pmatrix} \cos\theta & -\sin\theta \\ \sin\theta & \cos\theta \end{pmatrix}\begin{pmatrix} z \\ x \end{pmatrix}. \tag{1.38}$$

From the above, we see that rotations around two orthogonal directions can lead to a rotation around the third axis, as is indicated by the Lie algebra of the three-dimensional rotation group given in equation (1.5). Likewise, we see that the Lorentz boost does not apply to only one direction, but it can be used along all directions, confirming the Lie algebra given in equations (1.5) and (1.10). Thus it follows that the three transformations given in equations (1.31), (1.34), and (1.37) yield the most general transformations in the four-dimensional Minkowski space; specifically, two rotations and one boost are sufficient to generate the six parameter group $SL(2, c)$, which has the same commutation relations as that of the Lorentz group.

1.2.2 Properties of the energy–momentum four-vector

We know from Einstein that the energy–momentum four-vector can be written in the same form as the Minkowski space–time four-vector. We wrote this vector in the two-by-two matrix formalism in equation (1.26). Therefore, we illustrate some transformations on this four-vector. If the particle moves along the z-direction, the energy–momentum four-vector two-by-two matrix becomes

$$P = \begin{pmatrix} E + p & 0 \\ 0 & E - p \end{pmatrix}, \tag{1.39}$$

where E and p are the energy and magnitude of the momentum, respectively. If we then consider a particle at rest, the energy–momentum four-vector can be written as

$$P_0 = \begin{pmatrix} m & 0 \\ 0 & m \end{pmatrix}. \tag{1.40}$$

If either the P or P_0 four-vector is rotated around the z- and y-directions separately, the four-vector remains invariant. For example, for P_0 we obtain

$$Z(\phi)P_0Z^{\dagger}(\phi) = R(\theta)P_0R^{\dagger}(\theta) = P_0, \tag{1.41}$$

where the two-by-two matrix for $Z(\phi)$ is given in equation (1.33) and that for $R(\theta)$ is given equation (1.36). Because the rotation matrices are Hermitian, this expression can also be written as

$$Z(\phi)P_0Z^{-1}(\phi) = R(\theta)P_0R^{-1}(\theta) = P_0. \tag{1.42}$$

It is evident from the form of the boost matrix of equation (1.30) that the Hermitian conjugate is not the inverse. Therefore, we have

$$\begin{aligned} B(\eta)P_0B^{\dagger}(\eta) &= \begin{pmatrix} e^{\eta/2} & 0 \\ 0 & e^{-\eta/2} \end{pmatrix}\begin{pmatrix} m & 0 \\ 0 & m \end{pmatrix}\begin{pmatrix} e^{\eta/2} & 0 \\ 0 & e^{-\eta/2} \end{pmatrix} \\ &= \begin{pmatrix} me^{\eta} & 0 \\ 0 & me^{-\eta} \end{pmatrix}. \end{aligned} \tag{1.43}$$

However, when we apply a similarity transformation, we obtain $B(\eta)\ P_0\ B^{-1}(\eta) = P_0$, which is invariant. We can see this explicitly from

$$\begin{pmatrix} e^{\eta/2} & 0 \\ 0 & e^{-\eta/2} \end{pmatrix}\begin{pmatrix} m & 0 \\ 0 & m \end{pmatrix}\begin{pmatrix} e^{-\eta/2} & 0 \\ 0 & e^{\eta/2} \end{pmatrix} = \begin{pmatrix} m & 0 \\ 0 & m \end{pmatrix}. \tag{1.44}$$

This difference occurs because not all the matrices of $SL(2, c)$ are Hermitian.

1.3 Conformal representation of the Lorentz group

In the two-by-two representation of the Lorentz group, we used the transformation matrix

$$\begin{pmatrix} \alpha & \beta \\ \gamma & \delta \end{pmatrix}, \tag{1.45}$$

with the condition that the determinant be one. Thus, there are six independent parameters. The multiplication of two of these matrices leads to

$$\begin{pmatrix} \alpha_2 & \beta_2 \\ \gamma_2 & \delta_2 \end{pmatrix}\begin{pmatrix} \alpha_1 & \beta_1 \\ \gamma_1 & \delta_1 \end{pmatrix} = \begin{pmatrix} \alpha_2\alpha_1 + \beta_2\gamma_1 & \alpha_2\beta_1 + \beta_2\delta_1 \\ \gamma_2\alpha_1 + \delta_2\gamma_1 & \gamma_2\beta_1 + \delta_2\delta_1 \end{pmatrix}. \tag{1.46}$$

It is interesting to note that this algebra can be achieved by a conformal transformation of the form [1]

$$z' = \frac{\alpha z + \beta}{\gamma z + \delta} \tag{1.47}$$

where z and z' are complex numbers. In order to see whether this leads to the algebra of the conformal transformation, let us write

$$z' = \frac{\alpha_1 z + \beta_1}{\gamma_1 z + \delta_1} \quad \text{and} \quad z'' = \frac{\alpha_2 z' + \beta_2}{\gamma_2 z' + \delta_2}. \tag{1.48}$$

Then

$$z'' = \frac{(\alpha_2\alpha_1 + \beta_2\gamma_1)z + (\alpha_2\beta_1 + \beta_2\delta_1)}{(\gamma_2\alpha_1 + \delta_2\gamma_1)z + (\gamma_2\beta_1 + \delta_2\delta_1)}. \tag{1.49}$$

There are many two-by-two matrices in the Lorentz group, but not all of them are able to be derived from the Pauli spin matrices. The group $SL(2, c)$, with six independent parameters is able to accommodate all of these matrices. It is also interesting that the two-by-two representation of the $SL(2, c)$ can be represented by the conformal representation given in this section.

1.4 Representations of the Poincaré group

The study of the Poincaré group also known as the inhomogeneous Lorentz group, starts with the group of inhomogeneous Lorentz transformations on the four-dimensional Minkowski space in the form

$$x'^{\mu} = a_{\nu}^{\mu} x^{\nu} + b^{\nu}. \tag{1.50}$$

The matrix performing the above linear transformation can be constructed by considering five-by-five matrices of the form

$$A = \begin{pmatrix} \text{4 − by − 4} & 0 \\ \text{Lorentz} & 0 \\ \text{Transformation} & 0 \\ \text{Matrix} & 0 \\ 0\ 0\ 0\ 0 & 1 \end{pmatrix} \quad \text{and} \quad B = \begin{pmatrix} 1 & 0 & 0 & 0 & b_0 \\ 0 & 1 & 0 & 0 & b_3 \\ 0 & 0 & 1 & 0 & b_1 \\ 0 & 0 & 0 & 1 & b_2 \\ 0 & 0 & 0 & 0 & 1 \end{pmatrix} \tag{1.51}$$

applicable to the five-element column vector $(t, z, x, y, 1)$. We can thus write the inhomogeneous Lorentz transformation of equation (1.50) as

$$x' = B(b)A(a)x, \tag{1.52}$$

where x, x', a, and b are abbreviations of their respective quantities with indices. It is then evident that this matrix is the representation of the inhomogeneous Lorentz transformation.

From what was discussed in section 1.1, we see that the A matrix has the same properties as the four-by-four Lorentz transformation matrices. It can be understood as well that the B matrix represents translations in the four-dimensional Minkowski space. The B matrix is Abelian (the commutative law for group multiplication holds) and is an invariant subgroup of the inhomogeneous Lorentz group. We can then write

$$B(b')B(b) = B(b' + b) \quad \text{and} \quad A(a)B(b) = B(aba^{-1})A(a). \tag{1.53}$$

Matrices of the form BA have the algebraic property:

$$[B(b')A(a')][B(b)A(a)] = B(b' + a'ba'^{-1})A(a'a). \tag{1.54}$$

From this, we see that the Poincaré group is a semi-direct product of the Lorentz group $A(a)$ and the translation group $B(b)$.

Since B is an invariant subgroup, matrices of the form AB are equally adequate for representing the group as are matrices of the form BA. It is possible then to conveniently use this representation when we study the space–time symmetries of a relativistic particle with a given energy–momentum four-vector p. Here the state vector has the momentum dependence in the form

$$\exp(\pm ip \cdot x), \tag{1.55}$$

where

$$p \cdot x = p_\mu x^\mu = p_0 t - p_z z - p_x x - p_y y. \tag{1.56}$$

We know from equation (1.19) that this is a Lorentz invariant quantity. It is clear from the five-by-five matrix form of B and also from the above expression that the generator of translations is given by

$$P_\mu = i\left(\frac{\partial}{\partial x^\mu}\right). \tag{1.57}$$

Using the discussions given in section 1.1, the most general form for the generators of the Lorentz transformation, given equation (1.14), can be written as:

$$M_{\mu\nu} = L^*_{\mu\nu} + L_{\mu\nu}. \tag{1.58}$$

Here the $L^*_{\mu\nu}$ part is applicable to coordinate variables, while $L_{\mu\nu}$ applies to all other variables which are not affected by $L^*_{\mu\nu}$. For massive Dirac particles, the $L_{\mu\nu}$ operators generate Lorentz transformations on the Dirac spinor. For massless photons, it is necessary to construct $L_{\mu\nu}$ so that it applies to the polarization vector. When we have finite-size particles with internal motion of constituent particles, such as a hadron composed of two or three quarks, the $L_{\mu\nu}$ operator has to generate Lorentz transformations on the internal space–time coordinate variables. The internal coordinate variables, in this instance, are independent of the particle coordinates.

In section 6.3, we show that the coupled harmonic oscillator equation can be represented by the Poincaré group. There the hadron is considered to be composed of two quarks. We see from equation (6.21) that the translation operator P_μ acts only on the external hadron coordinates. However, the Lorentz transformations, which include boosts and rotations generated by equation (1.58), have the explicit form given in equation (6.23) as

$$L_{\mu\nu}^{*} = i\left(X_\mu \frac{\partial}{\partial X^\nu} - X_\nu \frac{\partial}{\partial X^\mu}\right) \quad \text{and} \quad L_{\mu\nu} = i\left(x_\mu \frac{\partial}{\partial x^\nu} - x_\nu \frac{\partial}{\partial x^\mu}\right). \tag{1.59}$$

Here $L_{\mu\nu}^{*}$ represents only the external hadron coordinates while $L_{\mu\nu}$ represents the internal coordinates of the space–time separation of the quarks.

We can also write the commutation relations among the generators of the Poincaré group. These take the explicit form

$$\begin{gathered} [P_\mu, P_\nu] = 0, \qquad [M_{\mu\nu}, P_\rho] = -i(\eta_{\mu\rho}P_\nu - \eta_{\nu\rho}P_\mu), \\ [M_{\mu\nu}, M_{\rho\sigma}] = -\, i(\eta_{\mu\rho}M_{\nu\sigma} - \eta_{\nu\rho}M_{\mu\sigma} + \eta_{\mu\sigma}M_{\rho\nu} - \eta_{\nu\sigma}M_{\rho\mu}), \end{gathered} \tag{1.60}$$

where $\eta_{\mu\nu}$ are the components of

$$\eta_{\mu\nu} = \begin{pmatrix} 1 & 0 & 0 & 0 \\ 0 & -1 & 0 & 0 \\ 0 & 0 & -1 & 0 \\ 0 & 0 & 0 & -1 \end{pmatrix}. \tag{1.61}$$

Since P_μ does not commute with $M_{\mu\nu}$, it is clear that the Poincaré group is not a direct product of the Lorentz and translation groups.

1.5 Representations of the Lorentz group based on harmonic oscillators

The matrix representations of the Lorentz group in section 1.2 are primarily for coordinate transformations. The question then is how we can transform functions. This problem has a stormy history. For plane waves, the form

$$\exp(ip \cdot x) \tag{1.62}$$

is widely used in the literature. Since from equation (1.19) we know that

$$p \cdot x = p_0 t - p_z z - p_x x - p_y y \tag{1.63}$$

is a Lorentz-invariant quantity, there is no problem from the mathematical point of view.

However, for standing waves, we have to consider boundary conditions. The issue is then how to transform these conditions. One way to circumvent this difficulty is to study harmonic oscillators with built-in boundary conditions.

Indeed, Dirac [7, 8], Yukawa [20], and Feynman [10] struggled with this problem using harmonic oscillator wave functions. Later, it was shown to be possible to construct the representation of the Poincaré group for relativistic extended particles based on harmonic oscillators [14, 15]. This representation serves useful purposes for understanding high-speed hadrons. We shall discuss these problems systematically in chapters 6 and 7.

1.5.1 Harmonic oscillators and Fock space

In quantum mechanics, as well as in quantum field theory, creation and annihilation operators are considered to constitute an essential component. They act on the elements of the space of quantum states of various types of particles, called the Fock space [11]. Particularly, a creation operator (or step-up operator) increases the number of particles in a given state by one, while its adjoint, the annihilation operator (step-down operator), decreases the number of particles by one. They satisfy the commutation relations:

$$[a, a^{\dagger}] = 1. \tag{1.64}$$

It is possible to consider the step-down and step-up operators defined above as

$$\begin{aligned} a_i &= \frac{1}{\sqrt{2}}(x_i + ip_i) = \frac{1}{\sqrt{2}}\left(x_i + \frac{\partial}{\partial x_i}\right), \\ a_i^{\dagger} &= \frac{1}{\sqrt{2}}(x_i - ip_i) = \frac{1}{\sqrt{2}}\left(x_i - \frac{\partial}{\partial x_i}\right). \end{aligned} \tag{1.65}$$

Then

$$\left[a_i, a_j^{\dagger}\right] = \delta_{ij} \quad \text{and} \quad [a_i, a_j] = 0. \tag{1.66}$$

To further illustrate the use of these operators, we will consider the one-dimensional harmonic oscillator and drop the subscripts i and j.

In this instance, the operators of equation (1.65) become

$$\begin{aligned} a &= \frac{1}{\sqrt{2}}(x + ip) = \frac{1}{\sqrt{2}}\left(x + \frac{\partial}{\partial x}\right), \\ a^{\dagger} &= \frac{1}{\sqrt{2}}(x - ip) = \frac{1}{\sqrt{2}}\left(x - \frac{\partial}{\partial x}\right). \end{aligned} \tag{1.67}$$

The harmonic oscillator differential equation can be written as:

$$\frac{1}{2}\left\{-\left(\frac{\partial}{\partial x}\right)^2 + x^2\right\}\chi_n(x) = \left(n + \frac{1}{2}\right)\chi_n(x). \tag{1.68}$$

The solution to equation (1.68) is then the harmonic oscillator wave function in the nth excited state which takes the form

$$\chi_n(x) = \frac{1}{\sqrt{\pi 2^n n!}} H_n(x) e^{-x^2/2}, \tag{1.69}$$

where $H_n(x)$ is the Hermite polynomial. In the Fock space [11] where the number n serves as the number of particles or photons in a given state, we are only interested in the quantum number n in equation (1.69). Hence it is more convenient to write this wave function as $|n\rangle$. Then we can write

$$a|n\rangle = \sqrt{n}\ |n-1\rangle \quad \text{and} \quad a^\dagger|n\rangle = \sqrt{n+1}\ |n+1\rangle. \tag{1.70}$$

It is then apparent that a and $a^\dagger$ are step-down and step-up (annihilation and creation) operators for the harmonic oscillator wave functions. The operator a reduces the number of particles by one, and $a^\dagger$ adds one. These operators are a basic mathematics device for many procedures in quantum optics.

1.5.2 Bogoliubov transformation

In physics, we often use the harmonic oscillators for Heisenberg brackets. We use a and $a^\dagger$ for step-down and step-up operators. Remembering that they satisfy the commutation relation

$$[a, a^\dagger] = 1, \tag{1.71}$$

we can then consider the transformation

$$b = \alpha a + \beta a^\dagger \quad \text{and} \quad b^\dagger = \alpha^* a^\dagger + \beta^* a. \tag{1.72}$$

If we demand the commutation relation

$$[b, b^\dagger] = 1, \tag{1.73}$$

this leads to

$$\alpha\alpha^* - \beta\beta^* = 1. \tag{1.74}$$

This transformation is widely known as the Bogoliubov transformation in the literature [5, 9]. To relate this to the Lorentz group, we can consider that α and β should take the form

$$\alpha = e^{i\phi} \cosh\eta \quad \text{and} \quad \beta = e^{i\xi} \sinh\eta. \tag{1.75}$$

As for these angles, it is only their difference that matters. This means that there are only two independent variables, namely η and θ, where

$$\theta = \phi - \xi. \tag{1.76}$$

We can then consider a three-dimensional space with the orthogonal x-,y-, and z-axes. Then

$$\eta = \sqrt{x^2 + y^2 + z^2} \quad \text{and} \quad \tan\theta = \frac{y}{x}. \tag{1.77}$$

Using equation (1.13), the rotation around the z-axis is generated by

$$J_3 = -i\left(x\frac{\partial}{\partial y} - y\frac{\partial}{\partial x}\right). \tag{1.78}$$

In addition, we can consider the generators:

$$K_1 = -i\left(z\frac{\partial}{\partial x} + x\frac{\partial}{\partial z}\right) \quad \text{and} \quad K_2 = -i\left(z\frac{\partial}{\partial y} + y\frac{\partial}{\partial z}\right). \tag{1.79}$$

It is interesting to note that these generators are like the boost generators given in equation (1.13) with t replaced by z. The transformation generated by these three generators moves the point (x, y, z) within the circular cone which makes a 45° angle with the z-axis, and the projection of the cone into the xz- and yz-plane.

These three generators satisfy the commutation relations:

$$[K_1, K_2] = -iJ_3, \quad [K_2, J_3] = iK_1, \quad \text{and} \quad [K_1, J_3] = -iK_2. \tag{1.80}$$

This closed set of commutation relations is exactly the Lie algebra for the $O(2, 1)$ Lorentz group (discussed in detail in section 2.4) which is like (isomorphic to) the $Sp(2)$ symplectic group (discussed in detail in section 8.1). This is precisely the symmetry of the Bogoliubov transformation.

Without loss of generality, we can fix the value of z at $z = 1$, and define η as

$$\eta = \sqrt{x^2 + y^2} \quad \text{and} \quad \tan\theta = \frac{y}{x}. \tag{1.81}$$

Then

$$K_1 \rightarrow P_1 = -i\frac{\partial}{\partial x} \quad \text{and} \quad K_2 \rightarrow P_2 = -i\frac{\partial}{\partial y}, \tag{1.82}$$

with

$$[P_1, P_2] = 0, \quad [P_2, J_3] = iP_1, \quad \text{and} \quad [P_1, J_3] = -iP_2. \tag{1.83}$$

This is the Lie algebra of the two-dimensional Euclidean group (discussed in detail in section 2.3; the contraction of $O(2, 1)$ to $E(2)$ is detailed in section 3.3), where P_1 and P_2 are translation generators along the x- and y-direction, respectively. It is interesting to note that the Bogoliubov transformation also has the $E(2)$ symmetry.

1.5.3 Dirac's two-oscillator system

Dirac published a paper in 1963 [8] in which he considered two oscillators in terms of the annihilation and creation operators as a_1, $a_1^\dagger$ and a_2, $a_2^\dagger$. He then organized them into the following sixteen bilinear forms:

$$a_i a_j, \quad a_i^\dagger a_j^\dagger, \quad a_i^\dagger a_j, \quad \text{and} \quad a_i a_j^\dagger. \tag{1.84}$$

This two-oscillator system is important as we shall see in chapter 9, where we use combinations of these bilinear forms to generate the two-photon coherent state [19].

Furthermore, starting with the Schrödinger equation for two harmonic oscillators, Dirac also observed in the same paper that unitary transformations applicable to the ground-state wave function could be generated by ten combinations of these bilinear forms He also noted that these ten combinations form the 3 + 2 de Sitter or $O(3, 2)$ group. This group, applicable to three space-like and two time-like coordinates, can be regarded as two coupled Lorentz groups. It was shown later [13] that fifteen combinations of these bilinear forms lead to the Lie algebra of the $O(3, 3)$ group or the Lorentz group applicable to three space-like and three time-like coordinates. We discuss these groups in detail in chapters 8 and 9, together with their relationship to the Dirac matrices.

1.6 Wigner functions for the Lorentz group

In quantum mechanics, the position and momentum variables are *q-numbers*. In the Wigner phase-space representation, they are both *c-numbers*. The Wigner function thus allows us to interpret Heisenberg's uncertainty principle in terms of a closed area in the phase space of x and p.

The Wigner phase-space distribution function is defined as [18]

$$W(x, p) = \frac{1}{\pi} \int \psi(x + x')\psi^*(x - x')\exp(2ipx')\, dx'. \tag{1.85}$$

This Wigner function is normalized as

$$\int W(x, p)\, dxdp = 1. \tag{1.86}$$

Furthermore, the density distribution $|\psi(x)|^2$ is recovered from the Wigner function as

$$|\psi(x)|^2 = \psi^*(x)\psi(x) = \int W(x, p)\, dp. \tag{1.87}$$

While this Wigner function is an important formula for quantum mechanics, it contains basic symmetry properties of the Lorentz group. Let us consider the Gaussian wave function

$$\psi(x) = \left(\frac{1}{\pi}\right)^{1/4} \exp(-x^2/2). \tag{1.88}$$

The Wigner function for the wave function then becomes

$$W(x, p) = \frac{1}{\pi} \exp(-[x^2 + p^2]). \tag{1.89}$$

This function defines a circle in the plane of x and p.

Heisenberg's uncertainty principle allows us to squeeze along the x-direction by the Lorentz boost along the z-axis, thus forming an ellipse. It is also possible to squeeze along the p-direction by the Lorentz boost along the x-axis, again forming an ellipse. The area of the ellipse remains the same as that of the circle even though the width of the distribution in both x and p becomes extended. Then, according to the rules of canonical transformations, it is possible to rotate this ellipse. This is also an area-preserving transformation. The group of these transformations is called the two-dimensional symplectic group, or the group $Sp(2)$. The generators of this group satisfy the closed set of commutation relations that is identical with that of the Lorentz group applicable to two space-like dimensions and one time-like dimension.

If the single oscillator contains this much content of the Lorentz group, two oscillators with couplings will give much richer contents of the Lorentz group. We shall discuss this issue in chapter 8.

References

[1] Bargmann V 1947 Irreducible unitary representations of the Lorentz group *Ann. Math.* **48** 568–640

[2] Başkal S, Kim Y S, and Noz M E 2014 Wigner's space–time symmetries based on the two-by-two matrices of the damped harmonic oscillators and the Poincaré sphere *Symmetry* **6** 473–515

[3] Başkal S, Kim Y S, and Noz M E 2019 *Mathematical Devices for Optical Sciences* (Bristol, UK: IOP Publishing) (ISBN 978-0-7503-1612-5, OCLC: 1034620988)

[4] Beresteckij V B, Lifshitz E M, Pitaevskij L P, and Landau L D 1982 *Quantum Electrodynamics: Course of Theoretical Physics* 2nd edn vol 4 (Amsterdam: Butterworth-Heinemann) reprinted 2007 (ISBN 978-0-7506-3371-0, Originally published 1980; OCLC: 254580834; original authors: L D Landau and E M Lifshitz).

[5] Bogoljubov N N, Tolmachov V V, and Širkov D V 1958 A new method in the theory of superconductivity *Fortschr. Phys.* **6** 605–82

[6] Dirac P A M 1945 Application of quaternions to Lorentz transformations *Proc. R. Irish Acad. A: Math. Phys. Sci.* **50** 261–70

[7] Dirac P A M 1945 Unitary representations of the Lorentz group *Proc. R. Soc. A: Math. Phys. Eng. Sci.* **183** 284–95

[8] Dirac P A M 1963 A remarkable representation of the 3 + 2 de Sitter group *J. Math. Phys.* **4** 901–9

[9] Fetter A L and Walecka J D 2003 *Quantum Theory of Many-Particle Systems* (Mineola, NY: Dover Publications) (ISBN 978-0-486-42827-7, Originally published: New York: McGraw-Hill, 1971)

[10] Feynman R P, Kislinger M, and Ravndal F 1971 Current matrix elements from a relativistic quark model *Phys. Rev.* D **3** 2706–32

[11] Fock V 1934 Quanten elecktrodynamik *Phys. Zeitschr. Sovietunion* **6** 425–69

[12] Gilmore R 2005 *Lie Groups, Lie Algebras, and Some of Their Applications* (Mineola, NY: Dover Publications) (ISBN 978-0-486-44529-8, Originally published: New York: Wiley, 1974)

[13] Han D, Kim Y S, and Noz M E 1995 *O* (3,3)–like symmetries of coupled harmonic oscillators *J. Math. Phys.* **36** 3940–54

[14] Kim Y S and Noz M E 1986 *Theory and Applications of the Poincaré Group* (Dordrecht: Springer) (ISBN 978-94-010-8526-7, 978-94-009-4558-6)

[15] Kim Y S, Noz M E, and Oh S H 1979 Representations of the Poincaré group for relativistic extended hadrons *J. Math. Phys.* **20** 1341–4 See also, Physics Auxiliary Publication Service Document No. PAPS JMAPA-20-1336-12

[16] Naimark M A 1954 Linear representation of the Lorentz group *Usp. Mat. Nauk* **9** 19–93
Naimark M A 1957 Linear representation of the Lorentz group *Am. Math. Soc. Transl. Ser.* **2** 379–458 (Engl. transl.)
Naimark M A 1964 *Linear Representations of the Lorentz Group (International Series of Monographs in Pure and Applied Mathematics)* vol 63 (Oxford: Pergamon) (Engl. transl.)
[17] Wightman A S 1961 L'invariance dans la mecanique quantique relativiste *Relations de dispersion et particules elementaires* ed C DeWitt and R Omnes (New York: Wiley) pp 150–226 (ASIN: B003IGFWDI)
[18] Wigner E P 1932 On the quantum correction for thermodynamic equilibrium *Phys. Rev.* **40** 749–59
[19] Yuen H P 1976 Two-photon coherent states of the radiation field *Phys. Rev.* A **13** 2226–43
[20] Yukawa H 1953 Structure and mass spectrum of elementary particles. I. General considerations *Phys. Rev.* **91** 415–6

IOP Publishing

Physics of the Lorentz Group (Second Edition)

Beyond high-energy physics and optics

Sibel Başkal, Young S Kim, and Marilyn E Noz

Chapter 2

Wigner's little groups for internal space–time symmetries

When Einstein formulated his theory of special relativity, he was interested in point particles, without internal space–time structures. If a massive particle is at rest, the momentum is zero. On the other hand, the particle has spin with three different directions indicating an internal space–time structure. A light wave has electric and magnetic fields perpendicular to the direction of propagation which is translated into the photon polarization. Massless spin-half neutrinos are polarized. These are the issues concerning the internal space–time symmetries. The hydrogen atom is a bound state of the electron and proton with non-zero size but is small enough to be regarded as a point particle obeying Einstein's law of Lorentz transformations. However, the hydrogen atom has a rich internal space–time structure, rich enough to provide the foundation of quantum mechanics. The question then is how these particles would look to a moving observer.

To answer this question, we study in this chapter, the internal space–time symmetries defined by Wigner in his 1939 paper [21]. Wigner noted that the most important quantities for a given particle are the mass and momentum associated with it; thus, the energy–momentum four-vector. He then considered the subgroups of the Lorentz group whose transformations leave the particle energy–momentum four-vector invariant. He named them *little groups*. Thus these little groups can transform the internal space–time structure of particles. Since the particle momentum is fixed and remains invariant, we can assume that the particle momentum is along the z-direction.

This momentum is then invariant under rotations around the z-axis. In addition, these rotations commute with the Lorentz boost along the z-axis. From the Lie algebra of equation (1.10), we have

$$[J_3, K_3] = 0. \tag{2.1}$$

With these preparations, we can simplify the problem using the Euler angles, which can be used to parameterize three-dimensional rotations [8]. Euler formulated his

doi:10.1088/978-0-7503-3607-9ch2

kinematics to understand spinning tops in classical mechanics. In quantum mechanics, we study this problem by constructing representations of the rotation group, such as the spherical harmonics and Pauli matrices. Now the pressing issue is what happens if the system is Lorentz-boosted.

In section 2.1, we discuss how to use Euler angles to decompose Wigner's little groups. In section 2.2, we boost the rotation group applicable to the massive particle and examine the $O(3)$-like little group for massive particles. In section 2.3, the $E(2)$-like little group is used to study what happens when the massive particle becomes massless. We consider the four potential of a photon whose momentum is along the z-axis. We use, in section 2.3.1, the geometry of a plane tangential to the north pole of a large sphere to describe the $E(2)$-like little group. Another method to accomplish this by using a cylinder tangent to the equatorial plane of a similarly large sphere is given in section 2.3.2. In section 2.4, we discuss the $O(2, 1)$-like little group for imaginary-mass particles, which, although they do not exist in our world, play a theoretical role in quantum field theory. In section 2.5, we give some additional comments on what we understand about Wigner's little groups. In section 2.6, we discuss Wigner's little groups in Dirac's light-cone coordinate system.

2.1 Euler decomposition of Wigner's little group

The Euler angles constitute a convenient parameterization of the three-dimensional rotations [8]. The Euler kinematics consists of two rotations around the z-axis with one rotation around the y-axis between them. These three operations also cover the rotation around the x-axis, thanks to the commutation relation

$$[J_2, J_3] = iJ_1. \tag{2.2}$$

In this way, it is possible to study the essential features of three-dimensional rotations using the two-dimensional space of z and x. This aspect is well known.

The first question is, what happens if we add a Lorentz boost along the z-direction to this traditional procedure [10]? Since the rotation around the z-axis is not affected by the boost along the same axis, we are asking what happens to the rotation around the y-axis if we apply a Lorentz boost along the z-direction.

2.2 $O(3)$-like little group for massive particles

Wigner observed first that a massive particle at rest has three rotational degrees of freedom leading to the concept of spin. Thus, the little group for this massive particle is like $O(3)$. How about this massive particle moving in the z-direction? Since the particle has a positive mass, when it is in the Lorentz frame in which it is at rest, the particle's energy–momentum four-vector is proportional to

$$P = (1, 0, 0, 0). \tag{2.3}$$

This momentum remains invariant under rotations.

The three generators of this little group are J_1, J_2, and J_3, satisfying the Lie algebra of equation (1.5). The dynamical variables associated with these Hermitian operators are known to be particle spins.

The system can be boosted along the z-axis, with the boost matrix $B(\eta)$ given in equation (1.6). If this matrix is applied to the energy–momentum four-vector of equation (2.3), it becomes

$$P' = (\cosh\eta, \sinh\eta, 0, 0). \tag{2.4}$$

The generators become

$$J_i' = B(\eta)J_iB^{-1}(\eta). \tag{2.5}$$

Under this boost operation, J_3 remains invariant, but J_2' becomes

$$J_2' = (\cosh\eta)J_2 - (\sinh\eta)K_1. \tag{2.6}$$

As for J_1, the boost results in

$$J_1' = (\cosh\eta)J_1 + (\sinh\eta)K_2. \tag{2.7}$$

However, we can obtain the same result by rotating J_2' by $-90°$ around the z-axis, thanks to the Euler angles effect discussed in section 2.1.

Although the generators J_i' satisfy the same Lie algebra as J_i, they are not the same. Thus, we shall call the operators J_i' generators of the $O(3)$-like little group for the massive particle with a non-zero momentum.

An interesting issue is what happens when η becomes infinite, and $\sinh\eta = \cosh\eta$. We shall discuss this problem in section 4.6.

2.3 $E(2)$-like little group for massless particles

Wigner also observed that a massless particle could not be brought to its rest frame. Still, he showed that the little group for the massless particle also has three degrees of freedom and that this little group is like $E(2)$ or the two-dimensional Euclidean group. Hence, the generators of this little group share the same Lie algebra with the two-dimensional Euclidean group with one rotational and two translational degrees of freedom. It is not difficult to associate the rotational degree of freedom of $E(2)$ to the helicity of the massless particle. However, what is the physics of those two translational degrees of freedom? Wigner did not provide the answer to this question in his original paper [21].

If the particle is massless, its energy–momentum four-vector is proportional to

$$P = (1, 1, 0, 0). \tag{2.8}$$

This expression is, of course, invariant under rotations around the z-axis. Wigner [21] observed that it is also invariant under the transformation

$$D(\gamma, \phi) = \exp[-i\gamma(N_1\cos\phi + N_2\sin\phi)] \tag{2.9}$$

with

$$N_1 = K_1 - J_2 = \begin{pmatrix} 0 & 0 & i & 0 \\ 0 & 0 & i & 0 \\ i & -i & 0 & 0 \\ 0 & 0 & 0 & 0 \end{pmatrix} \quad \text{and} \quad N_2 = J_1 + K_2 = \begin{pmatrix} 0 & 0 & 0 & i \\ 0 & 0 & 0 & i \\ 0 & 0 & 0 & 0 \\ i & -i & 0 & 0 \end{pmatrix}. \tag{2.10}$$

As a consequence,

$$D(\gamma, \phi) = \begin{pmatrix} 1 + \gamma^2/2 & -\gamma^2/2 & \gamma\cos\phi & \gamma\sin\phi \\ \gamma^2/2 & 1 - \gamma^2/2 & \gamma\cos\phi & \gamma\sin\phi \\ \gamma\cos\phi & -\gamma\cos\phi & 1 & 0 \\ \gamma\sin\phi & -\gamma\sin\phi & 0 & 1 \end{pmatrix}. \tag{2.11}$$

Thus the generators of this little group are N_1, N_2, and J_3. They satisfy the following set of commutation relations,

$$[N_1, N_2] = 0, \quad [J_3, N_1] = iN_2, \quad \text{and} \quad [J_3, N_2] = -iN_1. \tag{2.12}$$

As Wigner notes, this Lie algebra is the same as that of the two-dimensional Euclidean group, with

$$[P_1, P_2] = 0, \quad [J_3, P_1] = iP_2, \quad \text{and} \quad [J_3, P_2] = -iP_1, \tag{2.13}$$

where P_1 and P_2 generate translations along the x- and y-directions, respectively. Thus Wigner called this the $E(2)$-like little group.

In addition, Kim and Wigner [16, 17] considered the following operators,

$$Z_1 = -ix\frac{\partial}{\partial z} \quad \text{and} \quad Z_2 = -iy\frac{\partial}{\partial z}, \tag{2.14}$$

with $x^2 + y^2 =$ constant. The Z_i generate translations along the z-direction on the surface of a circular cylinder, as described in figure 2.1. As with the generators P_i, the Z_i satisfy the following commutation relations:

$$[Z_1, Z_2] = 0, \quad [J_3, Z_1] = iZ_2, \quad \text{and} \quad [J_3, Z_2] = -iZ_1. \tag{2.15}$$

We can say that this is the Lie algebra for the *cylindrical group*.

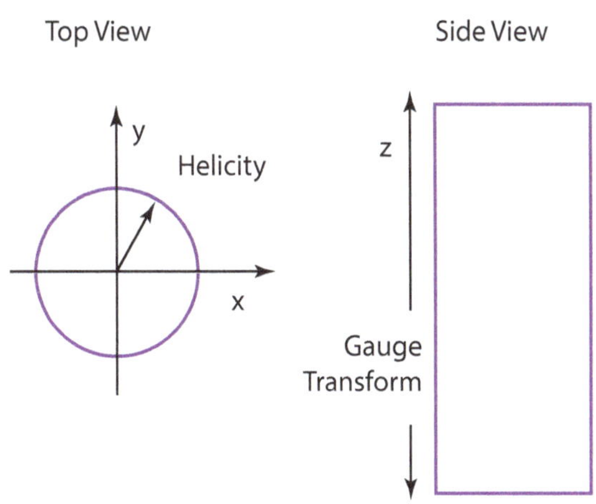

Figure 2.1. The internal space–time structure of photons in the cylindrical picture. In the side view of this cylinder the top-down translation corresponds to a gauge transformation. The top view is a circle that shows the rotational degree of freedom corresponds to the helicity of the photon [17].

Let us consider a photon whose momentum is along the z-direction. It has the four-potential

$$(A_0, A_z, A_x, A_y). \tag{2.16}$$

According to the Lorentz condition, $A_0 = A_z$, and the four-potential is

$$(A_0, A_0, A_x, A_y). \tag{2.17}$$

If we apply the $D(\gamma, \phi)$ of equation (2.11) to this four-potential:

$$\begin{pmatrix} 1 + \gamma^2/2 & -\gamma^2/2 & \gamma\cos\phi & \gamma\sin\phi \\ \gamma^2/2 & 1 - \gamma^2/2 & \gamma\cos\phi & \gamma\sin\phi \\ \gamma\cos\phi & -\gamma\cos\phi & 1 & 0 \\ \gamma\sin\phi & -\gamma\sin\phi & 0 & 1 \end{pmatrix} \begin{pmatrix} A_0 \\ A_0 \\ A_x \\ A_y \end{pmatrix}, \tag{2.18}$$

the result is

$$\begin{pmatrix} A_0 + \gamma\left(A_x\cos\phi + A_y\sin\phi\right) \\ A_0 + \gamma\left(A_x\cos\phi + A_y\sin\phi\right) \\ A_x \\ A_y \end{pmatrix}. \tag{2.19}$$

The little group transformation $D(\gamma, \phi)$ leaves the transverse components A_x and A_y invariant but provides an addition to A_0. This form is a cylindrical transformation. In the language of physics, it is a gauge transformation. These results are summarized in table 2.1.

If we boost the energy–momentum four-vector of equation (2.8) along the z-direction, the energy–momentum four-vector becomes

$$P' = e^{\eta}(1, 1, 0, 0), \tag{2.20}$$

and N_1 and N_2 become $e^{\eta}N_1$ and $e^{\eta}N_2$, respectively. J_3 remains invariant.

Table 2.1. Here the covariance of the energy–momentum relation is shown together with the covariance of the internal space–time symmetry. The J_3 component of the angular momentum remains invariant under the Lorentz boost along the z-direction. It is known as the helicity. In contrast, the transverse J_1 and J_2 components collapse into a gauge transformation. In the four-by-four matrix formulation, earlier papers [9, 10, 17] have studied the γ parameter for the massless particle.

	Massive Slow	Lorentz Covariance	Massless Fast
Energy-Momentum	$E = p^2/2m$	Einstein's $E = [p^2 + m^2]^{1/2}$	$E = cp$
Spin, Helicity	J_3	Wigner's	Helicity
Gauge Transformation	J_1, J_2	Little Group	Gauge Transformation

2.3.1 Geometric aspects of the *E*(2)-like little group

The Euclidean group, often known as $E(2)$, consists of a rotation and translations in a two-dimensional Euclidean plane. The coordinate transformations applicable to this plane are familiar to us. They are:

$$x' = x\cos\phi - y\sin\phi + a \quad \text{and} \quad y' = x\sin\phi + y\cos\phi + b. \tag{2.21}$$

Here the a and b represent the translations in the xy-plane. Writing this in matrix form using our four-dimensional coordinate system, we obtain:

$$\begin{pmatrix} 1 \\ 1 \\ x' \\ y' \end{pmatrix} = \begin{pmatrix} 1 & 0 & 0 & 0 \\ 0 & 1 & 0 & 0 \\ 0 & a & \cos\phi & -\sin\phi \\ 0 & b & \sin\phi & \cos\phi \end{pmatrix} \begin{pmatrix} 1 \\ 1 \\ x \\ y \end{pmatrix}. \tag{2.22}$$

Let us consider a small xyz-plane located at the north pole of a very large sphere. Then the x and y are very small as they are translations on this small xyz-plane. We can exponentiate the above matrix as

$$E(u, v, \phi) = \exp[-i(aP_1 + bP_2)]\exp(-i\phi J_3), \tag{2.23}$$

where J_3 is the generator of rotations about the z-axis given in equation (1.3) and P_1 and P_2 generate translations. These translation generators can be written in matrix form as:

$$P_1 = \begin{pmatrix} 0 & 0 & 0 & 0 \\ 0 & 0 & 0 & 0 \\ 0 & i & 0 & 0 \\ 0 & 0 & 0 & 0 \end{pmatrix} \quad \text{and} \quad P_2 = \begin{pmatrix} 0 & 0 & 0 & 0 \\ 0 & 0 & 0 & 0 \\ 0 & 0 & 0 & 0 \\ 0 & i & 0 & 0 \end{pmatrix}. \tag{2.24}$$

They can also be written as

$$P_1 = -i\frac{\partial}{\partial x} \quad \text{and} \quad P_2 = -i\frac{\partial}{\partial y}, \tag{2.25}$$

while the rotation generator J_3 takes the form

$$J_3 = -i\left(x\frac{\partial}{\partial y} - y\frac{\partial}{\partial x}\right). \tag{2.26}$$

These generators satisfy the commutation relations given in equation (2.13).

2.3.2 Cylindrical interpretation of the *E*(2)-like little group

Previously, we discussed the $E(2)$-like little group in terms of the cylindrical group. In cylindrical coordinates our transformation equations take the form

$$x' = x\cos\phi - y\sin\phi, \quad y' = x\sin\phi + y\cos\phi, \quad \text{and} \quad z' = z + ax + by. \tag{2.27}$$

This transformation leaves $(x^2 + y^2)^{\frac{1}{2}}$ invariant, while z can vary from $-\infty$ to $+\infty$. This linear transformation thus represents the cylindrical group, as illustrated in figure 2.1. Here x and y are transformations on the circular part of the cylinder, and z represents translations up and down the z-axis of the cylinder. Writing this in matrix form using our four-dimensional coordinate system, we obtain:

$$\begin{pmatrix} 1 \\ z' \\ x' \\ y' \end{pmatrix} = \begin{pmatrix} 1 & 0 & 0 & 0 \\ 0 & 1 & a & b \\ 0 & 0 & \cos\phi & -\sin\phi \\ 0 & 0 & \sin\phi & \cos\phi \end{pmatrix} \begin{pmatrix} 1 \\ z \\ x \\ y \end{pmatrix}. \tag{2.28}$$

As z can be much larger than x and y, we can exponentiate the above matrix as

$$C(u, v, \phi) = \exp[-i(aZ_1 + bZ_2)]\exp(-i\phi J_3). \tag{2.29}$$

As before, J_3 represents rotations around z in the xyz-plane and the Z_i can be written in matrix form as:

$$Z_1 = \begin{pmatrix} 0 & 0 & 0 & 0 \\ 0 & 0 & i & 0 \\ 0 & 0 & 0 & 0 \\ 0 & 0 & 0 & 0 \end{pmatrix} \quad \text{and} \quad Z_2 = \begin{pmatrix} 0 & 0 & 0 & 0 \\ 0 & 0 & 0 & i \\ 0 & 0 & 0 & 0 \\ 0 & 0 & 0 & 0 \end{pmatrix}. \tag{2.30}$$

The Z_i again satisfy the same commutation relations given in equation (2.15). It is interesting to note that the commutation relations in equations (2.13) and (2.15) are invariant under a sign change in P_i or Z_i. They are also invariant under Hermitian conjugation. Since J_3 is Hermitian, we can write the relationship between P_i or Z_i as:

$$Z_i = -(P_i)^\dagger \quad \text{and} \quad P_i = -(Z_i)^\dagger. \tag{2.31}$$

These two geometric interpretations are illustrated in figure 2.2.

2.4 *O*(2, 1)-like little group for imaginary-mass particles

We are now interested in transformations which leave the four-vector of the form

$$P = (0, 1, 0, 0) \tag{2.32}$$

invariant. Then $P^2 = -1$, and it is a negative number. We are accustomed to positive values of $P^2 = (\text{mass})^2$. A negative P^2 means that the particle mass is imaginary, and it moves faster than light. We are thus talking about a particle we cannot observe in the real world. On the other hand, these particles play a major theoretical role in quantum field theory, specifically in Feynman diagrams, and string theory [7, 11].

We are now interested in transformations that leave the four-vector of equation (2.32) invariant. Let us consider the Lorentz boost along the y-direction, with

$$S(\lambda) = \begin{pmatrix} \cosh\lambda & 0 & 0 & \sinh\lambda \\ 0 & 1 & 0 & 0 \\ 0 & 0 & 1 & 0 \\ \sinh\lambda & 0 & 0 & \cosh\lambda \end{pmatrix}, \tag{2.33}$$

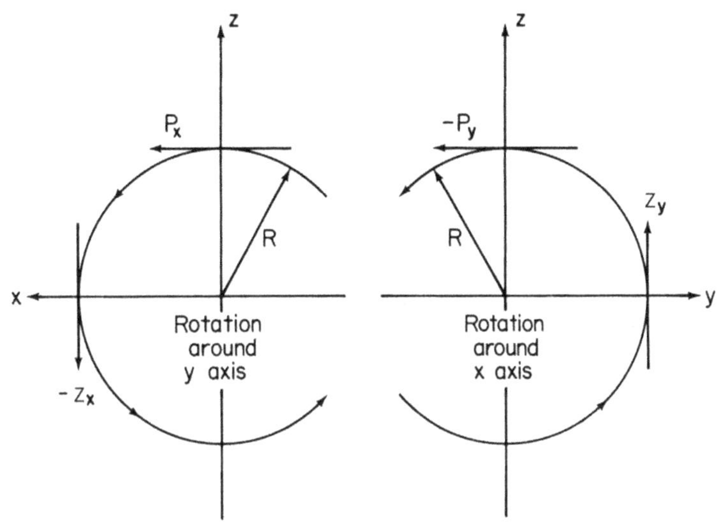

Figure 2.2. Geometry of the Euclidean and cylindrical groups. There can be a plane tangential at the north pole for a given sphere, as well as a cylinder tangential to the equatorial belt. The gauge transformation corresponds to the up-down z translation along the axis of the circular cylindrical surface [16].

which is generated by K_2. Likewise, it is invariant under the boost along the x-direction. Thus, we can consider the set of commutation relations

$$[J_3, K_1] = iK_2, \quad [J_3, K_2] = -iK_1, \quad \text{and} \quad [K_1, K_2] = -iJ_3. \tag{2.34}$$

These commutation relations form the Lie algebra of the Lorentz group applicable to two space and one time dimensions. This group is known in the literature as $O(2, 1)$.

If we boost the energy–momentum four-vector of equation (2.32) along the z-direction, it becomes

$$(\sinh\eta, \cosh\eta, 0, 0). \tag{2.35}$$

While J_3 remains invariant, K_1 and K_2 become

$$K_1' = (\cosh\eta)K_1 - (\sinh\eta)J_2 \quad \text{and} \quad K_2' = (\cosh\eta)K_2 + (\sinh\eta)J_1, \tag{2.36}$$

respectively. The generators K_1', K_2', and J_3 satisfy the same Lie algebra as that of equation (2.34).

If we interchange $\sinh\eta$ and $\cosh\eta$, these generators become those of equation (2.6). If η becomes very large, the energy–momentum four-vector takes the same form for the massive, massless, and imaginary-mass particles. The generators of the little groups also become the same. We shall discuss this issue in section 4.6.

Even though we are talking about imaginary-mass particles in this section, this little group's mathematics applies to many branches of physics [1, 2, 14].

2.4.1 $O(2, 1)$-like little group in a fish tank—Cherenkov radiation

In Wigner's original paper of 1939 [21], the little group for the space-like momentum is like $O(2, 1)$, namely the Lorentz group applicable to two space-like and one time-like directions. Unlike the $O(3)$-like little group for massive particles, or the $E(2)$-like little group for massless particles, this $O(2, 1)$-like little group is seldom discussed in the literature, because we do not observe particles with imaginary mass.

On the other hand, if the speed of light is less than c, the situation could be different [18]. Indeed, the speed is reduced by the index of refraction n. For water, the value of n is 1.33. Thus the speed is $c/1.33$. On the other hand, massive particles can move faster than this speed. If a particle reaches this speed in water, it causes a release of radiation that is widely known as Cherenkov radiation [5, 13].

If the particle moves faster than light, the four-momentum becomes space-like, which means the energy of the particle is smaller than the momentum. In that case, Wigner's $O(2, 1)$-like little group is applicable.

Let us look at figure 2.3. We start with a massive particle A with the energy–momentum relation $E = \sqrt{p^2 + m^2}$. This particle can be boosted to the rest state with $p = 0$, and $E = m$, as specified by point B in the figure. On the other hand, if this point crosses the momentum-energy dividing line, it becomes space-like. Then it can be brought to the state C where the four-momentum becomes purely space-like. For this space-like four-vector, Wigner's $O(2, 1)$-like little group is applicable.

Let us consider a light wave traveling in a vacuum along the z-direction as it enters the fish tank. When that happens, ω remains unchanged, but the speed of light is reduced by the index of refraction. Thus, the expression for this wave becomes

$$\exp(i(kz - \omega t)), \tag{2.37}$$

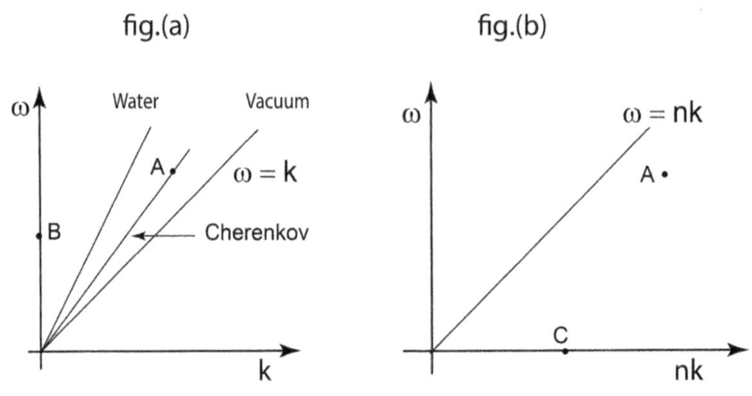

Figure 2.3. A particle moving in a fish tank. The particle moves with a time-like four-momentum A, and can be Lorentz-boosted to B without space-like component. Since the index of refraction is greater than one for water, the light-cone in the $k\omega$ plane is steeper than the 45° line, as shown in (a). When the line passes through the energy–momentum point A of a moving particle, it gives an observable effect known as Cherenkov radiation. The line can become steeper depending on the index of refraction. It is possible to rotate this steep line to 45° by expanding the horizontal axis, as shown in (b). The k axis is now expanded to nk. Then the particle four-momentum at A becomes space-like, and it can be boosted to C without a time-like component.

with $k = \omega/c$. If the beam enters the water with the index of refraction n, ω remains the same, but c is reduced to c/n. Thus, in the water, the light-wave becomes [18]

$$\exp(i(nkz - \omega t)). \tag{2.38}$$

This expression can be written as

$$\exp(i\omega(nz - t)), \tag{2.39}$$

in the unit system where $c = 1$.

In the water, a particle can move with speed greater than c/n. When it does, it causes a release of radiation, in the form of a γ-ray widely known as Cherenkov radiation [5]. It is like an airplane moving with a speed greater than that of sound—it produces a sonic boom.

Thus, to the observer in the particle frame, the light wave would move backward. If the particle moves with the same speed as the light-wave in the water, z and t should become as

$$z \to \frac{z + \beta t}{\sqrt{1 - \beta^2}}, \quad \text{and} \quad t \to \frac{t + \beta z}{\sqrt{1 - \beta^2}}, \tag{2.40}$$

where β is the speed of the particle. The exponential form of equation (2.39) then becomes

$$\exp\left\{ i\omega \left[\frac{(n - \beta)z - (n\beta - 1)t}{\sqrt{1 - \beta^2}} \right] \right\}. \tag{2.41}$$

If $\beta = 1/n$, this expression becomes

$$\exp\left\{ i\left(\omega\sqrt{n^2 - 1}\right)z \right\}. \tag{2.42}$$

This is a time-independent standing wave.

Let us go to figure 2.3. The energy–momentum dividing line makes a 45° angle with respect to each axis in vacuum, as shown in (a). If the light speed becomes less than c, the line becomes steeper. The four-momentum A of the particle is between the two lines. This four-momentum can be brought to the rest state at B in the figure.

In order to make this picture into the Lorentzian system, we can expand the horizontal axis from k to nk, as shown in (b) where the separation line makes a 45° angle. This figure allows us to use the language of Wigner's little group. The four-momentum A is now space-like, and it can be boosted to C in the same figure. This four-momentum does not have the energy component. If we use the notation (p_0, p_z, p_x, p_y), the energy–momentum four-vector at C can be written as $(0, 1, 0, 0)$. This energy–momentum four-vector must be invariant under rotations around z-axis. It should also be invariant under the boosts along the x and y directions. This is known as the $O(2, 1)$-like little group for the space-like four-momentum.

We can mention the same thing using the two-by-two representation of the Lorentz group. In this representation, the energy–momentum four-vector takes the form

$$\begin{pmatrix} p_0 + p_z & p_x - ip_y \\ p_x + ip_y & p_0 - p_x \end{pmatrix}, \tag{2.43}$$

with the determinant:

$$p_0^2 - p_z^2 - p_x^2 - p_y^2. \tag{2.44}$$

The space-like four-vector at C can be written as

$$P = \begin{pmatrix} 1 & 0 \\ 0 & -1 \end{pmatrix}. \tag{2.45}$$

This momentum matrix is invariant under the transformation

$$M \;\, P \;\, M^{\dagger} = P, \tag{2.46}$$

if the transformation matrix takes the form

$$\begin{pmatrix} \exp(i\phi/2) & 0 \\ 0 & \exp(-i\phi/2) \end{pmatrix}, \quad \begin{pmatrix} \cosh(\eta/2) & \sinh(\eta/2) \\ \sinh(\eta/2) & \cosh(\eta/2) \end{pmatrix}, \quad \text{or} \\ \begin{pmatrix} \cosh(\lambda/2) & i\sinh(\lambda/2) \\ -i\sinh(\lambda/2) & \cosh(\lambda/2) \end{pmatrix}. \tag{2.47}$$

These matrices perform a rotation around the z-axis, boost along the x-direction, and boost along the y-direction, respectively. They are generated by

$$J_3 = \frac{1}{2}\sigma_2, \quad K_1 = \frac{i}{2}\sigma_1, \quad \text{and} \quad K_2 = \frac{i}{2}\sigma_2, \tag{2.48}$$

respectively. These generators satisfy the Lie algebra of the $O(2, 1)$ Lorentz group.

While the transformation matrices given in equation (2.47) leave the four-momentum invariant, they can change the spin of the spin-$\frac{1}{2}$ particles. Let us start with a particle with the spin along the z-direction which is the direction of the momentum. Then its spinor is

$$\begin{pmatrix} 1 \\ 0 \end{pmatrix}. \tag{2.49}$$

This spinor remains invariant under the rotation around the z-axis. However, if we apply the second matrix, which corresponds to the boost along the x-direction,

$$\begin{pmatrix} \cosh(\eta/2) & \sinh(\eta/2) \\ \sinh(\eta/2) & \cosh(\eta/2) \end{pmatrix}\begin{pmatrix} 1 \\ 0 \end{pmatrix} = \begin{pmatrix} \cosh(\eta/2) \\ \sinh(\eta/2) \end{pmatrix}. \tag{2.50}$$

This is not a unitary transformation, and it requires further interpretation in quantum mechanics. Indeed, the generator of the boost is not Hermitian, and thus we cannot associate it with observable dynamical variables, unlike the rotation generators corresponding to spins. Additionally, this transformation has not been observed in laboratories.

2.5 Further properties of Wigner's little groups

In sections 2.2, 2.3, and 2.4, we discussed Wigner's little groups whose transformations leave the energy–momentum four-vector for the massive, massless, and imaginary-mass particle invariant. We saw that as the rotation around the z-axis commutes with the boost along the same direction, it is sufficient to consider the transformations in the zx-plane. We, therefore, discussed the Lorentz group applicable to two space and one time dimensions, namely to the (t, x, y) coordinates. However, the rotation around the z-axis will also extend the representation to the full (t, z, x, y) space. With this point in mind, we list in table 2.2 the energy–momentum four-vectors and transformation matrices that leave their respective four-vectors invariant. Since Wigner [21] constructed these four-vectors and four-by-four matrices, we can call them Wigner four-vectors and Wigner transformation matrices. The Lorentz frames where they take these forms can be called Wigner frames.

According to table 2.1, we are encouraged to look for one covariant expression for all three types of particles listed in table 2.2. We shall discuss this problem in section 4.6.

The $E(2)$-like little group for massless particles was formulated by Wigner in 1939 [21], but the physical interpretation was not completely settled until 1990 [17]. This problem had a stormy history of forty-one years, and a comprehensive

Table 2.2. Wigner energy–momentum four-vectors and Wigner transformation matrices applicable to two space-like and one time-like dimensions. The y component remains invariant. Each Wigner energy–momentum four-vector remains invariant under the application of its Wigner transformation matrix.

Mass	Wigner Four-vector	Wigner Transformation Matrix
Massive	$(1, 0, 0, 0)$	$\begin{pmatrix} 1 & 0 & 0 & 0 \\ 0 & \cos\theta & -\sin\theta & 0 \\ 0 & \sin\theta & \cos\theta & 0 \\ 0 & 0 & 0 & 1 \end{pmatrix}$
Massless	$(1, 1, 0, 0)$	$\begin{pmatrix} 1+\gamma^2/2 & -\gamma^2/2 & \gamma & 0 \\ \gamma^2/2 & 1-\gamma^2/2 & \gamma & 0 \\ \gamma & -\gamma & 1 & 0 \\ 0 & 0 & 0 & 1 \end{pmatrix}$
Imaginary-mass	$(0, 1, 0, 0)$	$\begin{pmatrix} \cosh\lambda & 0 & \sinh\lambda & 0 \\ 0 & 1 & 0 & 0 \\ \sinh\lambda & 0 & \cosh\lambda & 0 \\ 0 & 0 & 0 & 1 \end{pmatrix}$

list of earlier papers on this subject was given by Kim and Wigner in their 1987 paper [16]. The debate continues, though, with more recent papers on this subject [4, 19, 20]. To gain insight into the $E(2)$-like little group, we discussed the geometry of the $E(2)$ or Euclidean group in terms of a large sphere with a tangential plane attached to the north pole and studied the geometry of the cylindrical group in terms of a tangential cylinder attached to the equatorial plane of a similarly large sphere.

As for the Lorentz group applicable to the (2 + 1)-dimensional space, physical applications seldom go beyond transformations in the t, x, and y coordinates. This smaller group is called the $O(2, 1)$-like little group. Calculations in high-energy physics involving Lorentz transformations are mostly based on this $O(2, 1)$-like little group [14]. Additionally, this group serves as one of the basic languages in classical and modern optics [2, 15]. We shall discuss these physical applications in later chapters of this book.

2.6 Little groups in the light-cone coordinate system

To gain further insight, we construct the little group for the massive, massless, and imaginary-mass particles in terms of Dirac's light-cone coordinate system [6]. The advantage of this is that the Lorentz boost is diagonal. It is also the natural language for the internal symmetry of the photon.

As we shall see in chapter 3, it was Inönü and Wigner [12] who first introduced the idea that the $E(2)$ group could be constructed as a contraction of the $O(3)$ group by considering a flat plane tangential to the north pole of a very large sphere. We saw in section 2.3.1 that the geometry of the $E(2)$ group could be illustrated just by taking into account such a plane. In section 2.3.2 we discussed the geometry of a cylinder tangential to the equatorial belt of that same sphere. We shall consider that the four-by-four representation of the Lorentz group can produce both the $E(2)$ and the cylindrical symmetries. This aspect is illustrated in (a) of figure 2.4.

We noted in sections 2.2 and 2.3 that the little groups for massive and massless particles are like $O(3)$ and $E(2)$, respectively. Let us start with a massive particle at rest. The little group is generated by J_1, J_2, and J_3, whose four-by-four forms are given in section 1.1.

For the massless particle, the little group is generated by N_1, N_2, and J_3. These generators satisfy the closed set of commutation relations which constitute the Lie algebra of $E(2)$ or the two-dimensional Euclidean group.

In this section, we approach the same problem using Dirac's light-cone coordinate system [6]. Here the variables z_+ and z_- are defined as

$$z_+ = \frac{t + z}{\sqrt{2}} \quad \text{and} \quad z_- = \frac{t - z}{\sqrt{2}}, \tag{2.51}$$

and we have to work with the four-vector (z_+, z_-, x, y). Likewise, for the energy–momentum four-vector, we can use

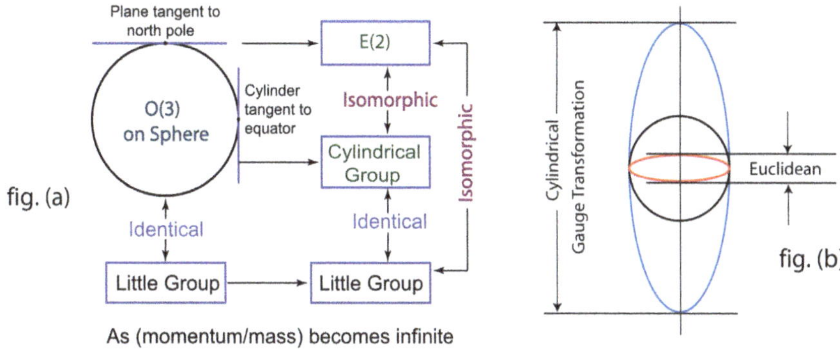

Figure 2.4. Contractions of the three-dimensional rotation group. This group can be illustrated by a sphere. This group can become the two-dimensional Euclidean group on a plane tangential to the north pole, as illustrated in (a). It was later noted that there is a cylinder tangential to the equatorial plane of this sphere, and the up and down translations on this cylinder correspond to the gauge transformation for photons [16]. As illustrated in (b), the four-dimensional representation of the Lorentz group contains both elongation and contraction of the z-axis, as the system is boosted along this direction. The elongation and contraction become the cylindrical and Euclidean groups, respectively [16, 17].

$$p_+ = \frac{p_0 + p_z}{\sqrt{2}} \quad \text{and} \quad p_- = \frac{p_0 - p_z}{\sqrt{2}}. \tag{2.52}$$

Thus the transformation to the light-cone system is the same for (t, z, x, y) and for (p_0, p_z, p_x, p_y). For the momentum, the transformation takes the explicit form:

$$\begin{pmatrix} p_+ \\ p_- \\ p_x \\ p_y \end{pmatrix} = \begin{pmatrix} 1/\sqrt{2} & 1/\sqrt{2} & 0 & 0 \\ 1/\sqrt{2} & -1/\sqrt{2} & 0 & 0 \\ 0 & 0 & 1 & 0 \\ 0 & 0 & 0 & 1 \end{pmatrix} \begin{pmatrix} p_0 \\ p_z \\ p_x \\ p_y \end{pmatrix}. \tag{2.53}$$

The major advantage of the light-cone variables is that the Lorentz boost is diagonal. Under the boost, the z_+ and z_- variables become

$$e^{\eta} z_+ = e^{\eta}\left(\frac{t + z}{\sqrt{2}}\right) \quad \text{and} \quad e^{-\eta} z_- = e^{-\eta}\left(\frac{t - z}{\sqrt{2}}\right), \tag{2.54}$$

while the boost matrix takes the form

$$e^{-i\eta K_3} = \begin{pmatrix} e^{\eta} & 0 & 0 & 0 \\ 0 & e^{-\eta} & 0 & 0 \\ 0 & 0 & 1 & 0 \\ 0 & 0 & 0 & 1 \end{pmatrix}. \tag{2.55}$$

In the light-cone coordinate system, the generators take the form

$$J_3 = \begin{pmatrix} 0 & 0 & 0 & 0 \\ 0 & 0 & 0 & 0 \\ 0 & 0 & 0 & -i \\ 0 & 0 & i & 0 \end{pmatrix}, \qquad K_3 = \begin{pmatrix} i & 0 & 0 & 0 \\ 0 & -i & 0 & 0 \\ 0 & 0 & 0 & 0 \\ 0 & 0 & 0 & 0 \end{pmatrix},$$

$$J_1 = \frac{1}{\sqrt{2}} \begin{pmatrix} 0 & 0 & 0 & i \\ 0 & 0 & 0 & i \\ 0 & 0 & 0 & 0 \\ -i & -i & 0 & 0 \end{pmatrix}, \qquad K_1 = \frac{1}{\sqrt{2}} \begin{pmatrix} 0 & 0 & i & 0 \\ 0 & 0 & -i & 0 \\ i & -i & 0 & 0 \\ 0 & 0 & 0 & 0 \end{pmatrix}, \tag{2.56}$$

$$J_2 = \frac{1}{\sqrt{2}} \begin{pmatrix} 0 & 0 & -i & 0 \\ 0 & 0 & -i & 0 \\ i & i & 0 & 0 \\ 0 & 0 & 0 & 0 \end{pmatrix}, \qquad K_2 = \frac{1}{\sqrt{2}} \begin{pmatrix} 0 & 0 & 0 & i \\ 0 & 0 & 0 & -i \\ 0 & 0 & 0 & 0 \\ i & -i & 0 & 0 \end{pmatrix}.$$

If a massive particle is at rest, its little group is generated by J_1, J_2, in addition to J_3. For a massless particle moving along the z-direction, the little group is generated by N_1, N_2, and J_3, where

$$N_1 = K_1 - J_2 \quad \text{and} \quad N_2 = K_2 + J_1. \tag{2.57}$$

The N_i matrices are then written in the light-cone coordinate system as

$$N_1 = \begin{pmatrix} 0 & 0 & i & 0 \\ 0 & 0 & 0 & 0 \\ 0 & -i & 0 & 0 \\ 0 & 0 & 0 & 0 \end{pmatrix} \quad \text{and} \quad N_2 = \begin{pmatrix} 0 & 0 & 0 & i \\ 0 & 0 & 0 & 0 \\ 0 & 0 & 0 & 0 \\ 0 & -i & 0 & 0 \end{pmatrix}. \tag{2.58}$$

For convenience, we have dropped the normalization factor of $\sqrt{2}$.

These matrices satisfy the same commutation relations as given in equation (2.12), namely:

$$[J_3, N_1] = iN_2, \qquad [J_3, N_2] = -iN_1, \qquad \text{and} \quad [N_1, N_2] = 0. \tag{2.59}$$

The transformation generated by the N_1 and N_2 matrices is

$$D(\gamma, \phi) = \exp\{-i\gamma[(\cos\phi)N_1 + (\sin\phi)N_2]\}. \tag{2.60}$$

Since $N_1^3 = N_2^3 = 0$, the Taylor expansion of this exponential form truncates, and it becomes

$$D(\gamma, \phi) = \begin{pmatrix} 1 & \gamma^2/2 & \gamma\cos\phi & \gamma\sin\phi \\ 0 & 1 & 0 & 0 \\ 0 & -\gamma\cos\phi & 1 & 0 \\ 0 & -\gamma\sin\phi & 0 & 1 \end{pmatrix}. \tag{2.61}$$

In this light-cone coordinate system, the Wigner four-vector for the massive particle is proportional to (1, 1, 0, 0), and, using the transformation matrix from equation (2.53), the Wigner transformation matrix takes the form

$$\begin{pmatrix} (1+\cos\theta)/2 & (1-\cos\theta)/2 & -\sin\theta/\sqrt{2} & 0 \\ (1-\cos\theta)/2 & (1+\cos\theta)/2 & \sin\theta/\sqrt{2} & 0 \\ \sin\theta/\sqrt{2} & -\sin\theta/\sqrt{2} & \cos\theta & 0 \\ 0 & 0 & 0 & 1 \end{pmatrix}. \tag{2.62}$$

For the massless particle, the Wigner four-vector becomes (1, 0, 0, 0) in the light-cone coordinate system. The Wigner transformation matrix becomes

$$\begin{pmatrix} 1 & \gamma^2/2 & \gamma & 0 \\ 0 & 1 & 0 & 0 \\ 0 & \gamma & 1 & 0 \\ 0 & 0 & 0 & 1 \end{pmatrix}. \tag{2.63}$$

There is a scale change of the $\sqrt{2}$ on the γ parameter from the expression given in equation (2.11). This change does not affect our reasoning.

For the imaginary-mass particle, the Wigner four-vector becomes (1, −1, 0, 0) and the Wigner transformation matrix is

$$\begin{pmatrix} (\cosh\lambda+1)/2 & (\cosh\lambda-1)/2 & \sinh\lambda/\sqrt{2} & 0 \\ (\cosh\lambda-1)/2 & (\cosh\lambda+1)/2 & \sinh\lambda/\sqrt{2} & 0 \\ \sinh\lambda/\sqrt{2} & \sinh\lambda/\sqrt{2} & \cosh\lambda & 0 \\ 0 & 0 & 0 & 1 \end{pmatrix}. \tag{2.64}$$

These three Wigner transformation matrices are tabulated in table 2.3.

Table 2.3. Wigner energy–momentum four-vectors and Wigner transformation matrices in the light-cone coordinate system.

Mass	Wigner Four-vector	Wigner Transformation Matrix
Massive	$\begin{pmatrix} 1 \\ 1 \\ 0 \\ 0 \end{pmatrix}$	$\begin{pmatrix} (1+\cos\theta)/2 & (1-\cos\theta)/2 & -\sin\theta/\sqrt{2} & 0 \\ (1-\cos\theta)/2 & (1+\cos\theta)/2 & \sin\theta/\sqrt{2} & 0 \\ \sin\theta/\sqrt{2} & -\sin\theta/\sqrt{2} & \cos\theta & 0 \\ 0 & 0 & 0 & 1 \end{pmatrix}$
Massless	$\begin{pmatrix} 1 \\ 0 \\ 0 \\ 0 \end{pmatrix}$	$\begin{pmatrix} 1 & \gamma^2/2 & \gamma & 0 \\ 0 & 1 & 0 & 0 \\ 0 & \gamma & 1 & 0 \\ 0 & 0 & 0 & 1 \end{pmatrix}$
Imaginary mass	$\begin{pmatrix} 1 \\ -1 \\ 0 \\ 0 \end{pmatrix}$	$\begin{pmatrix} (\cosh\lambda+1)/2 & (\cosh\lambda-1)/2 & \sinh\lambda/\sqrt{2} & 0 \\ (\cosh\lambda-1)/2 & (\cosh\lambda+1)/2 & \sinh\lambda/\sqrt{2} & 0 \\ \sinh\lambda/\sqrt{2} & \sinh\lambda/\sqrt{2} & \cosh\lambda & 0 \\ 0 & 0 & 0 & 1 \end{pmatrix}$

Among the three Wigner transformation matrices given there, the matrix for the massless particle is much simpler than the other two. Thus, we are interested in whether these two complicated matrices can be collapsed into the simple Wigner transformation matrix for the massless particle. We know from physics that the massive particle becomes like the massless particle in the large momentum/small mass limit. Let us then Lorentz-boost those two complicated matrices using the diagonal boost matrix $B(\eta)$ given in equation (2.55).

For the massive particle, the Wigner four-vector becomes $(e^{\eta}, e^{-\eta}, 0, 0)$. It becomes $e^{\eta}(1, 0, 0, 0)$ in the large-η limit. The boosted Wigner transformation matrix becomes

$$\begin{pmatrix} (1+\cos\theta)/2 & e^{2\eta}(1-\cos\theta)/2 & -e^{\eta}\sin\theta/\sqrt{2} & 0 \\ e^{-2\eta}(1-\cos\theta)/2 & (1+\cos\theta)/2 & e^{-\eta}\sin\theta/\sqrt{2} & 0 \\ e^{-\eta}\sin\theta/\sqrt{2} & -e^{\eta}\sin\theta/\sqrt{2} & \cos\theta & 0 \\ 0 & 0 & 0 & 1 \end{pmatrix}. \tag{2.65}$$

If this matrix is to remain finite, the angle θ has to be small, and $e^{\eta}\theta$ can remain finite. If we let

$$e^{\eta}\theta = -\sqrt{2}\gamma, \tag{2.66}$$

this boosted matrix becomes the Wigner transformation matrix of equation (2.63). This limiting process is illustrated by the mass hyperbola in figure 2.5. After this limit, the Wigner vector can come back to the finite-energy state through the light-cone track, also shown in figure 2.5.

For the imaginary-mass particle, the boosted Wigner transformation matrix takes the form

$$\begin{pmatrix} (\cosh\lambda+1)/2 & e^{2\eta}(\cosh\lambda-1)/2 & e^{\eta}\sinh\lambda/\sqrt{2} & 0 \\ e^{-2\eta}(\cosh\lambda-1)/2 & (\cosh\lambda+1)/2 & e^{-\eta}\sinh\lambda/\sqrt{2} & 0 \\ e^{-\eta}\sinh\lambda/\sqrt{2} & e^{\eta}\sinh\lambda/\sqrt{2} & \cosh\lambda & 0 \\ 0 & 0 & 0 & 1 \end{pmatrix}. \tag{2.67}$$

Here again, the parameter λ should become small when η becomes very large. If we let

$$e^{\eta}\lambda = \sqrt{2}\gamma, \tag{2.68}$$

this boosted matrix becomes the Wigner transformation matrix for the massless particle.

Figure 2.6 illustrates the collapse of these two complicated Wigner transformation matrices into one simple matrix for the massless particle. The $O(3)$-like little group for the massive particle is shown as the circle in that figure. The little group for imaginary-mass particles is illustrated as hyperbolic curves. Both the circle and

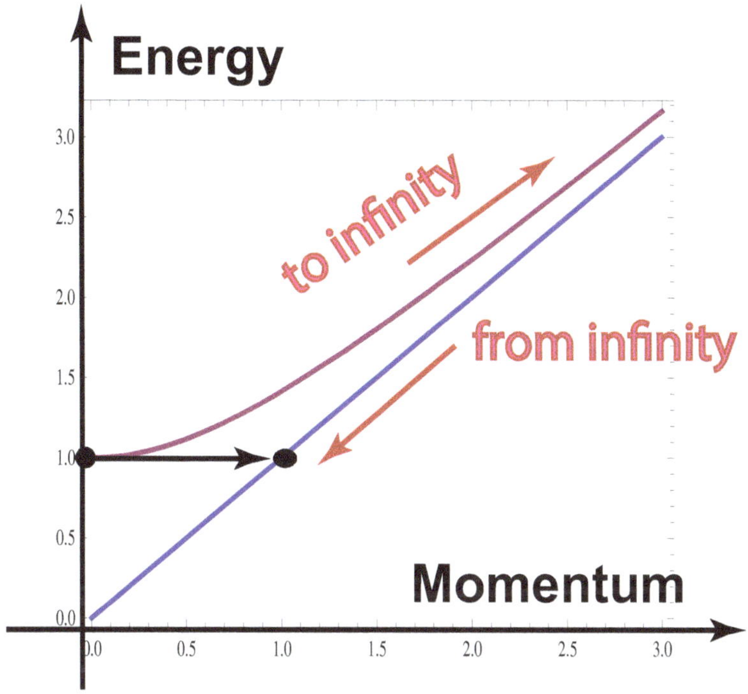

Figure 2.5. Wigner excursion. For the massless particle the Wigner transformation matrix can be obtained by making the exclusions specified here. For the massive particle, the Wigner transformation matrix can be boosted through the hyperbola to the infinite momentum limit. There it coincides with the light cone and can then come back through the light-cone line for the massless particle [3].

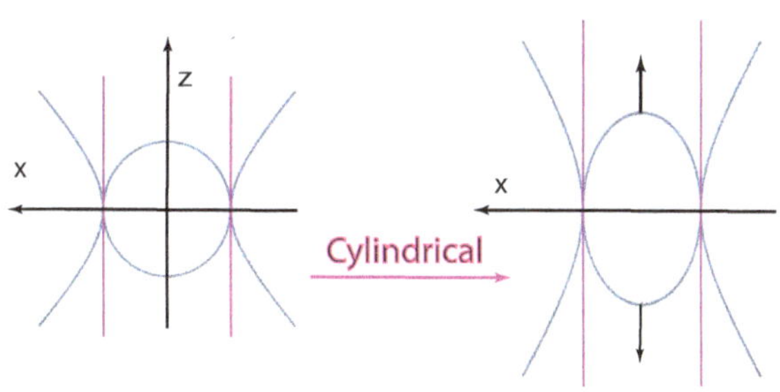

Figure 2.6. Wigner's three little groups as spherical, cylindrical, and hyperbolic surfaces for massive, massless, and imaginary-mass particles, respectively. These three of them become one cylindrical surface in the high-speed limit [3].

hyperbola collapse into two vertical lines corresponding to the cylindrical symmetry of the $E(2)$-like little group for the massless particle.

Since the Wigner transformation matrix for the massless particle is much simpler than the other two, we can go back to equation (2.63) and rotate it around the z-axis. Then the matrix becomes

$$D(\gamma, \phi) = \begin{pmatrix} 1 & \gamma^2/2 & \gamma\cos\theta & \gamma\sin\theta \\ 0 & 1 & 0 & 0 \\ 0 & \gamma\cos\theta & 1 & 0 \\ 0 & \gamma\sin\theta & 0 & 1 \end{pmatrix}. \tag{2.69}$$

This matrix is the $D(\gamma, \phi)$ matrix given in equation (2.11) written in the light-cone coordinate system. There is a scale change $\sqrt{2}$ for γ. The D matrix generates gauge transformations when applied to the photon four-vector.

In the light-cone coordinate system, the four-potential of a photon with momentum along the z-axis takes the form

$$(A_+, A_-, A_x, A_y) \tag{2.70}$$

with

$$A_+ = \frac{A_0 + A_z}{\sqrt{2}} \quad \text{and} \quad A_- = \frac{A_0 - A_z}{\sqrt{2}}. \tag{2.71}$$

If boosted along the hyperbolic route in figure 2.5, this four-vector becomes

$$(e^{\eta}A_+, 0, A_x, A_y) \tag{2.72}$$

for large values of η.

If the system is boosted back along the light-cone, the four-vector becomes

$$(A_+, 0, A_x, A_y). \tag{2.73}$$

During this process, A_- vanishes and leads to $A_0 = A_z$. The excursion given in figure 2.5 is equivalent to imposing the Lorentz boost. As for the Wigner transformation matrix, the η dependence is gone, and it remains invariant during this return trip.

When the D matrix is applied to this photon four-vector,

$$\begin{pmatrix} 1 & \gamma^2/2 & \gamma\cos\phi & \gamma\sin\phi \\ 0 & 1 & 0 & 0 \\ 0 & \gamma\cos\phi & 1 & 0 \\ 0 & \gamma\sin\phi & 0 & 1 \end{pmatrix} \begin{pmatrix} A_+ \\ 0 \\ A_x \\ A_y \end{pmatrix} = \begin{pmatrix} A_+ + \gamma(A_x\cos\phi + A_y\sin\phi) \\ 0 \\ A_x \\ A_y \end{pmatrix}, \tag{2.74}$$

the transverse components A_x and A_y are not affected. The transformation changes only the A_0 and A_z components and thus, is a gauge transformation.

This cylindrical transformation has been discussed in section 2.3. The light-cone coordinate system gives a more transparent mathematics. Indeed, this coordinate system is the natural language for the internal symmetry of the photon.

As another illustration, we use the energy–momentum four-vector for a massive particle

$$p = (p_+, p_-, p_x, p_y). \tag{2.75}$$

If the particle is at rest, this should represent the three-dimensional symmetry, with three independent coordinate variables. This means $p_+ = p_-$ in the light-cone coordinate system. If we boost the system, this four-vector becomes

$$(e^{\eta}p_+, e^{-\eta}p_-, p_x, p_y). \tag{2.76}$$

For large values of η, the $e^{-\eta}p_-$ component vanishes. In the language of photons, it is called the Lorentz condition or $A_0 = A_z$. The interesting problem is what happens between those two limits.

To study this problem, we can start with two spheres, both with a unit horizontal radius. One of them has a vertical radius e^{η} and the other, $e^{-\eta}$. Since

$$e^{\eta} = \left[\frac{\sqrt{p^2 + m^2} + p}{\sqrt{p^2 + m^2} - p}\right]^{1/2}, \tag{2.77}$$

this expression is one when the particle is at rest with $p = 0$. In the limit of large values of p, this number is infinite, while $e^{-\eta}$ becomes zero.

As p increases, both go through elliptic deformations, as illustrated in (b) of figure 2.4. In the large-η limit, one of them becomes a cylinder, and the other a flat surface. This picture is equivalent to the tangential plane and cylinder described in (a) of figure 2.4.

References

[1] Barci D G, Bollini C G, and Rocca M 1995 Quantization of a six-dimensional Wess–Zumino model *Il Nuovo Cimento* A **108** 797–808

[2] Başkal S and Kim Y S 2013 Lorentz group in ray and polarization optics *Mathematical Optics: Classical, Quantum and Computational Methods* ed V Lakshminarayanan, M L Calvo, and T Alieva (Boca Raton, FL: Taylor and Francis) pp 303–49 (ISBN 978-1-4398-6961-1)

[3] Başkal S, Kim Y S, and Noz M E 2019 *Mathematical Devices for Optical Sciences* (Bristol, UK: IOP Publishing) (ISBN 978-0-7503-1612-5, OCLC: 1034620988)

[4] Caban P and Rembieliński J 2003 Photon polarization and Wigner's little group *Phys. Rev.* A **68**

[5] Cherenkov P A 1934 Visible emission of clean liquids by action of γ radiation *Dokl. Akad. Nauk SSSR* **2** 451–5

[6] Dirac P A M 1949 Forms of relativistic dynamics *Rev. Mod. Phys.* **21** 392–9

[7] Feinberg G 1967 Possibility of faster-than-light particles *Phys. Rev.* **159** 1089–105

[8] Goldstein H 1980 *Classical Mechanics* (Addison-Wesley Series in Physics) 2nd edn (Reading, MA: Addison-Wesley Pub. Co.) (ISBN 978-0-201-02918-5)

[9] Han D, Kim Y S, and Son D 1982 E(2)-like little group for massless particles and neutrino polarization as a consequence of gauge invariance *Phys. Rev.* D **26** 3717–25

[10] Han D, Kim Y S, and Son D 1986 Eulerian parametrization of Wigner's little groups and gauge transformations in terms of rotations in two-component spinors *J. Math. Phys.* **27** 2228–35

[11] Headrick M, Minwalla S, and Takayanagi T 2004 Closed string tachyon condensation: an overview *Classical Quant. Grav.* **21** S1539–65

[12] Inönü E and Wigner E P 1953 On the contraction of groups and their representations *Proc. Natl. Acad. Sci.* 39 510–24

[13] Jackson J D 1999 *Classical Electrodynamics* 3rd edn (New York: Wiley) (ISBN 978-0-471-30932-1)

[14] Kim Y S and Noz M E 1986 *Theory and Applications of the Poincaré Group* (Dordrecht: Springer) (ISBN 978-94-010-8526-7 978-94-009-4558-6)

[15] Kim Y S and Noz M E 1991 *Phase Space Picture of Quantum Mechanics: Group Theoretical Approach* Number 40 in Lecture notes in physics series (Singapore; Hackensack, NJ: World Scientific Publishing Co.) (ISBN 978-981-02-0360-3 978-981-02-0361-0)

[16] Kim Y S and Wigner E P 1987 Cylindrical group and massless particles *J. Math. Phys.* **28** 1175–9

[17] Kim Y S and Wigner E P 1990 Space–time geometry of relativistic particles *J. Math. Phys.* **31** 55–60

[18] Kim Y S and Yeh L 1992 *O* (2,1)-like little group for spacelike four-momenta and localized light waves in continuous media *J. Math. Phys.* **33** 3190–6

[19] Lindner N H, Peres A, and Terno D R 2003 Wigner's little group and Berry's phase for massless particles *J. Phys. A: Math. General* **36** L449–54

[20] Scaria T and Chakraborty B 2002 Wigner's little group as a gauge generator in linearized gravity theories *Classical Quant. Grav.* **19** 4445–62

[21] Wigner E P 1939 On unitary representations of the inhomogeneous Lorentz group *Ann. Math.* **40** 149–204

IOP Publishing

Physics of the Lorentz Group (Second Edition)
Beyond high-energy physics and optics
Sibel Başkal, Young S Kim, and Marilyn E Noz

Chapter 3

Group contractions

In studying the Lorentz group, we often face transforming a given group into a simpler group. The surface of the earth is a case in point. Since its radius is so large, we feel that the surface is flat when we stay in a city. The group theory of spheres is that of the rotation group, and the two-dimensional Euclidean group applies to the flat surface. The technique of transforming one group to another is called group contraction.

The basic method for contracting groups was introduced to physics first by Inönü and Wigner [1] as a limiting process. In this chapter, we elucidate group contraction using squeeze transformations, which are relatively new in the physics literature. It is more rewarding to formulate the limiting process involved in group contraction in terms of this new mathematical technique. For example, we discuss the squeezed states of light in chapter 9.

Section 3.1, illustrates how a function on the xy-plane can be transformed into a straight line through squeeze transformations. In section 3.2, we consider a plane and a cylinder tangent to a sphere and formulate the contraction of the rotation group to the two-dimensional Euclidean group and to the cylindrical group. In section 3.3, we study the contraction of the three-parameter Lorentz group applicable to two space-like and one time-like dimensions. We note the contraction of this group results in the two-dimensional Euclidean group and the cylindrical group, as in the case of the rotation group.

In section 3.4, the contraction of the Lorentz group is discussed. The pressing issue is whether this group can be contracted into the Galilean group. While tangent planes are useful to curved surfaces, a tangent sphere could also be useful. Thus, in section 3.5, we examine how the tangent sphere could be constructed.

3.1 Contraction with squeeze transformations

Let us consider the xy-plane. We are familiar with rotations and translations on the plane. On the other hand, the concept of squeeze is relatively new. The point is that

doi:10.1088/978-0-7503-3607-9ch3

we can expand the x-axis while contracting the y-axis so that the product xy remains constant. Thus the squeeze is an area-preserving transformation.

The squeeze transformation matrix takes the same form as given in table 1.1:

$$B(\eta) = \begin{pmatrix} \exp(\eta/2) & 0 \\ 0 & \exp(-\eta/2) \end{pmatrix}. \tag{3.1}$$

The inverse of this squeeze matrix is given by

$$B(\eta)^{-1} = \begin{pmatrix} \exp(-\eta/2) & 0 \\ 0 & \exp(\eta/2) \end{pmatrix}, \tag{3.2}$$

with $B^{-1}B = I$ where I is, of course, the unit matrix.

If we apply the squeeze matrix to the $x\ y$ column matrix, the transformation takes the form:

$$\begin{pmatrix} \exp(\eta/2) & 0 \\ 0 & \exp(-\eta/2) \end{pmatrix} \begin{pmatrix} x \\ y \end{pmatrix} = \begin{pmatrix} [\exp(\eta/2)]\, x \\ [\exp(-\eta)]\, y \end{pmatrix}. \tag{3.3}$$

However, in the limit of large η, this matrix may make the lower element very small, and we have the *freedom to change or not to change* this small quantity by another small quantity. If we change, the column vector could be different from the original form. We call this changed vector the *contracted vector*.

Let us consider $y = f(x)$, and the column matrix

$$\begin{pmatrix} x \\ f(x) \end{pmatrix}. \tag{3.4}$$

If we apply the squeeze matrix $B(\eta)$ to this column vector, we obtain:

$$\begin{pmatrix} \exp(\eta/2) & 0 \\ 0 & \exp(-\eta/2) \end{pmatrix} \begin{pmatrix} x \\ f(x) \end{pmatrix} = \begin{pmatrix} [\exp(\eta/2)]\, x \\ [\exp(-\eta/2)] f(x) \end{pmatrix}. \tag{3.5}$$

In the limit of large η, the lower element of the resulting column matrix becomes zero. Thus, we can replace $f(x)$ with any convenient quantity we choose. We can choose this quantity to be one. Thus, the column matrix takes the form

$$\begin{pmatrix} \exp(\eta/2) & 0 \\ 0 & \exp(-\eta/2) \end{pmatrix} \begin{pmatrix} x \\ 1 \end{pmatrix} = \begin{pmatrix} x\exp(\eta/2) \\ \exp(-\eta/2) \end{pmatrix}. \tag{3.6}$$

Let us then apply the inverse of the squeeze matrix to this column matrix. Then

$$\begin{pmatrix} \exp(-\eta/2) & 0 \\ 0 & \exp(\eta/2) \end{pmatrix} \begin{pmatrix} x\exp(\eta/2) \\ \exp(-\eta/2) \end{pmatrix}. \tag{3.7}$$

The resulting column matrix is

$$\begin{pmatrix} x \\ 1 \end{pmatrix}. \tag{3.8}$$

For example, if $f(x)$ is $\sin(x)$, then $y = \sin(x)$. The contraction procedure above is illustrated in figure 3.1. The expression given in equation (3.8) is the contracted x y column vector.

Let us next squeeze the transformation applied to the rotation matrix. The rotation in this two-dimensional space is generated by σ_2, as we saw in equation (1.16). The rotation matrix is

$$\exp(-i\alpha\,\sigma_2) = \begin{pmatrix} \cos\alpha & -\sin\alpha \\ \sin\alpha & \cos\alpha \end{pmatrix}. \tag{3.9}$$

If we apply the squeeze matrix $B(\eta)$ of equation (3.2) to σ_2,

$$B\sigma_2 B^{-1} = \begin{pmatrix} 0 & -i\exp(\eta) \\ i\exp(-\eta) & 0 \end{pmatrix}. \tag{3.10}$$

In the limit of large η, this expression can be written as

$$\begin{pmatrix} 0 & -i\exp(\eta) \\ 0 & 0 \end{pmatrix}. \tag{3.11}$$

If we apply the inverse of the squeeze transformation, the result is

$$\begin{pmatrix} 0 & -i \\ 0 & 0 \end{pmatrix}, \tag{3.12}$$

and the resulting transformation matrix is

$$\exp\left\{-i\alpha\begin{pmatrix} 0 & -i \\ 0 & 0 \end{pmatrix}\right\} = \begin{pmatrix} 1 & -\alpha \\ 0 & 1 \end{pmatrix}. \tag{3.13}$$

This matrix is the contracted rotation matrix.

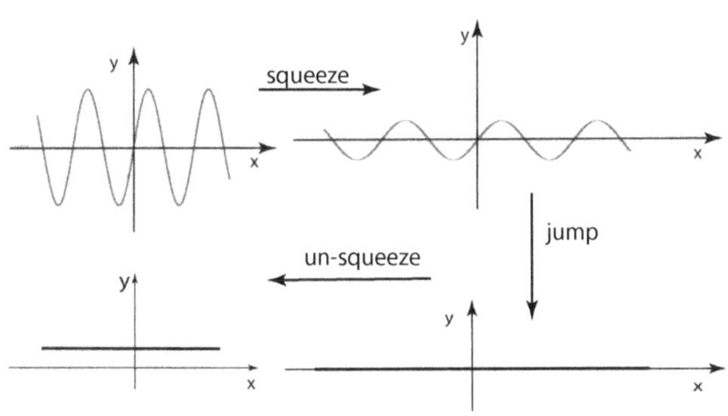

Figure 3.1. Contraction of the *sine* curve. First, the left side is expanded along the x-direction while it is contracted along the y-direction until it becomes a horizontal line along the x-axis. This horizontal line is un-contracted until it becomes a horizontal line at a chosen value of y.

If we apply this contracted matrix to the $x\ y$ column vector in equation (3.3),

$$\begin{pmatrix} 1 & -\alpha \\ 0 & 1 \end{pmatrix}\begin{pmatrix} x \\ y \end{pmatrix} = \begin{pmatrix} x - \alpha y \\ y \end{pmatrix}. \tag{3.14}$$

Here, the y variable remains invariant, while the translation αy along the x-direction depends on the y variable.

We also have freedom to choose the contracted column vector to be the same form as equation (3.8). The result will be

$$\begin{pmatrix} 1 & -\alpha \\ 0 & 1 \end{pmatrix}\begin{pmatrix} x \\ 1 \end{pmatrix} = \begin{pmatrix} x - \alpha \\ 1 \end{pmatrix}. \tag{3.15}$$

Here, the y variable remains invariant and is 1, but the x variable goes through a transition solely depending on the α parameter.

3.2 Contractions of the $O(3)$ rotation group

The three-dimensional rotation group is very familiar to us. We can start with a three-dimensional sphere. The rotation group on this spherical surface is well known. In this section, we are interested in a plane tangent at the north pole and a cylindrical surface touching the equatorial belt, as shown in figure 3.2. We propose to obtain the transformation group on these tangential surfaces using the squeeze procedure introduced in section 3.1.

The rotation group is generated by J_1, J_2, and J_3, and they satisfy the Lie algebra defined in equation (1.5):

$$[J_i, J_j] = i\epsilon_{ijk}J_k. \tag{3.16}$$

In section 1.1 we defined the rotation generators in the four dimensional Minkowski space of $(t,\ z,\ x,\ y)$. In this section we are not interested the full four-dimensional

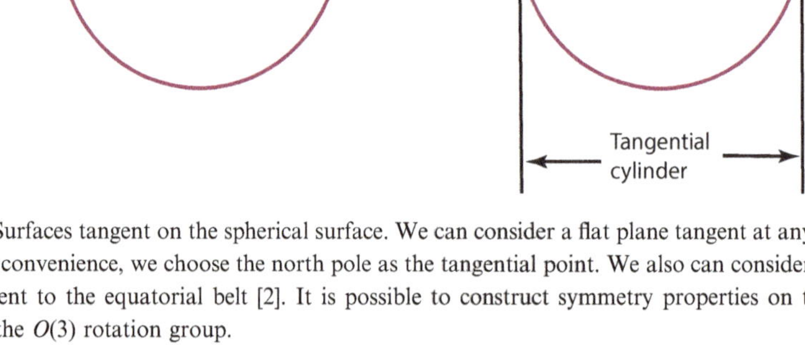

Figure 3.2. Surfaces tangent on the spherical surface. We can consider a flat plane tangent at any point on the surface. For convenience, we choose the north pole as the tangential point. We also can consider a cylindrical surface tangent to the equatorial belt [2]. It is possible to construct symmetry properties on these surfaces contracting the $O(3)$ rotation group.

space, so we shall eliminate the first row and column from equations (1.3) and (1.4). Consequently, the generators can take the more familiar form of three-by-three matrices applicable to the three-dimensional (z, x, y) space:

$$J_3 = \begin{pmatrix} 0 & 0 & 0 \\ 0 & 0 & -i \\ 0 & i & 0 \end{pmatrix}, \quad J_1 = \begin{pmatrix} 0 & 0 & i \\ 0 & 0 & 0 \\ -i & 0 & 0 \end{pmatrix}, \quad \text{and} \quad J_2 = \begin{pmatrix} 0 & -i & 0 \\ i & 0 & 0 \\ 0 & 0 & 0 \end{pmatrix}. \tag{3.17}$$

In terms of the differential operators, the rotation generators can be written as

$$J_3 = -i\left(x\frac{\partial}{\partial y} + y\frac{\partial}{\partial x}\right), \quad J_1 = -i\left(y\frac{\partial}{\partial z} + z\frac{\partial}{\partial y}\right), \quad \text{and} \quad J_2 = -i\left(z\frac{\partial}{\partial x} + x\frac{\partial}{\partial z}\right). \tag{3.18}$$

These generators satisfy the commutation relations given in (3.16).

Let us go to figure 3.3 and consider the spherical surface of the earth. We can consider first a flat plane tangent at the north pole. We can also consider a cylinder tangent at the equatorial belt. It is possible to achieve this purpose by considering the squeeze process which will shorten the z-axis, while expanding the xy-plane. This process will lead to the plane tangent at the north pole. It is also possible to squeeze in the opposite direction, namely to expand the z-axis while shrinking the xy-plane. This will lead to the single line along the z axis.

On the plane tangent at the north pole, transformations are generated by the rotation around the z-direction and translations along the x- and y-directions. The rotation generator, J_3, is given in equation (3.17). In equation (2.25) a set of translation generators were defined which represented translations in the x- and y-axis as:

$$P_1 = -i\frac{\partial}{\partial x} \quad \text{and} \quad P_2 = -i\frac{\partial}{\partial y}. \tag{3.19}$$

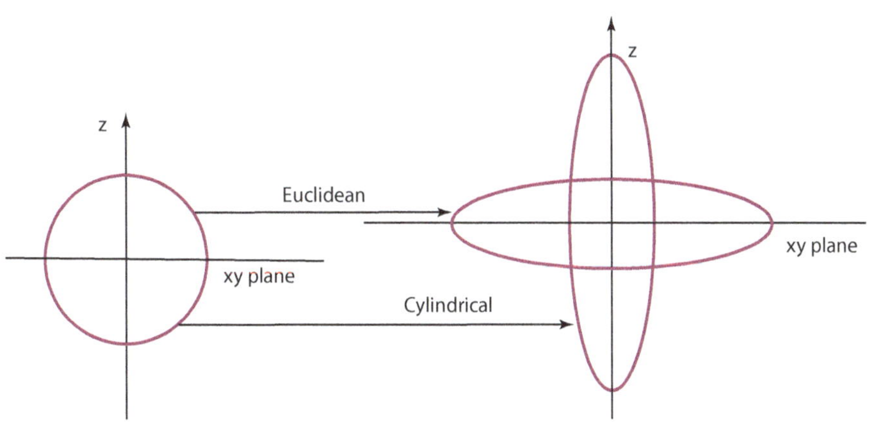

Figure 3.3. Squeezing the sphere. It is possible to obtain the flat plane by the squeezing process. By contracting the z-axis while expanding the xy-plane we obtain the tangential plane at the pole. It is also possible to obtain the tangential cylindrical surface by expanding the z-axis while contracting the radius of the xy-plane.

Together with J_3, they satisfy the following closed set of commutation relations

$$[P_1, P_2] = 0, \quad [P_1, J_3] = -iP_2, \quad \text{and} \quad [P_2, J_3] = iP_1. \tag{3.20}$$

These commutation relations form the Lie algebra for the two-dimensional Euclidean group, $E(2)$.

In the language of three-by-three matrices, we can use the squeeze matrix, similar to the one in section 3.1, as

$$B(\eta) = \begin{pmatrix} \exp(-\eta/2) & 0 & 0 \\ 0 & \exp(\eta/2) & 0 \\ 0 & 0 & \exp(\eta/2) \end{pmatrix}. \tag{3.21}$$

If we use the same technique as in equations (3.10), (3.11), and (3.12), J_3 is not affected by this transformation, but $J_2 \to P_1$ and $J_1 \to -P_2$. Thus these generators take the form

$$J_3 = \begin{pmatrix} 0 & 0 & 0 \\ 0 & 0 & -i \\ 0 & i & 0 \end{pmatrix}, \quad P_1 = \begin{pmatrix} 0 & 0 & 0 \\ i & 0 & 0 \\ 0 & 0 & 0 \end{pmatrix}, \quad \text{and} \quad P_2 = \begin{pmatrix} 0 & 0 & 0 \\ 0 & 0 & 0 \\ i & 0 & 0 \end{pmatrix}. \tag{3.22}$$

The transformation matrix which performs translations after rotation takes the form

$$\exp\left(-iaP_x - ibP_y\right)\exp(-i\phi J_3) = \begin{pmatrix} 1 & 0 & 0 \\ a & \cos\phi & -\sin\phi \\ b & \sin\phi & \cos\phi \end{pmatrix}. \tag{3.23}$$

If this matrix is applied to the contracted vector $(1,\ x,\ y)$,

$$\begin{pmatrix} 1 & 0 & 0 \\ a & \cos\phi & -\sin\phi \\ b & \sin\phi & \cos\phi \end{pmatrix}\begin{pmatrix} 1 \\ x \\ y \end{pmatrix} = \begin{pmatrix} 1 \\ a + x\cos\phi - y\sin\phi \\ b + x\sin\phi + y\cos\phi \end{pmatrix}. \tag{3.24}$$

We can take another limiting process. Consider a ring around the equator of this sphere, as seen in figure 3.3. In this case, z is much larger than the radius $R = \sqrt{x^2 + y^2}$. The rotation generator J_3 remains invariant, but as we did in equation (2.14) we could take

$$Z_1 = \frac{1}{z}J_2 \quad \to -ix\frac{\partial}{\partial z} \quad \text{and} \quad Z_2 = \frac{1}{z}J_1 \quad \to -iy\frac{\partial}{\partial z}. \tag{3.25}$$

These operators also satisfy a closed set of commutation relations similar to those given equation (3.20):

$$[Z_1, Z_2] = 0, \quad [Z_1, J_3] = -iZ_2, \quad \text{and} \quad [Z_2, J_3] = iZ_1. \tag{3.26}$$

These commutation relations form the Lie algebra for the cylindrical group. To obtain the explicit form for Z_1 and Z_2, we use the inverse of the procedure above as

$$\begin{aligned} B(\eta)^{-1}J_1B(\eta) &\to J_1' \to B(\eta)J_1'B(\eta)^{-1} \to Z_2, \\ B(\eta)^{-1}J_2B(\eta) &\to J_2' \to B(\eta)J_2'B(\eta)^{-1} \to -Z_1. \end{aligned} \tag{3.27}$$

Thus we have

$$J_3 = \begin{pmatrix} 0 & 0 & 0 \\ 0 & 0 & -i \\ 0 & i & 0 \end{pmatrix}, \quad Z_1 = \begin{pmatrix} 0 & i & 0 \\ 0 & 0 & 0 \\ 0 & 0 & 0 \end{pmatrix}, \quad \text{and} \quad Z_2 = \begin{pmatrix} 0 & 0 & i \\ 0 & 0 & 0 \\ 0 & 0 & 0 \end{pmatrix}, \tag{3.28}$$

where again J_3 remains the same, but $-J_2 \to Z_1$ and $J_1 \to Z_2$. This algebra is the same as that for the two-dimensional Euclidean group given in equation (3.20). Thus, the cylindrical group is locally isomorphic to the Euclidean group. Transformations in these two groups are illustrated in figure 3.4.

3.3 Contraction of the $O(2, 1)$ Lorentz group

This group consists of Lorentz boosts along the x- and y-directions and the rotation around the z-direction. The transformations of this group apply to the column matrix of (t, z, x, y). Here we use the transformation matrices given in table 1.1. The rotation around z-axis, therefore, takes the form:

$$\begin{pmatrix} 1 & 0 & 0 & 0 \\ 0 & 1 & 0 & 0 \\ 0 & 0 & \cos\phi & -\sin\phi \\ 0 & 0 & \sin\phi & \cos\phi \end{pmatrix} \begin{pmatrix} t \\ z \\ x \\ y \end{pmatrix} = \begin{pmatrix} t \\ z \\ x\cos\phi - y\sin\phi \\ x\sin\phi + y\sin\phi \end{pmatrix}. \tag{3.29}$$

The Lorentz boost along the x-direction is:

$$\begin{pmatrix} \cosh\lambda & 0 & \sinh\lambda & 0 \\ 0 & 1 & 0 & 0 \\ \sinh\lambda & 0 & \cosh\lambda & 0 \\ 0 & 0 & 0 & 1 \end{pmatrix} \begin{pmatrix} t \\ z \\ x \\ y \end{pmatrix} = \begin{pmatrix} t\cosh\lambda + x\sinh\lambda \\ z \\ t\sinh\lambda + x\cosh\lambda \\ y \end{pmatrix}. \tag{3.30}$$

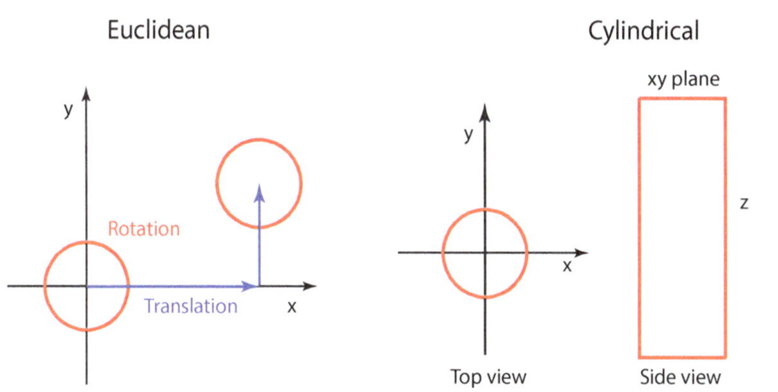

Figure 3.4. Euclidean and cylindrical groups. In the Euclidean group, we can make rotations around the z-axis and translations along the x- and y-axis. For the cylindrical group, we can make translations up and down on the z-axis and rotations in the xy-plane.

The generators of this group should consist of

$$J_3 = \begin{pmatrix} 0 & 0 & 0 & 0 \\ 0 & 0 & 0 & 0 \\ 0 & 0 & 0 & -i \\ 0 & 0 & i & 0 \end{pmatrix}, \quad K_1 = \begin{pmatrix} 0 & 0 & i & 0 \\ 0 & 0 & 0 & 0 \\ i & 0 & 0 & 0 \\ 0 & 0 & 0 & 0 \end{pmatrix}, \quad \text{and} \quad K_2 = \begin{pmatrix} 0 & 0 & 0 & i \\ 0 & 0 & 0 & 0 \\ 0 & 0 & 0 & 0 \\ i & 0 & 0 & 0 \end{pmatrix}, \tag{3.31}$$

where J_3 generates rotations around the z-axis, while K_1 and K_2 are the generators of the Lorentz boosts along the x- and y-directions, respectively.

For this four-dimensional problem, we can use the squeeze or contraction matrix:

$$C(\epsilon) = \begin{pmatrix} \epsilon & 0 & 0 & 0 \\ 0 & 1 & 0 & 0 \\ 0 & 0 & 1/\epsilon & 0 \\ 0 & 0 & 0 & 1/\epsilon \end{pmatrix} \quad \text{and} \quad C^{-1}(\epsilon) = \begin{pmatrix} 1/\epsilon & 0 & 0 & 0 \\ 0 & 1 & 0 & 0 \\ 0 & 0 & \epsilon & 0 \\ 0 & 0 & 0 & \epsilon \end{pmatrix}, \tag{3.32}$$

where ϵ can become very small as $e^{-\eta/2}$ became very small as η became large. Here, the z-coordinate does not participate in the transformations. With this matrix, we can contract the (t, z, x, y) column matrix as

$$\begin{pmatrix} t \\ z \\ x \\ y \end{pmatrix} \to \begin{pmatrix} 1 \\ 1 \\ x \\ y \end{pmatrix}. \tag{3.33}$$

It is now clear that the Euclidean contraction procedure will leave J_3 invariant, and

$$K_1 \to P_1 = \begin{pmatrix} 0 & 0 & 0 & 0 \\ 0 & 0 & 0 & 0 \\ i & 0 & 0 & 0 \\ 0 & 0 & 0 & 0 \end{pmatrix} \quad \text{and} \quad K_2 \to P_2 = \begin{pmatrix} 0 & 0 & 0 & 0 \\ 0 & 0 & 0 & 0 \\ 0 & 0 & 0 & 0 \\ i & 0 & 0 & 0 \end{pmatrix}. \tag{3.34}$$

This set of generators has the same form as that given in equation (3.22) for the contraction of the $O(3)$ rotation group to $E(2)$.

As for the cylindrical contraction,

$$K_1 \to Z_2 = \begin{pmatrix} 0 & 0 & 0 & i \\ 0 & 0 & 0 & 0 \\ 0 & 0 & 0 & 0 \\ 0 & 0 & 0 & 0 \end{pmatrix} \quad \text{and} \quad K_2 \to Z_1 = \begin{pmatrix} 0 & 0 & i & 0 \\ 0 & 0 & 0 & 0 \\ 0 & 0 & 0 & 0 \\ 0 & 0 & 0 & 0 \end{pmatrix}. \tag{3.35}$$

Again, this set of generators has the same form as that given in section 3.2 for the contraction of the $O(3)$ rotation group to the cylindrical group.

The tangential continuity is seldom mentioned in the physics literature, but the concept is straightforward. Let us look at the two curves: $y = x^2$ and $y = -x^2$, as shown in figure 3.5. It is possible to move continuously from one curve to the other at this point $x = y = 0$. Likewise, both the group $O(3)$ and the $O(2, 1)$ group can be contracted to the $E(2)$ Euclidean group or the cylindrical group. They are tangent to the $E(2)$ and cylindrical groups. Therefore, the $O(3)$ and $O(2, 1)$ groups are tangentially continuous to each other at the plane and at the cylinder.

3.4 Contraction of the Lorentz group

As we saw in chapter 1, the Lorentz group, the group of Lorentz transformations, is generated by four-by-four matrices applicable to the column vector of (t, z, x, y). In this space, the three rotation generators were given in equations (1.3) and (1.4) as:

$$J_1 = \begin{pmatrix} 0 & 0 & 0 & 0 \\ 0 & 0 & 0 & i \\ 0 & 0 & 0 & 0 \\ 0 & -i & 0 & 0 \end{pmatrix}, \quad J_2 = \begin{pmatrix} 0 & 0 & 0 & 0 \\ 0 & 0 & -i & 0 \\ 0 & i & 0 & 0 \\ 0 & 0 & 0 & 0 \end{pmatrix}, \quad \text{and} \quad J_3 = \begin{pmatrix} 0 & 0 & 0 & 0 \\ 0 & 0 & 0 & 0 \\ 0 & 0 & 0 & -i \\ 0 & 0 & i & 0 \end{pmatrix}. \tag{3.36}$$

The three boost generators were given in equations (1.8) and (1.9) as:

$$K_1 = \begin{pmatrix} 0 & 0 & i & 0 \\ 0 & 0 & 0 & 0 \\ i & 0 & 0 & 0 \\ 0 & 0 & 0 & 0 \end{pmatrix}, \quad K_2 = \begin{pmatrix} 0 & 0 & 0 & i \\ 0 & 0 & 0 & 0 \\ 0 & 0 & 0 & 0 \\ i & 0 & 0 & 0 \end{pmatrix}, \quad \text{and} \quad K_3 = \begin{pmatrix} 0 & i & 0 & 0 \\ i & 0 & 0 & 0 \\ 0 & 0 & 0 & 0 \\ 0 & 0 & 0 & 0 \end{pmatrix}. \tag{3.37}$$

We can now perform the contraction procedures using the four-by-four squeeze matrix of the form

$$B(\eta) = \begin{pmatrix} \exp(-\eta/2) & 0 & 0 & 0 \\ 0 & \exp(\eta/2) & 0 & 0 \\ 0 & 0 & \exp(\eta/2) & 0 \\ 0 & 0 & 0 & \exp(\eta/2) \end{pmatrix}, \tag{3.38}$$

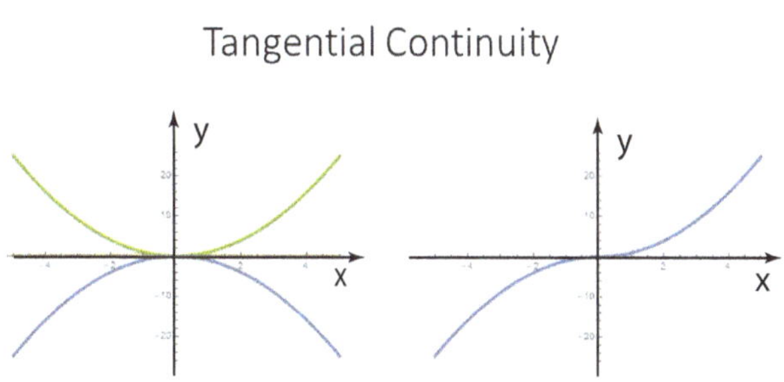

Figure 3.5. Tangential continuity of $y = -x^2$ to $y = x^2$. These two curves share the same tangential line at $x = y = 0$. Thus, it is possible to move continuously from one curve to the other at this point.

During this process, the rotation generators J_i remain unchanged, but the boost generators K_3, K_1, and K_2, become

$$K_3 \to P_3 = \begin{pmatrix} 0 & 0 & 0 & 0 \\ i & 0 & 0 & 0 \\ 0 & 0 & 0 & 0 \\ 0 & 0 & 0 & 0 \end{pmatrix}, \qquad K_1 \to P_1 = \begin{pmatrix} 0 & 0 & 0 & 0 \\ 0 & 0 & 0 & 0 \\ i & 0 & 0 & 0 \\ 0 & 0 & 0 & 0 \end{pmatrix},$$
$$K_2 \to P_2 = \begin{pmatrix} 0 & 0 & 0 & 0 \\ 0 & 0 & 0 & 0 \\ 0 & 0 & 0 & 0 \\ i & 0 & 0 & 0 \end{pmatrix}, \tag{3.39}$$

respectively. These matrices lead to the transformation of the form

$$\exp(-i[v_1P_1 + v_2P_2 + v_3P_3]) = \begin{pmatrix} 1 & 0 & 0 & 0 \\ v_3 & 1 & 0 & 0 \\ v_1 & 0 & 1 & 0 \\ v_2 & 0 & 0 & 1 \end{pmatrix}. \tag{3.40}$$

If this matrix is applied to the (t, z, x, y) column vector,

$$\begin{pmatrix} 1 & 0 & 0 & 0 \\ v_3 & 1 & 0 & 0 \\ v_1 & 0 & 1 & 0 \\ v_2 & 0 & 0 & 1 \end{pmatrix} \begin{pmatrix} t \\ z \\ x \\ y \end{pmatrix} = \begin{pmatrix} t \\ z + v_3t \\ x + v_1t \\ y + v_2t \end{pmatrix}. \tag{3.41}$$

The rotation matrix applicable to this four-dimensional space is

$$\begin{pmatrix} 1 & 0 & 0 & 0 \\ 0 & & \text{3-by-3} & \\ 0 & & \text{rotation} & \\ 0 & & \text{matrix} & \end{pmatrix}. \tag{3.42}$$

If this matrix is applied to the four-vector,

$$\begin{pmatrix} 1 & 0 & 0 & 0 \\ 0 & & \text{3-by-3} & \\ 0 & & \text{rotation} & \\ 0 & & \text{matrix} & \end{pmatrix} \begin{pmatrix} t \\ z \\ x \\ y \end{pmatrix} = \begin{pmatrix} t \\ z' \\ x' \\ y' \end{pmatrix}, \tag{3.43}$$

where z', x', and y' are the results of the three-dimensional rotation, and their forms are well known.

Thus the rotation followed by the Galilean boost matrix is

$$\begin{pmatrix} 1 & 0 & 0 & 0 \\ v_3 & 1 & 0 & 0 \\ v_1 & 0 & 1 & 0 \\ v_2 & 0 & 0 & 1 \end{pmatrix} \begin{pmatrix} 1 & 0 & 0 & 0 \\ 0 & & \text{3-by-3} & \\ 0 & & \text{rotation} & \\ 0 & & \text{matrix} & \end{pmatrix} = \begin{pmatrix} 1 & 0 & 0 & 0 \\ v_3 & & \text{3-by-3} & \\ v_1 & & \text{rotation} & \\ v_1 & & \text{matrix} & \end{pmatrix}. \tag{3.44}$$

Thus, the net effect of this transformation is

$$\begin{pmatrix} 1 & 0 & 0 & 0 \\ v_3 & & \text{3-by-3} & \\ v_1 & & \text{rotation} & \\ v_1 & & \text{matrix} & \end{pmatrix} \begin{pmatrix} t \\ z \\ x \\ y \end{pmatrix} = \begin{pmatrix} t \\ z' + v_3 t \\ x' + v_1 t \\ y' + v_2 t \end{pmatrix}. \tag{3.45}$$

It is also possible to apply this transformation matrix to the contracted four-vector with the elements $(1, z, x, y)$. If we apply the above transformation matrix to this contracted column matrix,

$$\begin{pmatrix} 1 & 0 & 0 & 0 \\ v_3 & & \text{3-by-3} & \\ v_1 & & \text{rotation} & \\ v_2 & & \text{matrix} & \end{pmatrix} \begin{pmatrix} 1 \\ z \\ x \\ y \end{pmatrix} = \begin{pmatrix} 1 \\ z' + v_3 \\ x' + v_1 \\ y' + v_2 \end{pmatrix}. \tag{3.46}$$

It is a Euclidean transformation in the three-dimensional space.

3.5 Tangential spheres

We have so far discussed a plane and a cylinder tangent to curved surfaces. In physics, there are cases where the curved surfaces can be simplified to a spherical surface. For simplicity, let us restrict ourselves to the surfaces which are invariant to rotations around one of the three axes. Then the problem reduces to a two-dimensional problem in the xy-plane. Consequently, the problem is reduced the circle sharing the tangent line at a given point, as shown in figure 3.6.

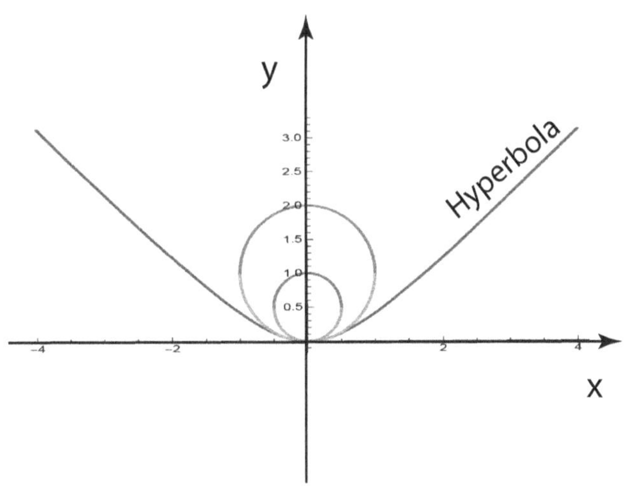

Figure 3.6. Two curves in contact. For a given hyperbola, we can consider circles in contact at $x = y = 0$, but the question is whether the radius of the circle coincides with that of the hyperbola. For small values of x, the hyperbola behaves like $y = x^2/2$, like a circle whose radius is 1, not 0.5.

The tangent line is the x-axis. However, is it enough? In this figure, the hyperbola is given by

$$(y + 1)^2 - x^2 = 1, \tag{3.47}$$

and the equations for the two circles are

$$x^2 + (y - 1)^2 = 1 \quad \text{and} \quad x^2 + (y - 0.5)^2 = 0.25, \tag{3.48}$$

having radii 1 and 0.5, respectively. For small values of x, the hyperbola can be written as

$$y = x^2/2 \tag{3.49}$$

and for the circles,

$$y = x^2/2 \quad \text{and} \quad y = x^2, \tag{3.50}$$

respectively. It is now clear which circle is a more accurate approximation to the hyperbola. In other words, the radius of the tangential sphere should coincide with the radius of curvature. This is indeed the definition of the radius of curvature. For the tangent line, the radius is infinite.

References

[1] Inönü E and Wigner E P 1953 On the contraction of groups and their representations *Proc. Natl. Acad. Sci.* **39** 510–24

[2] Kim Y S and Wigner E P 1987 Cylindrical group and massless particles *J. Math. Phys.* **28** 1175–9

Chapter 4

Two-by-two representations of Wigner's little groups

When Newton formulated his law of gravity, he did so for point particles. It took him another twenty years to make that law applicable to extended objects such as the Sun and moon. To do this, Newton had to develop a new mathematical method known as integral calculus. When Einstein formulated his special relativity theory, it was also for point particles, and the Lorentz group provided the mathematical tool. Shortly after the formulation of special relativity, Niels Bohr and others were concerned about the bound state of the hydrogen atom, which led to quantum mechanics. It was impossible at that time and still is, to accelerate the hydrogen atom to relativistic speeds. However, we now know that the proton and all hadrons are bound of states of more fundamental particles known as *quarks*. Hadrons can be accelerated to relativistic speeds. It is possible to study the internal space–time symmetry of these extended particles thanks to the mathematical tools provided by Wigner in his 1939 paper [8].

In chapter 2, we introduced the subgroups of the Lorentz group defined in that 1939 paper, whose transformations leave the particle momentum invariant, and we saw that a massive particle at rest has three rotational degrees of freedom leading to the concept of spin. Thus the little group for this massive particle is like the $O(3)$ group. What happens when this massive particle moves in the z-direction? Wigner did not comment on this. Wigner also observed that although a massless particle cannot be brought to its rest frame, the little group for the massless particle still has three degrees of freedom. This observation led to this little group being like the group $E(2)$ or the two-dimensional Euclidean group. It is not difficult to associate the rotational degree of freedom of $E(2)$ to the helicity of the massless particle. However, although we now know that those two translational degrees of freedom relate to gauge transformations, Wigner did not provide the answer to this question in his original paper [8]. In the present chapter, we will study Wigner's little groups in the two-by-two representation and then examine the transformation properties of

doi:10.1088/978-0-7503-3607-9ch4

these little groups. In this way we will attempt to close the gaps between Wigner's little groups and the physics world.

Section 4.1 defines the Wigner energy–momentum four-vector for the massive, massless, and imaginary-mass particle in the two-by-two matrix formalism. We further show how this four-vector is transformed. In section 4.2, we define the two-by-two transformation matrices, which leave the energy–momentum four-vector of these particles invariant when operating on it. In section 4.3, we complete our discussion of the transformation properties of the little groups by considering a boost along the z-direction.

In section 4.4, we discuss the Bargmann and Wigner decompositions. We illustrate space inversion (parity), time-reversal, and charge conjugation in terms of a loop diagram based on the Bargmann decomposition. Section 4.5 defines the dot conjugate representation, which applies to the opposite sign Lorentz boost defined in equation (1.12). In section 4.6, we regard the three little groups as one little group with three different branches, and we use this to explicitly see how the three branches become one. In section 4.7, we use the physics associated with the damped harmonic oscillator to illustrate the previous section.

4.1 Transformation properties of the energy–momentum four-vector

It was noted in section 1.2 that the Lorentz transformation of the energy–momentum four-vector can be represented by two-by-two matrices. An explicit form for the Lorentz transformation was given in equation (1.27) as [5, 7]

$$P' = G\, P\, G^{\dagger}, \tag{4.1}$$

where the two-by-two form of the G matrix is given in equation (1.18).

Let us use W as a subset of matrices which leaves the energy–momentum four-vector invariant. Then we can write

$$P = W\, P\, W^{\dagger}. \tag{4.2}$$

Of course, these matrices constitute Wigner's little group dictating the internal space–time symmetry of the particle.

From equation (1.39) we can derive the following forms for the energy–momentum four-vector. If the particle is massive, it can be brought to the system where it is at rest with $p = 0$. The energy–momentum four-vector in the two-by-two matrix representation is proportional to

$$P = \begin{pmatrix} 1 & 0 \\ 0 & 1 \end{pmatrix}. \tag{4.3}$$

For the massless particle, $E = p$. Thus the energy–momentum four-vector in the two-by-two matrix representation is proportional to

$$P = \begin{pmatrix} 1 & 0 \\ 0 & 0 \end{pmatrix}. \tag{4.4}$$

If the particle mass is imaginary, there is a Lorentz frame where the energy component vanishes. Thus, the energy–momentum four-vector becomes proportional to

$$P = \begin{pmatrix} 1 & 0 \\ 0 & -1 \end{pmatrix}. \tag{4.5}$$

For all three cases, the matrix of the form

$$Z(\phi) = \begin{pmatrix} e^{i\phi/2} & 0 \\ 0 & e^{-i\phi/2} \end{pmatrix} \tag{4.6}$$

will satisfy the Wigner condition of equation (4.2). This matrix corresponds to rotations around the z-axis as given in two-by-two matrix format in table 1.1.

4.2 Two-by-two representations of Wigner's little groups

For a massive particle, since the two-by-two energy–momentum matrix is proportional to the unit matrix, the W matrix should be Hermitian. The little group is the $SU(2)$ subgroup of the Lorentz group, namely the rotation subgroup. According to Euler, it is sufficient to consider rotations around the y-axis, using this rotation matrix from table 1.1

$$R(\theta) = \begin{pmatrix} \cos(\theta/2) & -\sin(\theta/2) \\ \sin(\theta/2) & \cos(\theta/2) \end{pmatrix}, \tag{4.7}$$

together with the rotation matrix $Z(\phi)$.

If we rotate equation (4.7) around the z-axis, it becomes

$$Z(\phi)R(\theta)Z^{\dagger}(\phi) = \begin{pmatrix} \cos(\theta/2) & -e^{i\phi}\sin(\theta/2) \\ e^{-i\phi}\sin(\theta/2) & \cos(\theta/2) \end{pmatrix}. \tag{4.8}$$

However, according to the Euler decomposition, it is sufficient to consider only the matrix $R(\theta)$ with $\phi = 0$ as the representation of Wigner's $O(3)$-like little group for massive particles.

If the particle is massless, the Wigner matrix is necessarily triangular and should take the form

$$T(\gamma) = \begin{pmatrix} 1 & -\gamma \\ 0 & 1 \end{pmatrix}. \tag{4.9}$$

This matrix has properties that are not too familiar to us. First of all, the matrix cannot be diagonalized. The inverse and Hermitian conjugate are

$$T^{-1}(\gamma) = \begin{pmatrix} 1 & \gamma \\ 0 & 1 \end{pmatrix} \quad \text{and} \quad T^{\dagger} = \begin{pmatrix} 1 & 0 \\ -\gamma & 1 \end{pmatrix}, \tag{4.10}$$

respectively. Since the inverse and Hermitian conjugate are not the same, T is not a Hermitian matrix.

Table 4.1. The Wigner energy–momentum four-vectors in the two-by-two matrix representation together with the corresponding transformation matrices. These energy–momentum four-vector matrices have determinants that are positive, zero, and negative for massive, massless, and imaginary-mass particles, respectively. The triangular matrix for the massless case is the most interesting expression here.

Particle mass	Wigner Four-vector	Wigner Transformation Matrix
Massive	$\begin{pmatrix} 1 & 0 \\ 0 & 1 \end{pmatrix}$	$\begin{pmatrix} \cos(\theta/2) & -\sin(\theta/2) \\ \sin(\theta/2) & \cos(\theta/2) \end{pmatrix}$
Massless	$\begin{pmatrix} 1 & 0 \\ 0 & 0 \end{pmatrix}$	$\begin{pmatrix} 1 & -\gamma \\ 0 & 1 \end{pmatrix}$
Imaginary-mass	$\begin{pmatrix} 1 & 0 \\ 0 & -1 \end{pmatrix}$	$\begin{pmatrix} \cosh(\lambda/2) & \sinh(\lambda/2) \\ \sinh(\lambda/2) & \cosh(\lambda/2) \end{pmatrix}$

If we rotate the T matrix around the z-axis, it becomes

$$Z(\phi)T(\gamma)Z^{\dagger}(\phi) = \begin{pmatrix} 1 & -e^{i\phi}\gamma \\ 0 & 1 \end{pmatrix}. \tag{4.11}$$

Thus, again taking $\phi = 0$, we shall use the triangular matrix of equation (4.9) as the representation of the $E(2)$-like little group.

For a particle with an imaginary-mass, we can choose the W matrix as

$$\exp(-i\lambda K_1) = \begin{pmatrix} \cosh(\lambda/2) & \sinh(\lambda/2) \\ \sinh(\lambda/2) & \cosh(\lambda/2) \end{pmatrix}, \tag{4.12}$$

where K_1 is given in table 1.1. This transformation leaves the energy–momentum four-vector of equation (4.5) invariant. If rotated around the z-axis, it becomes

$$\begin{pmatrix} \cosh(\lambda/2) & e^{i\phi}\sinh(\lambda/2) \\ e^{-i\phi}\sinh(\lambda/2) & \cosh(\lambda/2) \end{pmatrix}. \tag{4.13}$$

However, it is sufficient to choose the real matrix of equation (4.12) as the $O(2, 1)$-like little group for imaginary-mass particles.

Table 4.1 summarizes the transformation matrices for Wigner's little groups for massive, massless, and imaginary-mass particles.

4.3 Lorentz completion of the little groups

We are now interested in boosting Wigner's energy–momentum four-vectors and the representation matrices along the z-direction. The boosted energy–momentum four-vector is

$$P' = B(\eta)\, P\, B^{\dagger}(\eta), \tag{4.14}$$

with, as given in table 1.1,

$$B(\eta) = \begin{pmatrix} e^{\eta/2} & 0 \\ 0 & e^{-\eta/2} \end{pmatrix}. \tag{4.15}$$

The boosted energy–momentum four-vector should then take the form

$$\begin{pmatrix} e^{\eta} & 0 \\ 0 & e^{-\eta} \end{pmatrix}, \tag{4.16}$$

for the massive particle, and

$$\begin{pmatrix} e^{\eta} & 0 \\ 0 & 0 \end{pmatrix}, \quad \begin{pmatrix} e^{\eta} & 0 \\ 0 & -e^{-\eta} \end{pmatrix}, \tag{4.17}$$

respectively, for massless and imaginary-mass particles.

However, the boosted Wigner matrix becomes

$$W' = B(\eta)\, W\, B^{-1}(\eta). \tag{4.18}$$

It should be noted that $B^{\dagger}(\eta)$ is not the same as $B^{-1}(\eta)$. Hence, equation (4.18) is a similarity transformation, unlike the transformation of the energy–momentum four-vector given in equation (4.14).

The boosted W matrix becomes

$$W' = \begin{pmatrix} \cos(\theta/2) & -e^{\eta}\sin(\theta/2) \\ e^{-\eta}\sin(\theta/2) & \cos(\theta/2) \end{pmatrix}, \tag{4.19}$$

for the massive particle. For massless and imaginary-mass particles, they become

$$\begin{pmatrix} 1 & e^{\eta}\gamma \\ 0 & 1 \end{pmatrix} \quad \text{and} \quad \begin{pmatrix} \cosh(\lambda/2) & e^{\eta}\sinh(\lambda/2) \\ e^{-\eta}\sinh(\lambda/2) & \cosh(\lambda/2) \end{pmatrix}, \tag{4.20}$$

respectively.

These results are tabulated in table 4.2. In the limit of large η, all three Wigner four-vector matrices take the same form as that of the massless particle. If these

Table 4.2. Lorentz-boosted Wigner energy–momentum four-vectors and Wigner matrices in the two-by-two representation. They take the same form for infinite values of η, if the parameters θ, λ, and γ are made to decrease by $e^{-\eta}$.

Particle mass	Boosted Wigner Four-vector	Boosted Wigner Transformation Matrix
Massive	$\begin{pmatrix} e^{\eta} & 0 \\ 0 & e^{-\eta} \end{pmatrix}$	$\begin{pmatrix} \cos(\theta/2) & -e^{\eta}\sin(\theta/2) \\ e^{-\eta}\sin(\theta/2) & \cos(\theta/2) \end{pmatrix}$
Massless	$\begin{pmatrix} e^{\eta} & 0 \\ 0 & 0 \end{pmatrix}$	$\begin{pmatrix} 1 & -e^{\eta}\gamma \\ 0 & 1 \end{pmatrix}$
Imaginary-mass	$\begin{pmatrix} e^{\eta} & 0 \\ 0 & -e^{-\eta} \end{pmatrix}$	$\begin{pmatrix} \cosh(\lambda/2) & e^{\eta}\sinh(\lambda/2) \\ e^{-\eta}\sinh(\lambda/2) & \cosh(\lambda/2) \end{pmatrix}$

matrices are multiplied by e^{-n}, they all become the Wigner four-vector matrix of the massless particle in table 4.1. If the boosted Wigner matrices are multiplied by $e^{-\eta}$, they become $\sin(\theta/2)$, γ, and $\sinh(\lambda/2)$, respectively. Hence, we are led to look at the three little groups as three different branches of one little group. We shall discuss this problem more systematically in section 4.6.

4.4 Bargmann and Wigner decompositions

Let us restate the contents of section 4.3. In the case of a massive particle moving with the energy–momentum four-vector P' of equation (4.16), the representation of the little group is W' of equation (4.19), which is a Lorentz-boosted form of the W matrix. Indeed, it can be written as

$$B(\eta)WB(-\eta) = B(\eta)\left\{B(-\eta)\left[B(\eta)WB(-\eta)\right]B(\eta)\right\}B(-\eta), \tag{4.21}$$

meaning that the system is brought back to the frame in which the particle is at rest, a rotation is made without changing the momentum, and then the system is boosted back to a frame where it regains its original momentum. This three-step operation is illustrated on the left side in figure 4.1. Since this operation consists of three matrices, we shall call it the Wigner decomposition.

However, this is not the only momentum-preserving transformation. We can start with the momentum along the z-direction, as illustrated in the right side of figure 4.1. We can rotate this momentum around the y-axis, boost along the negative x-direction, and then rotate back to the original momentum along the z-direction. Then this operation can be written as

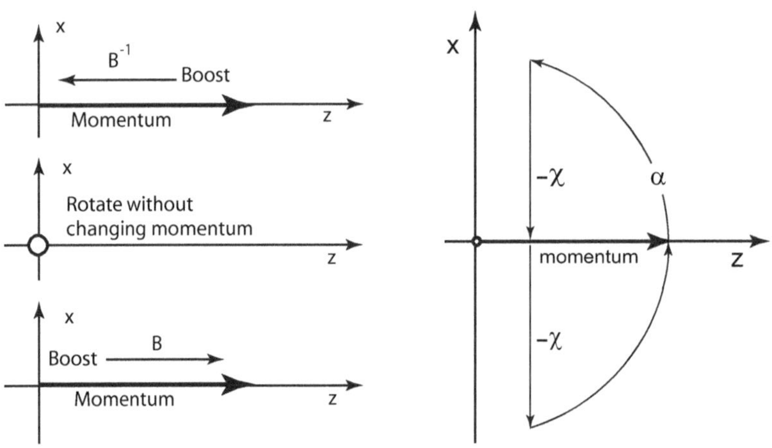

Figure 4.1. Wigner decomposition (left) and Bargmann decomposition (right). These figures illustrate momentum preserving transformations. In the Wigner transformation, a massive particle is brought to its rest frame. It can be rotated while the momentum remains the same. This particle is then boosted back to the frame with its original momentum. In the Bargmann decomposition, the momentum is rotated, boosted, and rotated to its original position [5].

$$R(\alpha)S(-2\chi)R(\alpha), \tag{4.22}$$

which is a product of one boost matrix sandwiched between two rotation matrices. This form is called the Bargmann decomposition [1].

The multiplication of these three matrices leads to

$$D(\alpha, \chi) = \begin{pmatrix} (\cos\alpha)\cosh\chi & -\sinh\chi - (\sin\alpha)\cosh\chi \\ -\sinh\chi + (\sin\alpha)\cosh\chi & (\cos\alpha)\cosh\chi \end{pmatrix}. \tag{4.23}$$

We can now compare this formula with the momentum preserving W' matrices given in table 4.2.

1. If the particle is massive, the off-diagonal elements should have opposite signs,

$$\cos(\theta/2) = (\cos\alpha)\cosh\chi, \qquad e^{2\eta} = \frac{(\sin\alpha)\cosh\chi - \sinh\chi}{\sinh\chi + (\sin\alpha)\cosh\chi}, \tag{4.24}$$

with $(\sin\alpha)\cosh\chi > \sinh\chi$.

2. If the particle is massless, one of the off-diagonal elements should vanish, and

$$\sinh\chi - (\sin\alpha)\cosh\chi = 0. \tag{4.25}$$

Thus, the diagonal elements become $(\cos\alpha)\cosh\chi = 1$, and the non-vanishing off-diagonal element becomes $2\sinh\chi = \gamma$. This procedure is commonly known as the Iwasawa decomposition [6] for the massless particle.

3. If the particle mass is imaginary, the off-diagonal elements have the same sign,

$$\cosh(\lambda/2) = (\cosh\chi)\cos\alpha \quad \text{and} \quad e^{-2\eta} = \frac{\sinh\chi - (\cosh\chi)\sin\alpha}{(\cosh\chi)\sin\alpha + \sinh\chi}. \tag{4.26}$$

It is now clear that the transformation matrices become the same triangular form in the limit of large η.

In the Wigner decomposition, the off-diagonal elements do not change their signs. We are assuming all the angles are positive and smaller than 90°. Thus, if we are interested in making a transition from the massive particle to the zero-mass, and then to the imaginary-mass particle, we have to make an excursion to the infinite value of η, and then come back through the appropriate route. This procedure is a singular operation in mathematics.

In the Bargmann decomposition, one of the off-diagonal elements can change its sign depending on the parameters α and χ. Thus, the transition from the massive to the massless and imaginary-mass particle is analytic. These mathematical properties are summarized in table 4.3.

Table 4.3. Bargmann and Wigner decompositions. Their mathematical properties are compared. In the Bargmann decomposition, one analytic expression covers the massive, massless, and imaginary-mass particles. As was noted in section 4.3, the Wigner decomposition is in the form of a similarity transformation which can serve many useful mathematical purposes.

Decompositions	Analytic	Similarity Transformation
Bargmann	Yes	No
Wigner	No	Yes

4.5 Conjugate transformations

The most general form of the $SL(2, c)$ matrix is given in equation (1.18), with six independent parameters. In terms of the six generators, this matrix can be written as

$$D = \exp\left\{-i\sum_{i=1}^{3}(\theta_i J_i + \eta_i K_i)\right\}, \tag{4.27}$$

where the J_i are the generators of rotations and the K_i are the generators of proper Lorentz boosts. They satisfy the Lie algebra given in equations (1.5) and (1.10). This set of commutation relations is invariant under the sign change of the boost generators K_i. Thus, using equation (1.12) we can consider the *dot conjugation* defined as

$$\dot{D} = \exp\left\{-i\sum_{i=1}^{3}(\theta_i J_i - \eta_i K_i)\right\}. \tag{4.28}$$

Since K_i are anti-Hermitian while J_i are Hermitian, the Hermitian conjugate of D is

$$D^{\dagger} = \exp\left\{-i\sum_{i=1}^{3}(-\theta_i J_i + \eta_i K_i)\right\}, \tag{4.29}$$

while the Hermitian conjugate of $\dot{D}$ is

$$\dot{D}^{\dagger} = \exp\left\{-i\sum_{i=1}^{3}(-\theta_i J_i - \eta_i K_i)\right\}. \tag{4.30}$$

Since we understand the rotation around the z-axis, we can now restrict the kinematics to the zt plane, and work with the $Sp(2)$ symmetry. Then recalling equation (4.23), the D matrices can be considered as Bargmann decompositions. First, the D and $\dot{D}$ matrices can be written as

$$\begin{aligned} D(\alpha, \chi) &= \begin{pmatrix} (\cos\alpha)\cosh\chi & -\sinh\chi - (\sin\alpha)\cosh\chi \\ -\sinh\chi + (\sin\alpha)\cosh\chi & (\cos\alpha)\cosh\chi \end{pmatrix}, \\ \dot{D}(\alpha, \chi) &= \begin{pmatrix} (\cos\alpha)\cosh\chi & \sinh\chi - (\sin\alpha)\cosh\chi \\ \sinh\chi + (\sin\alpha)\cosh\chi & (\cos\alpha)\cosh\chi \end{pmatrix}. \end{aligned} \tag{4.31}$$

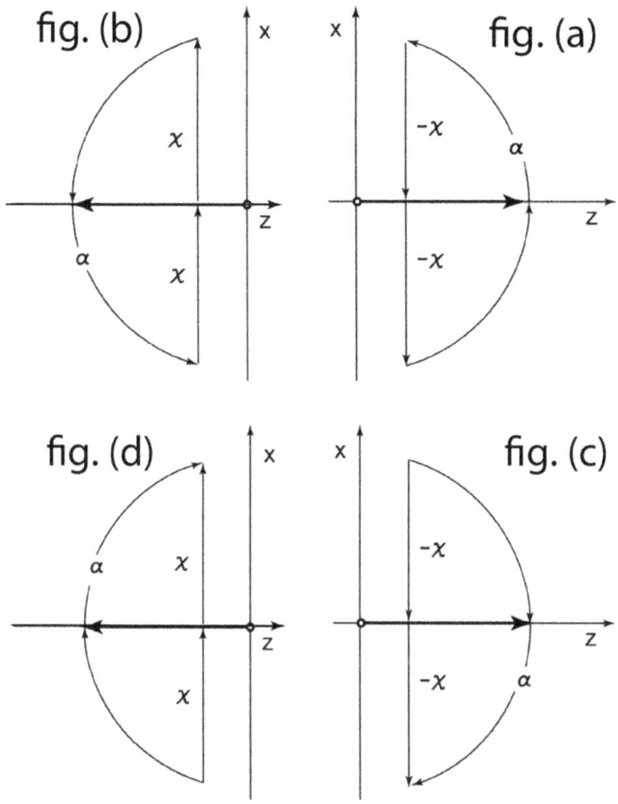

Figure 4.2. Four D-loops resulting from the Bargmann decomposition. If we start with (a), corresponding to the right side of figure 4.1 and reverse of the direction of the boost, the result is (d). From (a), if we reverse the direction of both the rotation and boost, we come back to (b). If we reverse the direction of the boost from (b), the result is (c). [5].

These matrices correspond to the *D loops* given in figure 4.2. The *dot conjugation* changes the direction of boosts and leads to the sign change in χ. This change in direction is the inversion of space, which is called the parity operation, as illustrated in (d).

We can also consider changing the direction of rotations, which results in using the Hermitian conjugates. We can write the Hermitian conjugate matrices as

$$D^{\dagger}(\alpha, \chi) = \begin{pmatrix} (\cos\alpha)\cosh\chi & -\sinh\chi + (\sin\alpha)\cosh\chi \\ -\sinh\chi - (\sin\alpha)\cosh\chi & (\cos\alpha)\cosh\chi \end{pmatrix},$$
$$\dot{D}^{\dagger}(\alpha, \chi) = \begin{pmatrix} (\cos\alpha)\cosh\chi & \sinh\chi + (\sin\alpha)\cosh\chi \\ \sinh\chi - (\sin\alpha)\cosh\chi & (\cos\alpha)\cosh\chi \end{pmatrix}. \quad (4.32)$$

The Hermitian conjugation of $\dot{D}$ leads to a change in direction of both the boost and the rotation, known as time-reversal, which corresponds to (b). If we then space invert $\dot{D}^{\dagger}$ we obtain the Hermitian conjugate of D. The combined transformation of

Table 4.4. Parity, time reversal, and charge conjugation in the loop representation.

Operation	Bargmann Decomposition	D Matrix
Start	$R(\alpha)S(-2\chi)R(\alpha)$	$D(\chi, \alpha)$
Parity–Space Inversion	$R(\alpha)S(2\chi)R(\alpha)$	$\dot{D}(\chi, \alpha)$
Time Reversal	$R(-\alpha)S(2\chi)R(-\alpha)$	$\dot{D}^{\dagger}(\chi, \alpha)$
Charge Conjugation	$R(-\alpha)S(-2\chi)R(-\alpha)$	$D^{\dagger}(\chi, \alpha)$

space inversion and time reversal is known as *charge conjugation* and corresponds to (c). These procedures are summarized in table 4.4.

The exponential expressions from equations (4.27) to (4.30), lead to

$$D^{\dagger} = \dot{D}^{-1} \quad \text{and} \quad \dot{D}^{\dagger} = D^{-1}, \tag{4.33}$$

and the Bargmann forms of equations (4.31) and (4.32) are consistent with these relations.

4.6 One little group with three branches

We have noted in sections 2.2 and 2.4 that J_2 and K_1 can serve as the starters of the representation of Wigner's little groups in their respective Wigner frames. They are for the massive and imaginary-mass particles, respectively. It was also noted that, when Lorentz boosted, they become

$$J_2' = (\cosh\eta)J_2 - (\sinh\eta)K_1 \quad \text{and} \quad K_1' = -(\sinh\eta)J_2 + (\cosh\eta)K_1. \tag{4.34}$$

These two equations can be combined into one formula, and we are led to consider the transformation matrix

$$D(x, y) = \exp[-i(yJ_2 - xK_1)] \tag{4.35}$$

as a function of the parameters x and y with J_2 and K_1 as given in table 1.1. We shall study this form of the Wigner matrix in detail.

During this study, there will be a problem of handling singularities that are not too familiar to us. In the next section we shall use the classical damped harmonic oscillator to study this singularity problem in detail, and conclude that this is not an analytic but can be called a *tangential* continuity.

Let us now write the D matrix of equation (4.35) as [2, 3, 5]

$$D(x, y) = \exp\left\{\begin{pmatrix} 0 & -(x+y) \\ -(x-y) & 0 \end{pmatrix}\right\}. \tag{4.36}$$

1. If $y > x$, we write

$$x + y = e^{\eta}\sqrt{y^2 - x^2} \quad \text{and} \quad x - y = e^{-\eta}\sqrt{y^2 - x^2}, \tag{4.37}$$

with

$$e^{\eta} = \sqrt{\frac{x+y}{|y-x|}}, \tag{4.38}$$

and $D(x, y)$ becomes

$$\exp\left\{\sqrt{y^2 - x^2}\begin{pmatrix} 0 & -e^{\eta} \\ e^{-\eta} & 0 \end{pmatrix}\right\} = \begin{pmatrix} \cos(\theta/2) & -e^{\eta}\sin(\theta/2) \\ e^{-\eta}\sin(\theta/2) & \cos(\theta/2) \end{pmatrix}, \tag{4.39}$$

with $\cos(\theta/2) = \cos\left(\sqrt{y^2 - x^2}\right)$. This expression is the same as that for the Lorentz-boosted W matrix given in equation (4.19) for massive particles.

2. If $x = y$, this expression becomes

$$D(x, y) = \exp\left\{\begin{pmatrix} 0 & -2x \\ 0 & 0 \end{pmatrix}\right\} = \begin{pmatrix} 1 & -2x \\ 0 & 1 \end{pmatrix}. \tag{4.40}$$

This form is for the little group for massless particles, as shown in $T(\gamma)$ of equation (4.9).

3. If $y < x$, we can write

$$x - y = e^{-\eta}\sqrt{x^2 - y^2} \quad \text{and} \quad x + y = e^{\eta}\sqrt{x^2 - y^2}. \tag{4.41}$$

Then the D matrix becomes

$$\exp\left\{\sqrt{x^2 - y^2}\begin{pmatrix} 0 & -e^{\eta} \\ -e^{-\eta} & 0 \end{pmatrix}\right\} = \begin{pmatrix} \cosh(\lambda/2) & -e^{\eta}\sinh(\lambda/2) \\ -e^{-\eta}\sinh(\lambda/2) & \cosh(\lambda/2) \end{pmatrix}, \tag{4.42}$$

with $\cosh(\lambda/2) = \cosh\left(\sqrt{x^2 - y^2}\right)$. This expression is the same as that for the Lorentz-boosted W matrix given in equation (4.20) for imaginary-mass particles, and e^{η} is given in equation (4.38).

Indeed, it is possible to derive three different forms of the W' matrix. The matrix

$$\begin{pmatrix} 0 & -(x+y) \\ -(x-y) & 0 \end{pmatrix} \tag{4.43}$$

is analytic in the x and y variables. However, this D matrix has three distinct branches. In order to understand the nature of this, let us look at what happens when $x - y$ is a small number

$$\varepsilon = x - y. \tag{4.44}$$

We can then write the D matrix as

$$D(x, \varepsilon) = \exp\left\{-\begin{pmatrix} 0 & 2x \\ \varepsilon & 0 \end{pmatrix}\right\}. \tag{4.45}$$

If ε is positive, the Taylor expansion leads to

$$D = \begin{pmatrix} \cosh\left(\sqrt{2x\varepsilon}\right) & -\left[\sqrt{2x/\varepsilon}\right]\sinh\sqrt{2x\varepsilon} \\ -\left[\sqrt{\varepsilon/2x}\right]\sinh\left(\sqrt{2x\varepsilon}\right) & \cosh\left(\sqrt{2x\varepsilon}\right) \end{pmatrix}. \tag{4.46}$$

If ε becomes zero, this expression becomes

$$\begin{pmatrix} 1 & -2x \\ 0 & 1 \end{pmatrix}. \tag{4.47}$$

If ε becomes negative,

$$\sqrt{2x\varepsilon} = i\sqrt{-2x\varepsilon}, \qquad \sqrt{\varepsilon/2x} = i\sqrt{-2x/\varepsilon}, \qquad \sqrt{2x/\varepsilon} = -i\sqrt{-2x/\varepsilon}, \tag{4.48}$$

if we take $\sqrt{-1} = i$. Thus, D becomes

$$D = \begin{pmatrix} \cos\left(\sqrt{-2x\varepsilon}\right) & -\left[\sqrt{-2x/\varepsilon}\right]\sin\left(\sqrt{-2x\varepsilon}\right) \\ \left[\sqrt{-\varepsilon/2x}\right]\sin\left(\sqrt{-2x\varepsilon}\right) & \cos\left(\sqrt{-2x\varepsilon}\right) \end{pmatrix}. \tag{4.49}$$

The result remains the same if we take $\sqrt{-1} = -i$.

This type of singularity is not common in literature. Let us study this point further using a physical example familiar to us.

4.7 Classical damped harmonic oscillator

Let us start with the second-order differential equation

$$\frac{d^2y}{dt^2} + 2\mu\frac{dy}{dt} + \omega^2 y = 0, \tag{4.50}$$

for a classical damped harmonic oscillator. If we introduce the function $\psi(t)$ as

$$\psi(t) = e^{-\mu t}y(t) \tag{4.51}$$

then $\psi(t)$ satisfies the simplified differential equation

$$\frac{d^2\psi(t)}{dt^2} + (\omega^2 - \mu^2)\psi(t) = 0. \tag{4.52}$$

This second-order differential equation has two independent solutions. Let us call them ψ_1 and ψ_2. They satisfy the first-order differential equations

$$\frac{d}{dt}\begin{pmatrix} \psi_1 \\ \psi_2 \end{pmatrix} = \begin{pmatrix} 0 & -(\omega + \mu) \\ (\omega - \mu) & 0 \end{pmatrix}\begin{pmatrix} \psi_1 \\ \psi_2 \end{pmatrix}. \tag{4.53}$$

This coupled equation leads to the second-order equation equation (4.52) for $\psi_1(t)$ and $\psi_2(t)$. The physical solution is an appropriate linear combination of these two wave functions.

The solution of this first-order differential equation is

$$\begin{pmatrix} \psi_1 \\ \psi_2 \end{pmatrix} = \exp\left\{\begin{pmatrix} 0 & -(\omega+\mu)t \\ (\omega-\mu)t & \end{pmatrix}\right\}\begin{pmatrix} C_1 \\ C_2 \end{pmatrix}, \tag{4.54}$$

where $C_1 = \psi_1(0)$ and $C_2 = \psi_2(0)$. We can then obtain the solutions by following the procedure developed in section 4.6.

1. If $\omega > \mu$, the solution becomes

$$\begin{pmatrix} \psi_1 \\ \psi_2 \end{pmatrix} = \begin{pmatrix} \cos(\omega' t) & -e^{\eta}\sin(\omega' t) \\ e^{-\eta}\sin(\omega' t) & \cos(\omega' t) \end{pmatrix}\begin{pmatrix} C_1 \\ C_2 \end{pmatrix}, \tag{4.55}$$

with

$$\omega' = \sqrt{\omega^2 - \mu^2} \quad \text{and} \quad e^{\eta} = \sqrt{\frac{\omega+\mu}{|\omega-\mu|}}. \tag{4.56}$$

2. If $\omega = \mu$, the solution becomes

$$\begin{pmatrix} \psi_1 \\ \psi_2 \end{pmatrix} = \begin{pmatrix} 1 & -2\omega t \\ 0 & 1 \end{pmatrix}\begin{pmatrix} C_1 \\ C_2 \end{pmatrix}. \tag{4.57}$$

3. If $\mu > \omega$, the solution matrix becomes

$$\begin{pmatrix} \cosh(\mu' t) & e^{\eta}\sinh(\mu' t) \\ e^{-\eta}\sinh(\mu' t) & \cosh(\mu' t) \end{pmatrix}, \tag{4.58}$$

with e^{η} given in equation (4.56), and

$$\mu' = \sqrt{\mu^2 - \omega^2}. \tag{4.59}$$

Let us now turn to the main issue of what happened when μ is close to ω. If ω is sufficiently close to μ, we can let

$$\mu - \omega = 2\mu\varepsilon, \quad \text{and} \quad \mu + \omega = 2\omega. \tag{4.60}$$

If ω is greater than μ, then ε defined in equation (4.60) becomes negative. The solution matrix becomes

$$\begin{pmatrix} 1 - (-\varepsilon/2)(2\omega t)^2 & -2\omega t \\ -(-\varepsilon)(2\omega t) & 1 - (-\varepsilon/2)(2\omega t)^2 \end{pmatrix}, \tag{4.61}$$

which can be written as

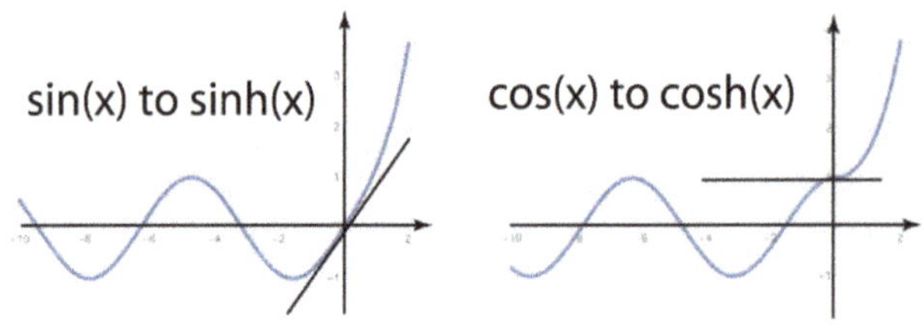

Figure 4.3. At $x = 0$, the transitions from sine to sinh and from cosine to cosh, respectively, are continuous. The first derivatives are continuous; the second derivatives are not. This means that the transitions are only tangentially continuous and not analytically continuous. [4].

Table 4.5. Here we see that the space–time symmetries and damped oscillators are based on the same set of two-by-two matrices.

Trace	Damped Oscillator	Particle Symmetry
Smaller than 2	Oscillation Mode	Massive Particles
Equal to 2	Transition Mode	Massless Particles
Larger than 2	Damping Mode	Imaginary-mass Particles

$$\begin{pmatrix} 1 - (1/2)[(2\omega\sqrt{-\varepsilon})t]^2 & -2\omega t \\ -\sqrt{-\varepsilon}[(2\omega\sqrt{-\varepsilon})t] & 1 - (1/2)[(2\omega\sqrt{-\varepsilon})t]^2 \end{pmatrix}. \tag{4.62}$$

If ε is positive, equation (4.58) can be written as

$$\begin{pmatrix} 1 + (1/2)[(2\omega\sqrt{\varepsilon})t]^2 & -2\omega t \\ -\sqrt{\varepsilon}[(2\omega\sqrt{\varepsilon})t] & 1 + (1/2)[(2\omega\sqrt{\varepsilon})t]^2 \end{pmatrix}. \tag{4.63}$$

The transition from equations (4.62) to (4.63) is continuous as they become identical when $\varepsilon = 0$. As ε changes its sign, the diagonal elements of the above matrices tell us how $\cos(\omega' t)$ becomes $\cosh(\mu' t)$. The upper-right element remains as $\sin(\omega' t)$ during this transitional process. The lower left element becomes $\sinh(\mu' t)$. This non-analytic continuity is illustrated in figure 4.3.

During this continuation process, the function remains the same. So does its first derivative, but the second derivative does not. Thus, the two functions share the same tangential line. It is indeed a *tangential continuity*. The continuity from one little group to another was discussed in section 4.6. This mathematical similarity is summarized in table 4.5.

References

[1] Bargmann V 1947 Irreducible unitary representations of the Lorentz group *Ann. Math.* **48** 568–640

[2] Başkal S and Kim Y S 2010 One analytic form for four branches of the *ABCD* matrix *J. Mod. Opt.* **57** 1251–9

[3] Başkal S and Kim Y S 2013 Lorentz group in ray and polarization optics *Mathematical Optics: Classical, Quantum and Computational Methods* ed V Lakshminarayanan, M L Calvo, and T Alieva (Boca Raton, FL: Taylor and Francis) pp 303–49 (ISBN 978-1-4398-6961-1)
[4] Başkal S and Kim Y S 2014 Lens optics and the continuity problems of the *ABCD* matrix *J. Mod. Opt.* **61** 161–6
[5] Başkal S, Kim Y S, and Noz M E 2014 Wigner's space–time symmetries based on the two-by-two matrices of the damped harmonic oscillators and the Poincaré sphere *Symmetry* **6** 473–515
[6] Iwasawa K 1949 On some types of topological groups *Ann. Math.* **50** 507
[7] Kim Y S and Noz M E 1986 *Theory and Applications of the Poincaré Group* (Dordrecht: Springer) (ISBN 978-94-010-8526-7 978-94-009-4558-6)
[8] Wigner E P 1939 On unitary representations of the inhomogeneous Lorentz group *Ann. Math.* **40** 149–204

IOP Publishing

Physics of the Lorentz Group (Second Edition)

Beyond high-energy physics and optics

Sibel Başkal, Young S Kim, and Marilyn E Noz

Chapter 5

Relativistic spinors and polarization of photons and neutrinos

When Wigner's 1939 paper [28] was published, it was not clear to everyone that the little groups could explain the Dirac equation for spin-$\frac{1}{2}$ particles or the Maxwell electromagnetic four-potential and four-tensor. Therefore the physical interpretation of Wigner's paper was not apparent. Nevertheless, some physicists, including Steven Weinberg, attempted to connect this paper with the physical world [23–25]. In his third 1964 paper, Weinberg constructed gauge-invariant four-tensors for the electromagnetic field [25], starting from Wigner's 1939 paper. In this chapter, we further discuss the physical interpretation of Wigner's 1939 paper by applying it to spinors and photon and neutrino polarization.

We know that Einstein's photoelectric effect allows us to regard the massless photon as the particle manifestation of the electromagnetic wave. In the quantum world, the photon has helicity or angular momentum aligned parallel or anti-parallel to the direction of the momentum. The question becomes, how is the photon helicity translated into the electric and magnetic fields perpendicular to the direction of propagation? This problem can be solved if we consider the internal space–time symmetries defined by Wigner's little groups [28] for massive and massless particles.

In the Maxwell picture, we can define the electromagnetic four-potential, which becomes the four-vector. Additionally, the electric and magnetic fields are grouped into the four-by-four Maxwell electromagnetic four-tensor. These can be represented by

$$A_\mu = \begin{pmatrix} A_0 \\ A_z \\ A_x \\ A_y \end{pmatrix} \quad \text{and} \quad F_{\mu\nu} = \begin{pmatrix} 0 & -E_z & -E_x & -E_y \\ E_z & 0 & -B_y & B_x \\ E_x & B_y & 0 & -B_z \\ E_y & -B_x & B_z & 0 \end{pmatrix}. \tag{5.1}$$

It is apparent that these expressions belong to Einstein's Lorentz-covariant world. Although the issue of the electromagnetic four-potential with the gauge degree of freedom was much debated in the literature, it was settled in later papers [20, 21, 26]. The present authors dealt with the problem in their publications [3, 4].

In section 5.1, we discuss $SL(2, c)$, the two-by-two representation of the Lorentz group, and of two-by-two spinors. We show that the undotted and dotted representation of the Lorentz group leads to the four spin states, which result in sixteen spin-1 and spin-0 states in the Lorentz-covariant world. The Wigner matrices for the dotted representation are defined. It is then illustrated that the dotted boost generators perform boosts in the opposite direction with respect to the undotted ones.

We show in section 5.2, how the massive and massless particles are related through Wigner's little groups. In section 5.3, we discuss Dirac spinors for massive particles and derive massless particles as the large-momentum or small-mass limit of massive particles.

In section 5.4, the polarization of massless spin-$\frac{1}{2}$ neutrinos is shown to be a consequence of Wigner's little groups. In the last section 5.5, we derive explicitly the sixteen spin states that result when the dotted representation is considered in addition to the undotted representation. We then derive the electromagnetic four-potential and group the electric and magnetic fields into the four-by-four Maxwell tensor. We show how photon polarization can be traced from this electromagnetic tensor.

5.1 Two-component spinors

In section 1.2, we saw that the two-by-two representation of the Lorentz group can be generated by the three two-by-two Pauli spin matrices. In section 2.2, we saw that the rotation subgroup of the proper Lorentz group defines the spin of a massive particle at rest. In the two-by-two representation, this group is commonly called $SU(2)$, and applies to two-component spinors. For the spinors we use

$$u = \begin{pmatrix} 1 \\ 0 \end{pmatrix} \quad \text{and} \quad v = \begin{pmatrix} 0 \\ 1 \end{pmatrix}, \tag{5.2}$$

for the spin-up and spin-down states, respectively. The direct product of two spinors yields four states leading to three spin-1 states and one spin-0 state. With these spinors, we can construct the spin-1 states using the symmetric combinations:

$$uu, \quad \frac{1}{\sqrt{2}}(uv + vu), \quad \text{and} \quad vv, \tag{5.3}$$

for the z-component spin 1, 0, and −1, respectively. For the spin-0 state, we use the anti-symmetric combination

$$\frac{1}{\sqrt{2}}(uv - vu). \tag{5.4}$$

We also saw in section 1.2 that the group $SL(2, c)$ serves as the two-by-two representation of the group of Lorentz transformations. From the mathematical point of view, this chapter is about extending the group $SU(2)$ to $SL(2, c)$ within the world of two-by-two matrices. From the physical point of view, we are building a bridge between Wigner's 1939 paper [28] and Dirac's equation for four-component spinors as well as Maxwell's Lorentz-covariant electromagnetic fields.

From equation (1.12), we know that there is both an undotted and a dotted representation of the Lorentz group. In this Lorentz-covariant world, there are, therefore, two additional spin states, which means that in the $SL(2, c)$ regime, there are four spin states for each particle. Corresponding to the undotted and the dotted representations, we can construct one set of two-component spinors (spin-up and spin-down) for the undotted representation, and another set for the dotted representation. This is the reason why the Dirac spinor has four components [18]. These four spin states are shown in table 5.1. If four spinors (two undotted and two dotted) are coupled, there are sixteen (4×4) states, which can be partitioned into the spin-0 and spin-1 states. We shall explicitly derive these sixteen states in section 5.5.

In sections 4.1 and 4.2, we defined the Wigner energy–momentum four-vector and Wigner transformation matrices in the two-by-two representation and summarized the results in table 4.1. We noted that the determinants of these matrices are positive, zero, and negative, respectively, corresponding to the energy–momentum four-vector of massive, massless, and imaginary-mass particles. In the world of $SL(2, c)$, we have to consider the dotted representation as well. These transformation matrices are shown in table 5.2.

In his 1939 paper [28] Wigner constructed the subgroups of the Lorentz group, i.e., Wigner's little groups, whose transformations leave the energy–momentum four-vector invariant. In chapter 2, we derived these little groups in the four-by-four representation. Here we examine these subgroups in the two-by-two matrix format.

We noted in section 4.1 that this energy–momentum four-vector is invariant under rotations around the z-axis. Referring to table 1.1, this rotation matrix is:

$$Z(\phi) = \begin{pmatrix} e^{i\phi/2} & 0 \\ 0 & e^{-i\phi/2} \end{pmatrix}. \tag{5.5}$$

In addition, we considered the transformation within the zx-plane. The matrix for these rotations is also given in table 1.1. For a massive particle at rest, the energy–momentum four-vector is invariant under rotations around the y-axis, whose transformation matrix is

Table 5.1. Spinors in the $SL(2, c)$ representation. The spinors u and $\dot{u}$ represent the spin-up states, while the spinors v and $\dot{v}$ represent spin-down states.

	Undotted	Dotted
Spin up	u	$\dot{u}$
Spin down	v	$\dot{v}$

Table 5.2. The undotted and dotted Wigner transformation matrices are presented with the corresponding two-by-two matrix representation of Wigner energy–momentum four-vectors. The determinants of the two-by-two four-vector matrices for the massive, massless, and imaginary-mass particles are positive, zero, and negative, respectively.

Wigner Four-vector	Wigner Transformation Matrix	Dotted Wigner Transformation Matrix
$\begin{pmatrix} 1 & 0 \\ 0 & 1 \end{pmatrix}$	$\begin{pmatrix} \cos(\theta/2) & -\sin(\theta/2) \\ \sin(\theta/2) & \cos(\theta/2) \end{pmatrix}$	$\begin{pmatrix} \cos(\theta/2) & -\sin(\theta/2) \\ \sin(\theta/2) & \cos(\theta/2) \end{pmatrix}$
$\begin{pmatrix} 1 & 0 \\ 0 & 0 \end{pmatrix}$	$\begin{pmatrix} 1 & -\gamma \\ 0 & 1 \end{pmatrix}$	$\begin{pmatrix} 1 & 0 \\ \gamma & 1 \end{pmatrix}$
$\begin{pmatrix} 1 & 0 \\ 0 & -1 \end{pmatrix}$	$\begin{pmatrix} \cosh(\lambda/2) & \sinh(\lambda/2) \\ \sinh(\lambda/2) & \cosh(\lambda/2) \end{pmatrix}$	$\begin{pmatrix} \cosh(\lambda/2) & -\sinh(\lambda/2) \\ -\sinh(\lambda/2) & \cosh(\lambda/2) \end{pmatrix}$

$$R(\theta) = \begin{pmatrix} \cos(\theta/2) & -\sin(\theta/2) \\ \sin(\theta/2) & \cos(\theta/2) \end{pmatrix}. \tag{5.6}$$

This matrix, together with $Z(\phi)$, leads to the rotation also around the x-axis. Thus, Wigner's little group for the massive particle is the three-dimensional rotation subgroup of the Lorentz group.

For the massless particle, the triangular matrix of the form:

$$T(\gamma) = \begin{pmatrix} 1 & -\gamma \\ 0 & 1 \end{pmatrix} \quad \text{or} \quad \dot{T}(\gamma) = \begin{pmatrix} 1 & 0 \\ \gamma & 1 \end{pmatrix} \tag{5.7}$$

satisfies the Wigner condition of equation (4.2). To preserve the Lorentz properties of the boosted momentum, γ must be real. These matrices are not Hermitian, but rather

$$\begin{aligned} T^{-1}(\gamma) &= \begin{pmatrix} 1 & \gamma \\ 0 & 1 \end{pmatrix} \quad \text{and} \quad T^{\dagger}(\gamma) = \begin{pmatrix} 1 & 0 \\ -\gamma & 1 \end{pmatrix}; \\ \dot{T}^{-1}(\gamma) &= \begin{pmatrix} 1 & 0 \\ -\gamma & 1 \end{pmatrix} \quad \text{and} \quad \dot{T}^{\dagger}(\gamma) = \begin{pmatrix} 1 & \gamma \\ 0 & 1 \end{pmatrix}. \end{aligned} \tag{5.8}$$

If we allow rotations around the z-axis, these triangular matrices become

$$T(\gamma e^{i\phi}) = \begin{pmatrix} 1 & -\gamma e^{i\phi} \\ 0 & 1 \end{pmatrix} \quad \text{or} \quad \dot{T}(\gamma e^{-i\phi}) = \begin{pmatrix} 1 & 0 \\ \gamma e^{-i\phi} & 1 \end{pmatrix}. \tag{5.9}$$

The T matrix is generated by the two-by-two version of the matrices defined in equation (2.10):

$$N_1 = K_1 - J_2 = \begin{pmatrix} 0 & i \\ 0 & 0 \end{pmatrix} \quad \text{and} \quad N_2 = K_2 + J_1 = \begin{pmatrix} 0 & 1 \\ 0 & 0 \end{pmatrix}. \tag{5.10}$$

The $\dot{T}$ matrix is generated by

$$\dot{N}_1 = \dot{K}_1 - J_2 = \begin{pmatrix} 0 & 0 \\ -i & 0 \end{pmatrix} \quad \text{and} \quad \dot{N}_2 = \dot{K}_2 + J_1 = \begin{pmatrix} 0 & 0 \\ 1 & 0 \end{pmatrix}. \tag{5.11}$$

Thus, the little group is generated by J_3, N_1, and N_2, or by J_3, $\dot{N}_1$, and $\dot{N}_2$. Hence, there are two sets of boost generators involved. These generators satisfy the following sets of commutation relations:

$$[N_1, N_2] = 0, \qquad [N_1, J_3] = iN_2, \qquad \text{and} \qquad [N_2, J_3] = -iN_1, \tag{5.12}$$

and

$$[\dot{N}_1, \dot{N}_2] = 0, \qquad [\dot{N}_1, J_3] = i\dot{N}_2, \qquad \text{and} \qquad [\dot{N}_2, J_3] = -i\dot{N}_1. \tag{5.13}$$

As usual, J_3 is the generator of rotations and the N_i generate translation-like transformations where:

$$D(\alpha, \beta) = D(\alpha, 0)D(0, \beta) = D(0, \beta)D(\alpha, 0). \tag{5.14}$$

Wigner in 1939 [28] observed that the first set given in equation (5.12) is the same as that of the generators for the two-dimensional Euclidean group with one rotation and two translations. The physical interpretation of the rotation is the helicity of the massless particle. On the other hand, the physics of the N_1 and N_2 matrices was not obvious, but they were subsequently shown to be gauge transformations [10, 20, 21, 26].

For the imaginary-mass particle, a matrix of the form:

$$S(\lambda) = \begin{pmatrix} \cosh(\lambda/2) & \sinh(\lambda/2) \\ \sinh(\lambda/2) & \cosh(\lambda/2) \end{pmatrix}, \tag{5.15}$$

leaves the energy–momentum four-vector invariant. This matrix corresponds to the Lorentz boost along the x-direction generated by K_1, as shown in table 1.1. Because of the rotational symmetry around the z-axis, the Wigner condition is also satisfied by the boost along the y-axis. The little group is thus generated by J_3, K_1, and K_2. The commutators of these three generators,

$$[J_3, K_1] = iK_2, \qquad [J_3, K_2] = -iK_1, \qquad \text{and} \qquad [K_1, K_2] = -iJ_3, \tag{5.16}$$

form the two-by-two representation of the little group $O(2, 1)$, the Lorentz group applicable to two space and one time dimensions, as discussed in section 2.4. The dotted matrices satisfy the same set of commutation relations.

We know that in the $SL(2, c)$ world, the rotations are the same in the undotted and the dotted representation because the rotation generators are Hermitian. However, for the boost generators, we have to consider both forms of the boost generators. As $\dot{K}_i$ equals $-K_i$ as given in equation (1.12), the Lorentz boosts for the dotted representation are performed in the opposite direction, as shown in table 5.3.

There are many interesting three-parameter subgroups of the Lorentz group. In 1939 [28], Wigner considered only those subgroups whose transformations leave the energy–momentum four-vector of a given particle invariant. First of all, consider a

Table 5.3. Generators and two-by-two transformation matrices for both the dotted and undotted representation of the Lorentz group. Here we see that although the rotations take the same form for both representations, boosts are performed in opposite directions for dotted and undotted representations.

Generators	Transformation Matrix	Dotted Trans. Matrix
$J_1 = \frac{1}{2}\begin{pmatrix} 0 & 1 \\ 1 & 0 \end{pmatrix}$	$\begin{pmatrix} \cos(\theta/2) & i\sin(\theta/2) \\ i\sin(\theta/2) & \cos(\theta/2) \end{pmatrix}$	same
$K_1 = \frac{1}{2}\begin{pmatrix} 0 & i \\ i & 0 \end{pmatrix}$	$\begin{pmatrix} \cosh(\lambda/2) & \sinh(\lambda/2) \\ \sinh(\lambda/2) & \cosh(\lambda/2) \end{pmatrix}$	inverse
$J_2 = \frac{1}{2}\begin{pmatrix} 0 & -i \\ i & 0 \end{pmatrix}$	$\begin{pmatrix} \cos(\theta/2) & -\sin(\theta/2) \\ \sin(\theta/2) & \cos(\theta/2) \end{pmatrix}$	same
$K_2 = \frac{1}{2}\begin{pmatrix} 0 & 1 \\ -1 & 0 \end{pmatrix}$	$\begin{pmatrix} \cosh(\lambda/2) & -i\sinh(\lambda/2) \\ i\sinh(\lambda/2) & \cosh(\lambda/2) \end{pmatrix}$	inverse
$J_3 = \frac{1}{2}\begin{pmatrix} 1 & 0 \\ 0 & -1 \end{pmatrix}$	$\begin{pmatrix} e^{i\phi/2} & 0 \\ 0 & e^{-i\phi/2} \end{pmatrix}$	same
$K_3 = \frac{1}{2}\begin{pmatrix} i & 0 \\ 0 & -i \end{pmatrix}$	$\begin{pmatrix} e^{\eta/2} & 0 \\ 0 & e^{-\eta/2} \end{pmatrix}$	inverse

massive particle at rest. The momentum of this particle is invariant under rotations in three-dimensional space. What happens for the massless particle that cannot be brought to a rest frame? We shall study this problem for understanding the space–time symmetry of neutrinos and photons in sections 5.4 and 5.5.

5.2 Massive and massless particles

Since the massive particle at rest remains invariant under rotations, we can Lorentz-boost this particle along the z-direction. Using the undotted boost matrix given in table 1.1,

$$B(\eta) = \begin{pmatrix} e^{\eta/2} & 0 \\ 0 & e^{-\eta/2} \end{pmatrix}, \tag{5.17}$$

the momentum of the massive particle becomes

$$p_z = m\sinh(\eta). \tag{5.18}$$

Here we can relate the boost parameter η to the physical variables:

$$e^{\eta} = \frac{p_z + \sqrt{p_z^2 + m^2}}{m}. \tag{5.19}$$

The boost matrix $B(\eta)$ of equation (5.17) commutes with the rotation matrix $Z(\phi)$ given in equation (5.5).

Table 5.4. The transformations of $T(\gamma)$ and $\dot{T}(\gamma)$ on the spinors. The parity invariance of the Lie algebra of the Lorentz group leads us to consider not only the transformation of triangular matrices applicable to both u and v, but also those applicable to $\dot{u}$ and $\dot{v}$.

	$T(\gamma)$ with $+\eta$	$\dot{T}(\gamma)$ with $-\eta$
Spinors	$T(\gamma)u = u$	$\dot{T}(\gamma)u = u + \gamma v$
	$T(\gamma)v = v - \gamma u$	$\dot{T}(\gamma)v = v$
Dotted spinors	$T(\gamma)\dot{u} = \dot{u}$	$\dot{T}(\gamma)\dot{u} = \dot{u} + \gamma\dot{v}$
	$T(\gamma)\dot{v} = \dot{v} - \gamma\dot{u}$	$\dot{T}(\gamma)\dot{v} = \dot{v}$

If instead, using the matrix given in equation (5.6), we rotate around the y-axis, which is perpendicular to the z-axis, this matrix becomes boosted to

$$B(\eta)R(\theta)B^{-1}(\eta) = \begin{pmatrix} \cos(\theta/2) & -e^{\eta}\sin(\theta/2) \\ e^{-\eta}\sin(\theta/2) & \cos(\theta/2) \end{pmatrix}. \tag{5.20}$$

It is clear from equation (5.18) that the mass becomes smaller as η becomes infinite. To keep the quantities in equation (5.20) finite, it is necessary that the upper-right element $e^{\eta}\sin(\theta/2)$ remain finite. We'll call this element γ. The lower-left element, $e^{-\eta}\sin(\theta/2)$, vanishes as η becomes infinite. The boosted rotation matrix then becomes the triangular Wigner matrix given in equation (5.7) as the angle θ becomes zero. The results of applying the matrices in equation (5.7) to the undotted and dotted spinors given in table 5.1 are shown in table 5.4.

Because of the rotational degree of freedom around the z-axis, the matrix of equation (5.20) can be generalized into

$$\begin{pmatrix} e^{i\phi/2} & 0 \\ 0 & e^{-i\phi/2} \end{pmatrix}\begin{pmatrix} \cos(\theta/2) & -e^{\eta}\sin(\theta/2) \\ e^{-\eta}\sin(\theta/2) & \cos(\theta/2) \end{pmatrix}\begin{pmatrix} e^{-i\phi/2} & 0 \\ 0 & e^{i\phi/2} \end{pmatrix}. \tag{5.21}$$

This becomes

$$\begin{pmatrix} \cos(\theta/2) & -e^{i\phi}e^{\eta}\sin(\theta/2) \\ e^{-i\phi}e^{-\eta}\sin(\theta/2) & \cos(\theta/2) \end{pmatrix}. \tag{5.22}$$

In the large-η limit, this expression leads to the triangular matrix of equation (5.9).

When applied to energy–momentum four-vectors, the triangular matrices T and $\dot{T}$ generate gauge transformations. From table 5.4, it can now be shown that they perform gauge transformations on massless spin-half particles [11, 13] as well as on photons.

5.3 Dirac spinors and massless particles

Dirac's equation is usually thought of as applicable to massive particles, in particular electrons. However, it is possible to consider the massless particle as the large-momentum/zero-mass limit of the Dirac spinors for the massive particle.

If we start with the spin operators defined in equations (1.16), we can apply a boost to J_i along the z-direction:

$$J_i' = B(\eta)\, J_i\, B^{-1}(\eta). \tag{5.23}$$

This similarity transformation is accomplished using the boost matrix given by equation (5.17). The boost parameter then becomes

$$e^{\eta} = \left(\frac{E+p}{E-p}\right)^{1/2}. \tag{5.24}$$

For a massive particle in the large-(momentum/mass) limit, we obtain:

$$e^{\eta} \to \frac{2E}{m}. \tag{5.25}$$

If we now use the similarity transformation of equation (5.23), as usual, J_3 is invariant, but J_1 and J_2 take the form:

$$J_1' = \frac{1}{2}\begin{pmatrix} 0 & e^{\eta} \\ e^{-\eta} & 0 \end{pmatrix} \quad \text{and} \quad J_2' = \frac{i}{2}\begin{pmatrix} 0 & -e^{\eta} \\ e^{-\eta} & 0 \end{pmatrix}. \tag{5.26}$$

Using the large-(momentum/mass) limit for a massive particle, the N_i matrices of equation (5.11) can be written as:

$$N_1 = -\frac{m}{E}J_2' \quad \text{and} \quad N_2 = \frac{E}{m}J_1'. \tag{5.27}$$

We can now discuss the large-momentum/zero-mass limit of the Dirac equation. Since, in $SL(2, c)$, it is necessary to consider both signs of the boost generators, we obtain:

$$J_i = \frac{1}{2}\begin{pmatrix} \sigma_i & 0 \\ 0 & \sigma_i \end{pmatrix} \quad \text{and} \quad K_i = \frac{i}{2}\begin{pmatrix} \sigma_i & 0 \\ 0 & -\sigma_i \end{pmatrix}. \tag{5.28}$$

This is applicable to Dirac wave functions in the Weyl representation [1, 18]. From the N_i and the $\dot{N}_i$ matrices defined in equations (5.10) and (5.11) we can write

$$\begin{aligned} D(\alpha, \beta) &= \exp(-i[\alpha N_1 + \beta N_2]) = \begin{pmatrix} 1 & \alpha - i\beta \\ 0 & 1 \end{pmatrix}, \\ \dot{D}(\alpha, \beta) &= \exp(-i[\alpha \dot{N}_1 + \beta \dot{N}_2]) = \begin{pmatrix} 1 & 0 \\ -\alpha - i\beta & 1 \end{pmatrix}. \end{aligned} \tag{5.29}$$

Recall that there are two sets of spinors in $SL(2, c)$. We have designated the spinors whose boosts are generated by $K_i = (i/2)\sigma_i$ as u and v. For the boosts generated by $\dot{K}_i = (-i/2)\sigma_i$ we have used $\dot{u}$ and $\dot{v}$. These spinors are gauge-invariant in the sense that

$$D(\alpha, \beta)u = u \quad \text{and} \quad \dot{D}(\alpha, \beta)\dot{v} = \dot{v}. \tag{5.30}$$

However, these spinors are gauge-dependent in the sense that

$$D(\alpha, \beta)v = v + (\alpha - i\beta)u \quad \text{and} \quad \dot{D}(\alpha, \beta)\dot{u} = \dot{u} - (\alpha + i\beta)\dot{v}. \tag{5.31}$$

It is then possible to use the gauge transformation matrices from equation (5.14) to obtain:

$$D(\alpha, \beta) = \begin{pmatrix} D(\alpha, \beta) & 0 \\ 0 & \dot{D}(\alpha, \beta) \end{pmatrix}. \tag{5.32}$$

This matrix can be applied to the Dirac spinors. When the result of applying the D matrix from equation (5.32) to the eigen-spinors given in equation (5.2) is evaluated for the massive Dirac particle at rest, we obtain:

$$u(0) = \begin{pmatrix} u \\ \pm\,\dot{u} \end{pmatrix} \quad \text{and} \quad v(0) = \begin{pmatrix} \pm\, v \\ \dot{v} \end{pmatrix}. \tag{5.33}$$

Here we represent the positive and negative energy states by the + and − signs, respectively. If these spinors are boosted along the z-axis using the operator generated by K_3, they can be written as:

$$u(\eta) = \begin{pmatrix} e^{+\eta/2}u \\ \pm\, e^{-\eta/2}\dot{u} \end{pmatrix} \quad \text{and} \quad v(\eta) = \begin{pmatrix} \pm\, e^{-\eta/2}v \\ e^{+\eta/2}\dot{v} \end{pmatrix}. \tag{5.34}$$

From equation (5.34), we see that in the large-momentum/zero-mass limit, the large components, $e^{(+\eta/2)}$, are gauge invariant. However, the small components are gauge dependent. Therefore, we see that non-zero mass, spin-$\frac{1}{2}$ particles are not invariant under gauge transformations. We can also renormalize the spinors of equation (5.34) in this limit, and thus obtain:

$$u = \begin{pmatrix} u \\ 0 \end{pmatrix} \quad \text{and} \quad v = \begin{pmatrix} 0 \\ \dot{v} \end{pmatrix}. \tag{5.35}$$

Now it is apparent that the D transformation leaves these spinors invariant. As shown in the next section, This invariance is responsible for the polarization of neutrinos [11, 16].

5.4 Polarization of massless neutrinos

To apply this analysis to spin-$\frac{1}{2}$ particles, we use the group generated by equation (1.16)

$$J_i = \frac{1}{2}\sigma_i \quad \text{and} \quad K_i = \frac{i}{2}\sigma_i. \tag{5.36}$$

These are identical to those for the proper Lorentz group and have the same algebraic properties as the $SL(2, c)$ group. Additionally, the Lie algebra for these generators is invariant if the sign of the boost operators is changed. In the case of $SL(2, c)$, or spin-$\frac{1}{2}$ particles, it is necessary to consider both the undotted and the dotted boost generators. We showed that the group $SL(2, c)$ consists of non-singular two-by-two matrices, which have the form defined in equation (1.18). This matrix is applicable to spinors defined in equation (5.2).

Among the subgroups of $SL(2, c)$, there are $E(2)$-like little groups which correspond to massless particles. If we consider a massless particle moving along the z-direction, then the little group is generated by J_3, N_1, and N_2, defined in equation (2.10). These N_i operators are the generators of gauge transformations in the case of the photon, thus we will refer to them as the gauge transformation in the $SL(2, c)$ regime [11, 28]. We shall examine their role with respect to massless particles of spin-$\frac{1}{2}$.

Since the sign of the generators of this subgroup remains invariant under a sign change in K_i, these generators remain unambiguous when applied to the space–time coordinate variable and the photon four-vectors. Here we choose the J_i to be the generators of rotations, but, because of the sign change allowed for K_i, we must consider both N_i and $\dot{N}_i$, where N_i and $\dot{N}_i$ are given in equations (5.10) and (5.11), respectively. The transformation matrices are then written as in equation (5.29). From equations (5.30) and (5.31), we know that the spinors whose boosts are generated by K_i and $\dot{K}_i$ are gauge-invariant in the sense given in equation (5.30) and gauge-dependent in the sense given in equation (5.31). The gauge-invariant spinors of equation (5.30) appear as polarized neutrinos [11, 14]. In the massless limit,

$$D(\alpha, \beta) = \begin{pmatrix} 1 & \gamma \\ 0 & 1 \end{pmatrix} \quad \text{and} \quad \dot{D}(\alpha, \beta) = \begin{pmatrix} 1 & 0 \\ -\gamma & 1 \end{pmatrix}. \tag{5.37}$$

However, if we insist on gauge invariance in the world of particles and fields, massless spin-$\frac{1}{2}$ particles are polarized, as shown in figure 5.1. The dotted spinor becomes right-handed, while the undotted spinor becomes left-handed [11, 13, 14]. Indeed, this is what we observe in the real world. Massless neutrinos and anti-neutrinos are left- and right-handed, respectively. Yes, neutrinos have non-zero mass [17, 22], but the mass is so small compared with the momentum that the small mass can be regarded as a small correction to the massless state. In other words, the massless state of neutrinos still plays important roles in physics.

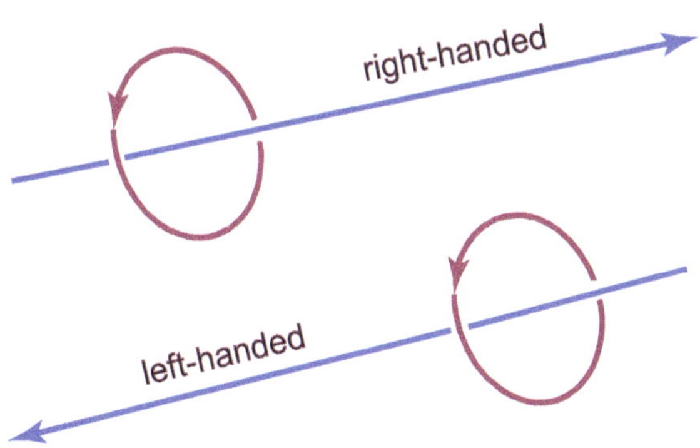

Figure 5.1. The polarization of massless neutrinos is a consequence of gauge invariance. As a result, what we see in nature is that the neutrino is left-handed while the anti-neutrino is right-handed [19].

5.5 Scalars, vectors, tensors, and the polarization of photons

We are quite familiar with the process of constructing three spin-1 states and one spin-0 state from two spin-$\frac{1}{2}$ spinors. In the Lorentz-covariant world, we have two undotted and two dotted states [5, 8, 18, 23], as we noted in section 5.1. If four of those two-state spinors are combined, there are sixteen states. In this section, we shall construct all sixteen states which can be partitioned as follows:

1. scalar with one state,
2. pseudo-scalar with one state,
3. four-vector with four states,
4. axial vector with four states,
5. second-rank tensor with six states.

For particles at rest, the addition of two spin-$\frac{1}{2}$ spinors results in spin-zero and spin-one states. When they are boosted, since we have two different sets of spinors, undotted and dotted, the spinors behave differently under the Lorentz boost, as shown in table 5.3. Around the z-direction, both sets of spinors are transformed by

$$Z(\phi) = \exp(-i\phi J_3) = \begin{pmatrix} e^{i\phi/2} & 0 \\ 0 & e^{-i\phi/2} \end{pmatrix}. \tag{5.38}$$

However, they are boosted by

$$\begin{aligned} B(\eta) &= \exp(-i\eta K_3) = \begin{pmatrix} e^{\eta/2} & 0 \\ 0 & e^{-\eta/2} \end{pmatrix}, \\ \dot{B}(\eta) &= \exp(i\eta K_3), = \begin{pmatrix} e^{-\eta/2} & 0 \\ 0 & e^{\eta/2} \end{pmatrix}, \end{aligned} \tag{5.39}$$

applicable to the undotted and dotted spinors, respectively. These two matrices commute with each other, and with the rotation matrix $Z(\phi)$ of equation (5.38). Since K_3 and J_3 commute with each other, we can work with the matrix $Q(\eta, \phi)$ defined as:

$$\begin{aligned} Q(\eta, \phi) = B(\eta)Z(\phi) &= \begin{pmatrix} e^{(\eta+i\phi)/2} & 0 \\ 0 & e^{-(\eta+i\phi)/2} \end{pmatrix}, \\ \dot{Q}(\eta, \phi) = \dot{B}(\eta)Z(\phi) &= \begin{pmatrix} e^{-(\eta-i\phi)/2} & 0 \\ 0 & e^{(\eta-i\phi)/2} \end{pmatrix}. \end{aligned} \tag{5.40}$$

When this combined matrix is applied to the spinors,

$$\begin{aligned} Q(\eta, \phi)u = e^{(\eta+i\phi)/2}u \quad &\text{and} \quad Q(\eta, \phi)v = e^{-(\eta+i\phi)/2}v; \\ \dot{Q}(\eta, \phi)\dot{u} = e^{-(\eta-i\phi)/2}\dot{u} \quad &\text{and} \quad \dot{Q}(\eta, \phi)\dot{v} = e^{(\eta-i\phi)/2}\dot{v}. \end{aligned} \tag{5.41}$$

If the particle is at rest, we can construct the combinations given in equations (5.3) and (5.4) to obtain the spin-one and spin-zero states. These are four bilinear states. In the $SL(2, c)$ regime, we must include both undotted and dotted spinors, resulting

Table 5.5. Bilinear combinations of the four $SL(2, c)$ spinors. Unlike $SU(2)$ which has two spinors leading to four bilinear forms, $SL(2, c)$ has two undotted and two dotted spinors leading to sixteen independent bilinear combinations.

Spin 1			Spin 0
uu,	$\frac{1}{\sqrt{2}}(uv + vu)$,	vv,	$\frac{1}{\sqrt{2}}(uv - vu)$
$\dot{u}\dot{u}$,	$\frac{1}{\sqrt{2}}(\dot{u}\dot{v} + \dot{v}\dot{u})$,	$\dot{v}\dot{v}$,	$\frac{1}{\sqrt{2}}(\dot{u}\dot{v} - \dot{v}\dot{u})$
$u\dot{u}$,	$\frac{1}{\sqrt{2}}(u\dot{v} + v\dot{u})$,	$v\dot{v}$,	$\frac{1}{\sqrt{2}}(u\dot{v} - v\dot{u})$
$\dot{u}u$,	$\frac{1}{\sqrt{2}}(\dot{u}v + \dot{v}u)$,	$\dot{v}v$,	$\frac{1}{\sqrt{2}}(\dot{u}v - \dot{v}u)$
After the operation of $Q(\eta, \phi)$ and $\dot{Q}(\eta, \phi)$			
$e^{i\phi}e^{\eta}uu$,	$\frac{1}{\sqrt{2}}(uv + vu)$,	$e^{-i\phi}e^{-\eta}vv$,	$\frac{1}{\sqrt{2}}(uv - vu)$
$e^{i\phi}e^{-\eta}\dot{u}\dot{u}$,	$\frac{1}{\sqrt{2}}(\dot{u}\dot{v} + \dot{v}\dot{u})$,	$e^{-i\phi}e^{\eta}\dot{v}\dot{v}$,	$\frac{1}{\sqrt{2}}(\dot{u}\dot{v} - \dot{v}\dot{u})$
$e^{i\phi}u\dot{u}$,	$\frac{1}{\sqrt{2}}(e^{\eta}u\dot{v} + e^{-\eta}v\dot{u})$,	$e^{-i\phi}v\dot{v}$,	$\frac{1}{\sqrt{2}}(e^{\eta}u\dot{v} - e^{-\eta}v\dot{u})$
$e^{i\phi}\dot{u}u$,	$\frac{1}{\sqrt{2}}(\dot{u}v + \dot{v}u)$,	$e^{-i\phi}\dot{v}v$,	$\frac{1}{\sqrt{2}}(e^{-\eta}\dot{u}v - e^{\eta}\dot{v}u)$

in sixteen independent bilinear combinations, as given in table 5.5. This table also gives the effect of the operation of $Q(\eta, \phi)$ and $\dot{Q}(\eta, \phi)$.

Among the bilinear combinations given in table 5.5, the following two equations are invariant under rotations and also under boosts:

$$S = \frac{1}{\sqrt{2}}(uv - vu) \quad \text{and} \quad \dot{S} = -\frac{1}{\sqrt{2}}(\dot{u}\dot{v} - \dot{v}\dot{u}). \tag{5.42}$$

They are thus scalars in the Lorentz-covariant world. Are they the same or different? Let us consider the following combinations:

$$S_{+} = \frac{1}{\sqrt{2}}(S + \dot{S}) \quad \text{and} \quad S_{-} = \frac{1}{\sqrt{2}}(S - \dot{S}). \tag{5.43}$$

Under the dot conjugation, S_{+} remains invariant, but S_{-} changes sign. The boost is performed in the opposite direction and therefore is the operation of space inversion. Thus S_{+} is a scalar while S_{-} is called a pseudo-scalar.

5.5.1 Four-vectors

In equation (1.20) we wrote the expression for the space–time four-vector as

$$\begin{pmatrix} t + z & x - iy \\ x + iy & t - z \end{pmatrix}. \tag{5.44}$$

Under the parity operation, this becomes

$$\begin{pmatrix} t - z & -x + iy \\ -x - iy & t + z \end{pmatrix}. \tag{5.45}$$

Here we see that off-diagonal elements undergo sign changes, and the diagonal elements become interchanged. Then the four-vectors in equation (5.44) and its dot conjugation can be constructed as

$$V \simeq \begin{pmatrix} \dot{v}u - u\dot{v} & \dot{v}v - v\dot{v} \\ u\dot{u} - \dot{u}u & v\dot{u} - \dot{u}v \end{pmatrix} \quad \text{and} \quad \dot{V} \simeq \begin{pmatrix} v\dot{u} - \dot{u}v & v\dot{v} - \dot{v}v \\ \dot{u}u - u\dot{u} & \dot{v}u - u\dot{v} \end{pmatrix}, \tag{5.46}$$

respectively. As in the above, the off-diagonal elements undergo sign changes, and the diagonal elements become interchanged.

If the expression of equation (5.44) is for an axial vector, the parity operation leads to

$$\begin{pmatrix} -t + z & x - iy \\ x + iy & -t - z \end{pmatrix}, \tag{5.47}$$

where only the sign of t is changed. The off-diagonal elements remain invariant, while the diagonal elements are interchanged with sign changes. The axial vector can be written as

$$A = \begin{pmatrix} u\dot{v} + \dot{v}u & v\dot{v} + \dot{v}v \\ u\dot{u} + \dot{u}u & -\dot{u}v - v\dot{u} \end{pmatrix}. \tag{5.48}$$

Here, the off-diagonal elements do not change their signs under dot conjugation, and the diagonal elements become interchanged with a sign change. This matrix thus represents an axial vector. Let us then write the electromagnetic axial vector or four-potential in the two-by-two vector format as

$$A = \begin{pmatrix} A_0 + A_z & A_x - iA_y \\ A_x + iA_y & A_0 - A_z \end{pmatrix}. \tag{5.49}$$

When this matrix is boosted along the z-direction, we obtain

$$A = \begin{pmatrix} (A_0 + A_z)e^{\eta} & A_x - iA_y \\ A_x + iA_y & (A_0 - A_z)e^{-\eta} \end{pmatrix}. \tag{5.50}$$

If we make then the Wigner excursion as illustrated in figure 2.5, this matrix for a massive particle at rest becomes transformed to that of a massless particle with the same energy. Therefore we obtain

$$A = \begin{pmatrix} A_0 + A_z & A_x - iA_y \\ A_x + iA_y & 0 \end{pmatrix}. \tag{5.51}$$

This results in $A_0 = A_z$ which is widely known as the Lorentz condition.

Let us now perform the $T(\gamma)$ transformation on u and v, while $\dot{T}(\gamma)$ is used on $\dot{u}$ and $\dot{v}$. Then the A matrix becomes

$$A + 2\gamma \begin{pmatrix} 0 & A_0 \\ A_0 & 0 \end{pmatrix}. \tag{5.52}$$

The result of this is that there is an addition of $2\gamma A_0$ to A_x, which is a translation in the plane of A_x and A_y.

On the other hand, if we perform the $\dot{T}(\gamma)$ transformation on u and v, while we use $T(\gamma)$ on $\dot{u}$ and $\dot{v}$, A becomes

$$A - 2\gamma\begin{pmatrix} A_x & 0 \\ 0 & 0 \end{pmatrix}. \tag{5.53}$$

The transformations for the triangular matrices can be found in table 5.4.

It now becomes a question of choosing between equations (5.52) and (5.53) for our transformation. To decide, we use the transformation rule given in equation (1.23) for the four-vector and apply the triangular matrix $T(\gamma)$ as the Lorentz transformation matrix, resulting in the matrix multiplication

$$\begin{pmatrix} 1 & -\gamma \\ 0 & 1 \end{pmatrix}\begin{pmatrix} A_0 + A_z & A_x - iA_y \\ A_x + iA_y & 0 \end{pmatrix}\begin{pmatrix} 1 & 0 \\ -\gamma & 1 \end{pmatrix}. \tag{5.54}$$

This then becomes

$$A - 2\gamma\begin{pmatrix} A_x & 0 \\ 0 & 0 \end{pmatrix}, \tag{5.55}$$

which has the same form as that given in equation (5.53). We are then able to choose the transformation of equation (5.53) as the $T(\gamma)$ and $\dot{T}(\gamma)$ transformations applicable to the spinors. We can see, since in both equations (5.53) and (5.55), A_x and A_y remain invariant while there is an addition to $A_0 = A_z$, that the $T(\gamma)$ matrices lead to a gauge transformation.

So far, everything we have done can be rotated around the z-axis. Then, γ is replaced by $\gamma e^{i\phi}$. The transformed A of equation (5.53) becomes

$$A - 2\gamma\begin{pmatrix} A_x \cos\phi + A_y \sin\phi & 0 \\ 0 & 0 \end{pmatrix}. \tag{5.56}$$

We can reach the same conclusion using the four-by-four formulation of the Lorentz group. The geometries contained in this larger representation lead to equations (5.53) and (5.56) [10, 20, 21, 26]. Thus, from equation (5.56) it is evident that $T(\gamma e^{i\phi})$ performs a gauge transformation.

Suppose we return to the limiting process discussed in section 2.3. Then we see that the transverse rotational degrees of freedom collapse into one gauge degree of freedom in the infinite-momentum or zero-mass limit. This gauge degree of freedom was first observed by Han *et al* in 1983 [12], and its geometry was given by Kim and Wigner in 1990 [20]. The most recent version of this geometry was given by the present authors in 2017 [3].

5.5.2 Second-rank tensor

In section 5.5.1, each bilinear spinor consisted of one dotted and one undotted spinor. There are also bilinear spinors which are both dotted or both undotted. We are interested in two sets of three quantities satisfying the $O(3)$ symmetry. They should therefore transform like

$$(x + iy)/\sqrt{2}, \qquad (x - iy)/\sqrt{2}, \qquad z, \tag{5.57}$$

which are like

$$uu, \qquad vv, \quad (uv + vu)/\sqrt{2}, \tag{5.58}$$

respectively, in the $O(3)$ regime. Since the dot conjugation is the parity operation, they are like

$$-\dot{u}\dot{u}, \quad -\dot{v}\dot{v}, \quad -(\dot{u}\dot{v} + \dot{v}\dot{u})/\sqrt{2}. \tag{5.59}$$

In other words,

$$\dot{(uu)} = -\dot{u}\dot{u}, \quad \text{and} \quad \dot{(vv)} = -\dot{v}\dot{v}. \tag{5.60}$$

To construct the z component of both equations (5.58) and (5.59) in this $O(3)$ space, let us first consider

$$f_z = \frac{1}{2}[(uv + vu) - (\dot{u}\dot{v} + \dot{v}\dot{u})] \quad \text{and} \quad g_z = \frac{1}{2i}[(uv + vu) + (\dot{u}\dot{v} + \dot{v}\dot{u})], \tag{5.61}$$

where f_z and g_z are, respectively, symmetric and antisymmetric under the dot conjugation or the parity operation. These quantities are invariant under the boost along the z-direction. They are also invariant under rotations around this axis, but they are not invariant under a boost along or a rotation around the x- or y-axis. They are different from the scalars given in equation (5.42).

Next, in order to construct the x and y components, we start with $f_\pm$ and $g_\pm$ as

$$\begin{aligned} f_+ &= \frac{1}{\sqrt{2}}(uu - \dot{u}\dot{u}), \qquad & g_+ &= \frac{1}{\sqrt{2}i}(uu + \dot{u}\dot{u}), \\ f_- &= \frac{1}{\sqrt{2}}(vv - \dot{v}\dot{v}), \qquad & g_- &= \frac{1}{\sqrt{2}i}(vv + \dot{v}\dot{v}). \end{aligned} \tag{5.62}$$

Then

$$\begin{aligned} f_x &= \frac{1}{\sqrt{2}}(f_+ + f_-) = \frac{1}{2}[(uu - \dot{u}\dot{u}) + (vv - \dot{v}\dot{v})], \\ f_y &= \frac{1}{\sqrt{2}i}(f_+ - f_-) = \frac{1}{2i}[(uu - \dot{u}\dot{u}) - (vv - \dot{v}\dot{v})], \end{aligned} \tag{5.63}$$

and

$$\begin{aligned} g_x &= \frac{1}{\sqrt{2}}(g_+ + g_-) = \frac{1}{2i}[(uu + \dot{u}\dot{u}) + (vv + \dot{v}\dot{v})], \\ g_y &= \frac{1}{\sqrt{2}i}(g_+ - g_-) = -\frac{1}{2}[(uu + \dot{u}\dot{u}) - (vv + \dot{v}\dot{v})]. \end{aligned} \tag{5.64}$$

Here f_x and f_y are symmetric under dot conjugation, while g_x and g_y are anti-symmetric.

Furthermore, f_z, f_x, and f_y of equations (5.61) and (5.63) transform like a three-dimensional vector. The same can be said for g_i of equations (5.61) and (5.64). Thus, they can be grouped into the second-rank tensor

$$T = \begin{pmatrix} 0 & -f_z & -f_x & -f_y \\ f_z & 0 & -g_y & g_x \\ f_x & g_y & 0 & -g_z \\ f_y & -g_x & g_z & 0 \end{pmatrix}, \tag{5.65}$$

whose Lorentz-transformation properties are well known. The g_i components change their signs under space inversion, while the f_i components remain invariant. They are like the electric and magnetic fields, respectively.

If the system is Lorentz-boosted, f_i and g_i can be computed from table 5.5. We are now interested in the symmetry of photons by taking the massless limit. According to the procedure developed in section 4.3, we can keep only the terms that become larger for larger values of η. Thus,

$$\begin{aligned} f_x &\to \frac{1}{2}(uu - \dot{v}\dot{v}), & f_y &\to \frac{1}{2i}(uu + \dot{v}\dot{v}), \\ g_x &\to \frac{1}{2i}(uu + \dot{v}\dot{v}), & g_y &\to -\frac{1}{2}(uu - \dot{v}\dot{v}), \end{aligned} \tag{5.66}$$

in the massless limit.

Then the tensor of equation (5.65) becomes

$$F = \begin{pmatrix} 0 & 0 & -E_x & -E_y \\ 0 & 0 & -B_y & B_x \\ E_x & B_y & 0 & 0 \\ E_y & -B_x & 0 & 0 \end{pmatrix}, \tag{5.67}$$

with

$$\begin{aligned} E_x &\simeq \frac{1}{2}(uu - \dot{v}\dot{v}), & E_y &\simeq \frac{1}{2i}(uu + \dot{v}\dot{v}), \\ B_x &\simeq \frac{1}{2i}(uu + \dot{v}\dot{v}), & B_y &\simeq -\frac{1}{2}(uu - \dot{v}\dot{v}). \end{aligned} \tag{5.68}$$

The electric and magnetic field components are perpendicular to each other. Furthermore,

$$B_x \simeq E_y, \quad B_y \simeq -E_x. \tag{5.69}$$

To address the symmetry of photons, let us go back to equation (5.62). In the massless limit,

$$B_+ \simeq E_+ \simeq uu, \qquad B_- \simeq E_- \simeq \dot{v}\dot{v}. \tag{5.70}$$

The gauge transformations applicable to u and $\dot{v}$ are the two-by-two matrices

$$T(\gamma) = \begin{pmatrix} 1 & -\gamma \\ 0 & 1 \end{pmatrix} \quad \text{and} \quad \dot{T}(\gamma) = \begin{pmatrix} 1 & 0 \\ \gamma & 1 \end{pmatrix}, \tag{5.71}$$

respectively, as given in equation (5.7). Both u and $\dot{v}$ are invariant under gauge transformations, while $\dot{u}$ and v are not as shown in table 5.4. The B_+ and E_+ are for the photon spin along the z-direction, while B_- and E_- are for the opposite direction. Although equation (5.70) is only for the field propagating along the z-direction, it can be rotated to suit an arbitrary direction [2].

It is important to note that the second-rank tensor in equation (5.65) is quite different from that given in equation (5.67). The former is for massive particles and forms a representation space for the $O(3)$-like little group. For massless particles, the only gauge-invariant products are uu and $\dot{v}\dot{v}$. Therefore, only these products are accommodated as the components of the Maxwell tensor given in equation (5.67).

Weinberg [25] constructed gauge-invariant state vectors for massless particles starting from Wigner's 1939 paper [28]. The bilinear spinors uu and $\dot{v}\dot{v}$ correspond to Weinberg's state vectors.

Most of the formulae presented in this section were given in our previous publication [3]. In the present section, we have given a more precise definition of Maxwell's tensor derived from the zero-mass limit using the Wigner excursion, as illustrated in figure 2.5. Thus we were able to give a more precise definition of the polarization of photons.

5.5.3 Higher spins

Wigner's original book of 1931 [27, 29], discussed the rotation group, without Lorentz transformations. Subsequently, this form of the rotation group has been extensively discussed in the literature [6, 8, 9]. The issue of how to construct the most general spin state from the two-component spinors for the spin-$\frac{1}{2}$ particle was most frequently the topic of interest.

As we know, four states can be constructed from two spin-$\frac{1}{2}$ spinors, which leads to three spin-1 states and one spin-0 state. If we have three spinors, four spin-$\frac{3}{2}$ states and two spin-$\frac{1}{2}$ states can be constructed, resulting in six states. Although this partition process is much more complicated [7, 15] for three spinors, it is possible for all higher spin states.

In the relativistic Lorentz-covariant world, there are four states for each spin-$\frac{1}{2}$ particle. If we match two spinors, we get the sixteen states tabulated in table 5.5. There should be 64 states for three spinors and for four spinors, 256 states. Also tabulated in table 5.5 was the Lorentz-boost applicable to those spinors. The transverse rotations were seen to be gauge transformations in the limit of zero-mass or infinite-η. The table given in figure 5.2 illustrates how the spins associated with massive and massless particles can be bundled together.

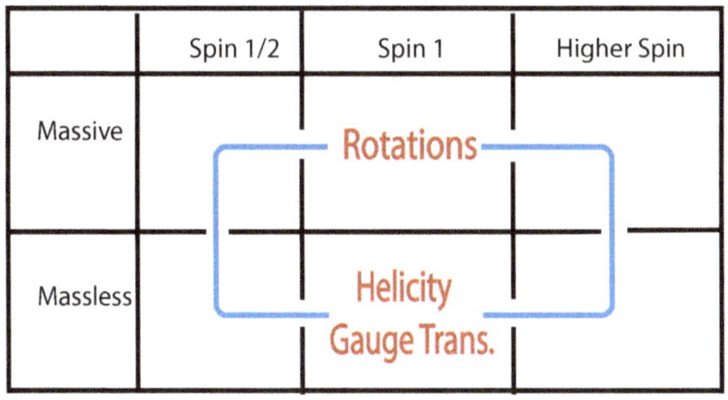

Figure 5.2. Massive and massless particles shown in a unified picture. Since the gauge transformation is a Lorentz-boosted rotation matrix, it is applicable to all massless particles. Therefore, in the Lorentz-covariant world, higher-spin states can be constructed from the four states of the spin-$\frac{1}{2}$ particle [3].

In this relativistic regime, photons and gravitons are of most interest. In sections 5.1 and 5.2, we saw that under gauge transformations, the observable components are invariant as well as being the terms that become largest for large values of η. We have additionally seen in section 5.5.2 that the photon states consisting of uu and $\dot{v}\dot{v}$ are for the spin parallel and anti-parallel to the momentum, respectively. It is then possible to conclude that for spin-2 gravitons, the states must be $uuuu$ and $\dot{v}\dot{v}\dot{v}\dot{v}$, respectively.

When Weinberg constructed his states for massless particles [24], especially photons and gravitons [25], he started with the conditions:

$$N_1|\text{state}\rangle = 0 \quad \text{and} \quad N_2|\text{state}\rangle = 0, \tag{5.72}$$

where N_1 and N_2 are defined in equation (5.10). As we saw in section 5.1, these are the generators of gauge transformations. Thus, Weinberg's states are gauge-invariant states. We can identify uu and $\dot{v}\dot{v}$ as Weinberg's states for photons. Then $uuuu$ and $\dot{v}\dot{v}\dot{v}\dot{v}$ can be identified as Weinberg's states for gravitons.

References

[1] Barut A O and McEwan J 1986 The four states of the massless neutrino with Pauli coupling by spin-gauge invariance *Lett. Math. Phys.* **11** 67–72

[2] Başkal S and Kim Y S 1997 Little groups and Maxwell-type tensors for massive and massless particles *Europhys. Lett.* **40** 375–80

[3] Başkal S, Kim Y S, and Noz M E 2017 Loop representation of Wigner's little groups *Symmetry* **9** 97–118

[4] Başkal S, Kim Y S, and Noz M E 2019 *Mathematical Devices for Optical Sciences* (Bristol, UK: IOP Publishing) (ISBN 978-0-7503-1612-5, OCLC: 1034620988)

[5] Beresteckij V B, Lifshitz E M, Pitaevskij L P, and Landau L D 1982 *Quantum Electrodynamics: Course of Theoretical Physics* (Amsterdam: Butterworth-Heinemann) 2nd edn vol 4 reprinted 2007 (ISBN 978-0-7506-3371-0, Originally published 1980; OCLC: 254580834; original authors: L D Landau and E M Lifshitz)

[6] Condon E U and Shortley G H 1979 *The Theory of Atomic Spectra* reprinted edition (Cambridge: Cambridge University Press) (ISBN 978-0-521-09209-8, Originally published 1951; OCLC: 258057861)
[7] Feynman R P, Kislinger M, and Ravndal F 1971 Current matrix elements from a relativistic quark model *Phys. Rev.* D **3** 2706–32
[8] Gel'fand I M, Minlos R A, and Shapiro Z Y 2018 *Representations of the Rotation and Lorentz Groups and Their Applications* (Mineola, NY: Dover Publications, Inc) (ISBN 978-0-486-82385-0, Originally published 1963; New York, NY USA: MacMillan)
[9] Hamermesh M 1989 *Group Theory and Its Application to Physical ProblemsDover Books on Physics and Chemistry* (New York: Dover Publications) (ISBN 978-0-486-66181-0, Originally published 1962, Reading, MA, USA: Addison-Wesley)
[10] Han D and Kim Y S 1981 Little group for photons and gauge transformations *Am. J. Phys.* **49** 348–51
[11] Han D, Kim Y S, and Son D 1982 E(2)-like little group for massless particles and neutrino polarization as a consequence of gauge invariance *Phys. Rev.* D **26** 3717–25
[12] Han D, Kim Y S, and Son D 1983 Gauge transformations as Lorentz-boosted rotations *Phys. Lett.* B **131** 327–9
[13] Han D, Kim Y S, and Son D 1986 Eulerian parametrization of Wigner's little groups and gauge transformations in terms of rotations in two-component spinors *J. Math. Phys.* **27** 2228–35
[14] Han D, Kim Y S, and Son D 1986 Photons, neutrinos, and gauge transformations *Am. J. Phys.* **54** 818–21
[15] Hussar P E, Kim Y S, and Noz M E 1980 Three-particle symmetry classifications according to the method of Dirac *Am. J. Phys.* **48** 1038–42
[16] Kim Y S 2002 Neutrino polarization as a consequence of gauge invariance *Czech. J. Phys.* **52** C353–60
[17] Kim Y S, Maguire G Q Jr, and Noz M E 2016 Do small-mass neutrinos participate in gauge transformations? *Adv. High Energy Phys.* **2016** 1–7
[18] Kim Y S and Noz M E 1986 *Theory and Applications of the Poincaré Group* (Dordrecht: Springer) (ISBN 978-94-010-8526-7 978-94-009-4558-6)
[19] Kim Y S and Noz M E 2018 *New perspectives on Einstein's* $E = mc^2$ (Singapore; Hackensack, NJ: World Scientific Publishing Co) (ISBN 978-981-323-770-4, OCLC: 1022620189)
[20] Kim Y S and Wigner E P 1990 Space–time geometry of relativistic particles *J. Math. Phys.* **31** 55–60
[21] Kupersztych J 1976 Is there a link between gauge invariance, relativistic invariance and electron spin? *Il Nuovo Cimento B Ser. 11* **31** 1–11
[22] Mohapatra R N and Smirnov A Y 2006 Neutrino mass and new physics *Annu. Rev. Nucl. Part. Sci.* **56** 569–628
[23] Weinberg S 1964 Feynman rules for any spin *Phys. Rev.* **133** B1318–32
[24] Weinberg S 1964 Feynman rules for any spin. II. Massless particles *Phys. Rev.* **134** B882–96
[25] Weinberg S 1964 Photons and gravitons in S-matrix theory: derivation of charge conservation and equality of gravitational and inertial mass *Phys. Rev.* **135** B1049–56
[26] Weinberg S 1995 *The Quantum Theory of Fields, Vol. 1 Foundations* (Cambridge; New York: Cambridge University Press) (ISBN 978-0-521-55001-7 978-0-521-55002-4 978-0-521-66000-6)

[27] Wigner E P 1931 *Gruppentheorie und ihre Anwendung auf die Quantenmechanik der Atomspektren* (Braunschweig: Springer)
[28] Wigner E P 1939 On unitary representations of the inhomogeneous Lorentz group *Ann. Math.* **40** 149–204
[29] Wigner E P 1959 *Group theory: and its application to the quantum mechanics of atomic spectra* (New York: Academic)

Chapter 6

Lorentz-covariant harmonic oscillators

Einstein and Bohr occasionally met, before and after 1927, to discuss physics. Einstein was interested in how things look to moving observers, while Bohr was interested in why the energy levels of the hydrogen atom are discrete. Then they must have talked about how the electron orbit of the hydrogen atom looks to a moving observer. There seem to be no written records to indicate how they sketched the orbits.

However, it is not uncommon to see in the literature the description of the Lorentz deformation, as described in figure 6.1. This figure became outdated in 1927. The electron orbit is now a standing wave. Thus, the question is how the standing wave appears when boosted in a given direction.

As is indicated in figure 6.1, the longitudinal component for the wave function of the electron orbit is affected while the transverse components remain unchanged. With this point in mind, we shall study harmonic oscillators. Since the wave equation for the three-dimensional oscillator is separable in the Cartesian coordinate system, it is sufficient to study the effect of the Lorentz boost only for the longitudinal component of the wave function.

Paul A M Dirac spent many years trying to construct a quantum mechanics consistent with special relativity. In his three papers published in 1927, 1945, and 1949 [8–10], Dirac presented all the essential ideas on how to construct harmonic oscillator wave functions that can be Lorentz transformed. However, Dirac's most serious problem was the lack of physical phenomena to which his ideas were applicable. The physical example did not exist until Gell-Mann published his paper on the quark model in 1964 [17], where the proton was described as a bound state of three quarks. The proton can be accelerated to speeds very close to that of light.

Dirac [10] said that the problem of Lorentz-covariant quantum mechanics is the same as that of constructing a representation of the Poincaré group. Thus, he was telling us to construct the representation based on harmonic oscillators. Since the standing wave is for a bound state, its symmetry is that of Wigner's little group, which dictates the internal space–time symmetry [40].

doi:10.1088/978-0-7503-3607-9ch6

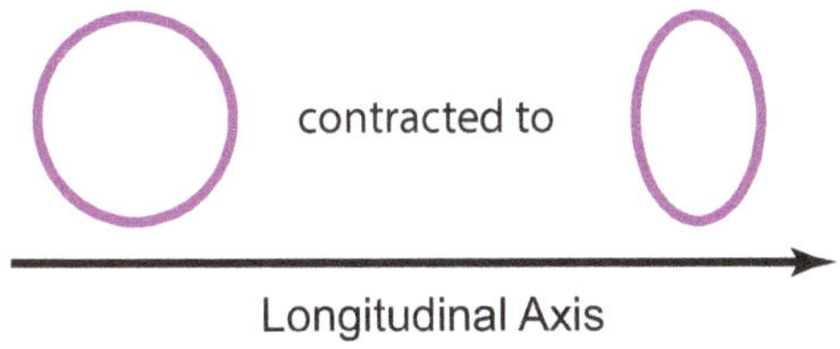

Figure 6.1. When John S Bell, in 1987, published the first edition of his book, 60 years after 1927, he included this classical picture of Lorentz contraction of the hydrogen atom's electron orbit as it would be viewed by an observer [4]. In this, Bell, while talking about quantum mechanics in his book, overlooked the fact that in 1927 a standing wave had replaced the electron orbit in the hydrogen atom. The remaining question then becomes how this standing wave orbit looks to moving observers.

Indeed, the most serious problem in Dirac's papers is that he did not specify the standing wave as a bound state. This problem was clarified later in the paper of Feynman, Kislinger, and Ravndal [15]. They start with a system of two quarks bound together by a harmonic oscillator force and write down a Lorentz-invariant differential equation. This differential equation has many different forms of solutions. It is possible to construct a set of solutions that constitute a representation of Wigner's little group. We shall thus start with the Lorentz-invariant differential equation of Feynman *et al.*

In section 6.1, we give a short introduction to Dirac's 1927 and 1945 papers in which he laid the foundation for constructing a Lorentz-covariant quantum mechanics. In section 6.2, we outline Dirac's points raised in his 1949 [10] paper in his pursuit of relativistic dynamics. We present in section 6.3, the basic material needed for constructing a bound state form of quantum mechanics. In section 6.4, using Wigner's little groups, we derive the solution for internal space–time coordinates of relativistic extended particles using the harmonic oscillator. This solution is presented both in the Cartesian and spherical coordinate systems consistent with Dirac's time–energy uncertainty relation. We detail in section 6.5 some further properties of the harmonic wave function derived in the previous sections. In section 6.6, we consider the overlapping wave function that results when the inner product of the ground state harmonic oscillator wave function is taken with a Lorentz boosted wave function. We show these results in a Lorentz contraction. Lastly, in section 6.7, we illustrate what happens when we include Feynman's rest of the Universe.

6.1 Dirac's plan to construct Lorentz-covariant quantum mechanics

The year 1927 is known for Heisenberg's uncertainty relation. In the same year, based on the line width and lifetime of excited energy levels, Dirac formulated his *c-number* time–energy uncertainty relation [8].

During World War II, Dirac was looking into the possibility of constructing representations of the Lorentz group using harmonic oscillator wave functions [9]. The Lorentz group is the language of special relativity, and the present form of quantum mechanics starts with harmonic oscillators. Therefore, he was interested in making quantum mechanics Lorentz-covariant by constructing representations of the Lorentz group using harmonic oscillators.

In his 1945 paper [9], Dirac considers the Gaussian form

$$\exp\left\{-\frac{1}{2}(x^2 + y^2 + z^2 + t^2)\right\}. \tag{6.1}$$

We note that this Gaussian form is in the $(x,\ y,\ z,\ t)$ coordinate variables. Thus, if we consider a Lorentz boost along the z-direction, we can drop the x and y variables and write the above equation as

$$\exp\left\{-\frac{1}{2}(z^2 + t^2)\right\}. \tag{6.2}$$

This expression is strange for those who believe in Lorentz invariance where $(z^2 - t^2)$ is an invariant quantity.

On the other hand, this expression is consistent with his earlier papers on the time–energy uncertainty relation [7, 8]. In those papers, Dirac observed there is a time–energy uncertainty relation, while there are no excitations along the time axis.

Let us look at figure 6.2 carefully. This figure is a pictorial representation of Dirac's equation (6.2), with localization in both space and time coordinates. Then Dirac's fundamental question would be how this figure appears to a moving observer.

6.2 Dirac's forms of relativistic dynamics

In 1949, *Reviews of Modern Physics* published a special issue to celebrate Einstein's 70th birthday. This issue contains Dirac's paper entitled *Forms of Relativistic Dynamics* [10]. In this paper, Dirac points out problems that have to be overcome in extending quantum mechanics to be relativistic. He considers possible constraints that can be imposed on the Lorentz transformations.

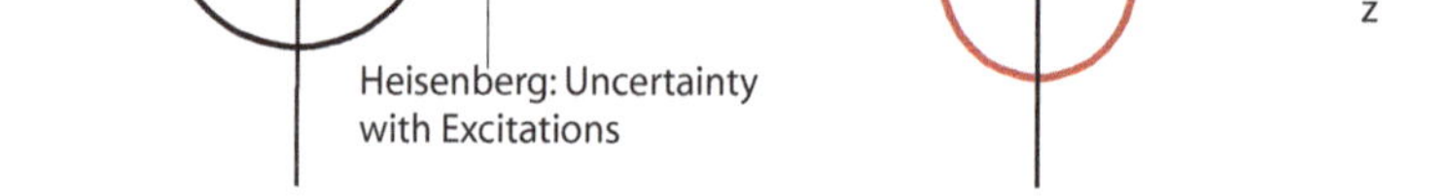

Figure 6.2. Space–time picture of quantum mechanics. In 1927, Dirac noted there is an uncertainty in the time and energy variables, but there are no excitations along the time axis. He called it the *c-number* time–energy relation. In 1945, Dirac attempted to construct a representation of the Lorentz group using harmonic oscillator wave functions. According to his 1927 paper, the Gaussian cut-off along the time axis represents the time–energy uncertainty relation. His c-number relation means that only the ground state is allowed along the time axis [23].

1. In this paper, Dirac's main problem was to deal with the time variable, and he writes down the equation

$$x_0 \approx 0. \tag{6.3}$$

Since Dirac offered no further explanations on his $\approx$ sign, we are free to interpret this as his c-number time–energy uncertainty relation. In terms of the harmonic oscillators, which he discussed in 1945, this condition means that the harmonic oscillator wave function in the time variable is always in the ground state.

2. Dirac wrote down ten generators of the Poincaré group and their commutation relations, which he called Poison brackets. The Poincaré group, defined in section 1.4, is the inhomogeneous Lorentz group with three rotation generators, three boost generators, and four generators of translations. Dirac notes that the rotation and translation generators are associated with angular momentum and space–time translations, respectively. On the other hand, there are no dynamical variables associated with the generators of boosts.

   Dirac then states that the construction of relativistic quantum mechanics is achieved through the construction of the representation of the Poincaré group. Even though Dirac did not tell us how to construct such a representation, he attempted in 1945 [9] to use harmonic oscillators to construct a representation of the Lorentz group. We should therefore use harmonic oscillator wave functions to construct the representation of the Poincaré group and the Lorentz-covariant quantum mechanics.

3. Also, in his 1949 paper, Dirac introduced the light-cone coordinate system, discussed in section 2.6. When the system is boosted along the z-direction, the transformation takes the form

$$\begin{pmatrix} z' \\ t' \end{pmatrix} = \begin{pmatrix} \cosh\eta & \sinh\eta \\ \sinh\eta & \cosh\eta \end{pmatrix} \begin{pmatrix} z \\ t \end{pmatrix}. \tag{6.4}$$

This transformation is not a rotation. This is a squeeze transformation. Dirac introduced his light-cone variables defined as

$$z_+ = (z + t)/\sqrt{2}, \quad \text{and} \quad z_- = (z - t)/\sqrt{2}. \tag{6.5}$$

Hence the boost transformation of equation (6.4) takes the form

$$z_+' = e^{\eta} z_+, \quad \text{and} \quad z_-' = e^{-\eta} z_-. \tag{6.6}$$

The z_+ variable becomes expanded while the z_- variable becomes contracted, as is illustrated in figure 6.3. The product

$$z_+ z_- = \frac{1}{2}(z + t)(z - t) = \frac{1}{2}(z^2 - t^2) \tag{6.7}$$

remains invariant. Indeed, in Dirac's picture, the Lorentz boost is an area-preserving squeeze transformation, as indicated in figure 6.3.

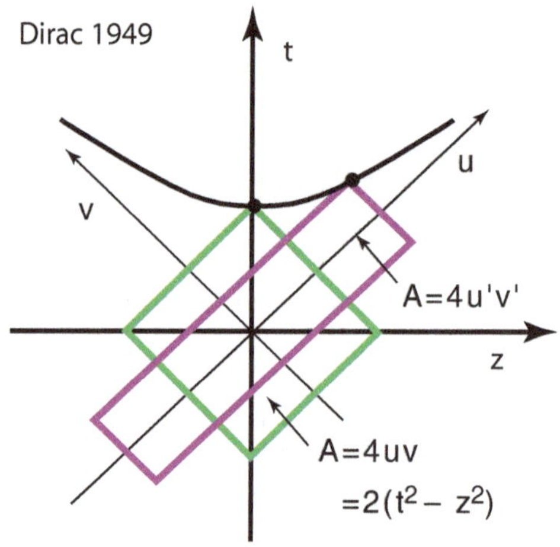

Figure 6.3. Lorentz boost in the light-cone coordinate system. The boost traces a point along the hyperbola. The boost also squeezes the square into a rectangle [23].

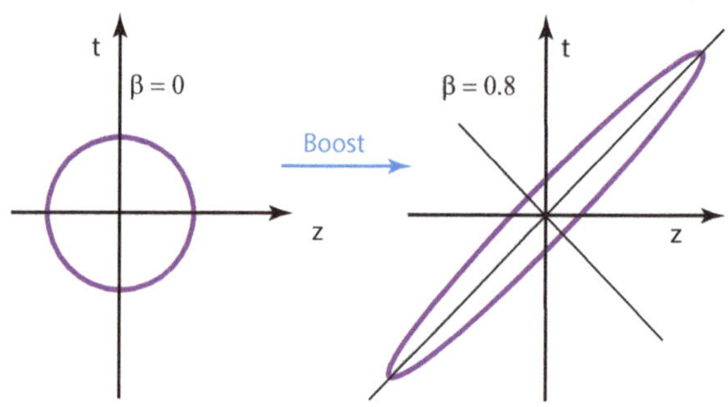

Figure 6.4. Lorentz-squeezed quantum mechanics. Starting from the Gaussian distribution, Dirac attempted to formulate a relativistic quantum mechanics, given in figure 6.2. In this figure the Gaussian distribution is pictured as it appears to an observer moving with the velocity parameter $\beta = v/c = 0.8$ [27].

Let us go back to the ground-state wave function of equation (6.2) whose Gaussian form is illustrated in figure 6.2. If the oscillator moves along the z-direction, it should appear as

$$\exp\left\{-\frac{1}{4}[e^{-2\eta}(z+t)^2 + e^{2\eta}(z-t)^2]\right\}, \tag{6.8}$$

according to the Lorentz-boost in the light-cone system given in equation (6.6). This is a Lorentz-squeezed Gaussian distribution, as shown in figure 6.4. Indeed, the Lorentz boost is a squeeze transformation [21].

6.3 Running waves and standing waves

We are quite familiar with the Klein–Gordon equation

$$\left[\left(\frac{\partial}{\partial x}\right)^2 + m^2\right]\phi(x) = 0, \tag{6.9}$$

for a single particle, where

$$\left(\frac{\partial}{\partial x}\right)^2 = \frac{\partial^2}{\partial t^2} - \frac{\partial^2}{\partial z^2} - \frac{\partial^2}{\partial x^2} - \frac{\partial^2}{\partial y^2}. \tag{6.10}$$

This equation and its solutions are well known in physics. Then the equation

$$\left[\left(\frac{\partial}{\partial x_a}\right)^2 + \left(\frac{\partial}{\partial x_b}\right)^2 + m_a^2 + m_b^2\right]\phi(x_a)\phi(x_b) = 0, \tag{6.11}$$

is for two independent particles whose coordinates are x_a and x_b, respectively. They are four-vectors, and we shall use the notation

$$x^2 \equiv t^2 - z^2 - x^2 - y^2 \tag{6.12}$$

for both x_a and x_b. The physics of this system for two free particles is also well known. The question is what happens when $(m_a^2 + m_b^2)$ is replaced by a term containing $(x_a - x_b)^2$.

Indeed, Feynman *et al* [15] wrote down the equation

$$\left\{2\left[\left(\frac{\partial}{\partial x_a}\right)^2 + \left(\frac{\partial}{\partial x_b}\right)^2\right] - \frac{1}{16}(x_a - x_b)^2 + m_0^2\right\}\phi(x_a, x_b) = 0, \tag{6.13}$$

for the bound state called the *hadron* consisting of two constituent particles called *quarks* bound together in a harmonic oscillator potential.

Let us introduce the coordinates

$$X = \frac{x_a + x_b}{2} \quad \text{and} \quad x = \frac{x_a - x_b}{2\sqrt{2}}. \tag{6.14}$$

The X coordinate is for the space–time specification of the hadron, while the x variable measures the relative space–time separation between the quarks. In terms of these variables, equation (6.13) can be written as

$$\left\{\left(\frac{\partial}{\partial X}\right)^2 + m_0^2 + \frac{1}{2}\left[\left(\frac{\partial}{\partial x}\right)^2 - x^2\right]\right\}\phi(X, x) = 0. \tag{6.15}$$

This equation is separable in the X and x variables and can be separated by using the equation:

$$\phi(X, x) = f(X)\psi(x), \tag{6.16}$$

where $f(X)$ and $\psi(x)$ satisfy the following differential equations, respectively:

$$\left\{\left(\frac{\partial}{\partial X}\right)^2 + m_0^2 + (\lambda + 1)\right\} f(X) = 0 \tag{6.17}$$

and

$$\frac{1}{2}\left[\left(\frac{\partial}{\partial x}\right)^2 - x^2\right]\psi(x) = (\lambda + 1)\psi(x). \tag{6.18}$$

The differential equation of equation (6.17) is a Klein–Gordon equation, and its solutions are well known. It takes the form

$$f(X) = \exp(\pm ip \cdot X) \tag{6.19}$$

with

$$p^2 = m_0^2 + (\lambda + 1), \tag{6.20}$$

where p is the energy–momentum four-vector of the hadron. Of course, p^2 is the $(\text{mass})^2$ of the hadron and is numerically constrained to take the values allowed by equation (6.20). The separation constant λ is determined from the solutions of the harmonic oscillator equation of equation (6.18).

Indeed, the wave function of equation (6.16) is for the hadron moving with the energy–momentum four-vector p_μ with the internal structure dictated by the oscillator equation, described in figure 6.5. Wigner's little group applies to the internal space–time symmetry dictated by the oscillator equation of equation (6.18) [22, 23, 27].

The space–time transformation of the total wave function of equation (6.16) is generated by the following ten generators of the Poincaré group. The operators

$$P_\mu = i\frac{\partial}{\partial X^\mu} \tag{6.21}$$

generate space–time translations. Lorentz transformations, which include boosts and rotations, are generated by

$$M_{\mu\nu} = L^*_{\mu\nu} + L_{\mu\nu} \tag{6.22}$$

where $L^*_{\mu\nu}$ and $L_{\mu\nu}$ are given by

$$\begin{aligned} L^*_{\mu\nu} &= i\left(X_\mu\frac{\partial}{\partial X^\nu} - X_\nu\frac{\partial}{\partial X^\mu}\right), \\ L_{\mu\nu} &= i\left(x_\mu\frac{\partial}{\partial x^\nu} - x_\nu\frac{\partial}{\partial x^\mu}\right). \end{aligned} \tag{6.23}$$

The translation operators P_μ, act only on the hadronic coordinate and do not affect the internal coordinate. The operators $L^*_{\mu\nu}$ and $L_{\mu\nu}$ Lorentz-transform the hadronic and internal coordinates, respectively. The above ten generators satisfy the commutation relations for the Poincaré group.

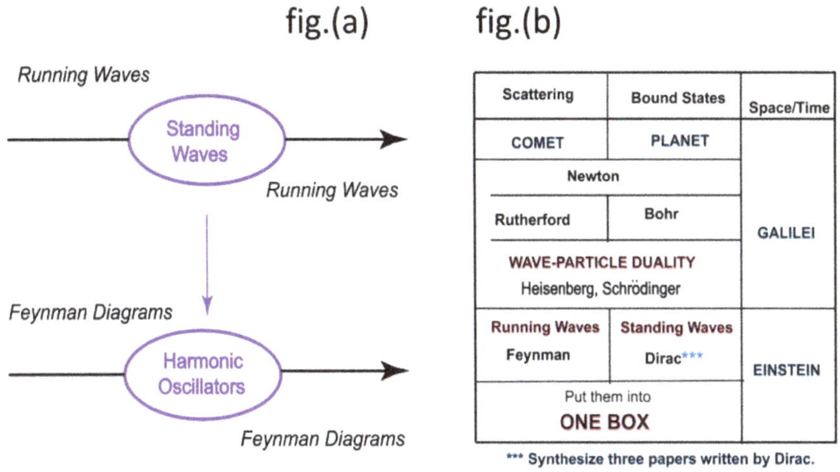

Figure 6.5. Feynman diagrams are satisfactory for running waves as they provide, in Einstein's world, a suitable interpretation for scattering states. However, they do not work for standing waves trapped inside a hadron. For these, Feynman [13] suggested that harmonic oscillators be used as the first step. In this way, Feynman suggested a way of combining quantum mechanics with special relativity. This is illustrated in (a). Based on Feynman's observations, we can construct the historical prospective, given in (b) that starts from the planets and comets. We discussed in section 6.1 and 6.2, Dirac's three papers. Additionally, as was shown by Han *et al* [19], Feynman diagrams for scattering problems and the covariant oscillator formalism share the same set of physical principles.

In order to consider irreducible representations of the Poincaré group, we have to construct wave functions which are diagonal in the invariant Casimir operators of the group, which commute with all the generators of equations (6.21) and (6.22). Here, the Casimir operators are

$$P^{\mu}P_{\mu} \quad \text{and} \quad W^{\mu}W_{\mu}, \tag{6.24}$$

where

$$W_{\mu} = \epsilon_{\mu\nu\alpha\beta}P^{\nu}M^{\alpha\beta}. \tag{6.25}$$

The eigenvalues of the above P^2 and W^2 represent, respectively, the mass and spin of the hadron.

The algebra of these generators becomes much simpler when the hadronic momentum is constant, as in Wigner's little group. While translation generators can be dropped from the algebra, the operator P^{ν} can be replaced by the four-vector

$$P = (E, p, 0, 0) \tag{6.26}$$

for the hadron momentum moving in the z-direction. As a consequence, the eigenvalues of the Casimir operators become $m^2 = (\text{mass})^2$ and $m^2\ell(\ell + 1)$, where ℓ is the total angular momentum of the oscillator. These eigenvalues are invariant under Poincaré or Lorentz transformations.

6.4 Little groups for relativistic extended particles

The harmonic oscillator equation of equation (6.18) is invariant under Lorentz transformations. For instance, if the system is boosted along the z-direction, according to equation (6.4), the differential equation takes the same form in the new coordinate variables. Thus, the solution also takes the previous form. With this point in mind, we can now study the solution of the differential equation in the Lorentz frame, where the hadron is at rest. Let us write the oscillator equation explicitly as:

$$\frac{1}{2}\left\{\left[x^2 + y^2 + z^2 - \left(\frac{\partial}{\partial x}\right)^2 - \left(\frac{\partial}{\partial y}\right)^2 - \left(\frac{\partial}{\partial z}\right)^2\right] - \left[t^2 - \left(\frac{\partial}{\partial t}\right)^2\right]\right\}\psi(x) \tag{6.27}$$
$$= (\lambda + 1)\psi(x).$$

According to Dirac's c-number time–energy uncertainty relation, the time component of the solution should always be in the ground state, and thus the solution takes the form

$$\psi(x) = \varphi(z, x, y)\left[\left(\frac{1}{\pi}\right)^{1/4} e^{-t^2/2}\right], \tag{6.28}$$

with

$$\frac{1}{2}\left[x^2 + y^2 + z^2 - \left(\frac{\partial}{\partial x}\right)^2 - \left(\frac{\partial}{\partial y}\right)^2 - \left(\frac{\partial}{\partial z}\right)^2\right]\varphi(z, x, y) = \left(\lambda + \frac{3}{2}\right)\varphi(z, x, y). \tag{6.29}$$

This equation is very familiar to us from textbooks. However, the equation carries the following additional interpretations.

1. The Cartesian variables z, x, and y are internal coordinate variables for the hadron.
2. This equation is separable both in the spherical and Cartesian coordinate systems. For the discussion of the Poincaré symmetry, we need the spherical coordinate system to construct the representation diagonal in the Casimir operators where the eigenvalue ℓ is needed.
3. When the system is boosted along the z-direction, the transverse x and y of the internal coordinate variables are not affected and can be separated from the differential equation of equation (6.27).
4. The spherical solutions can be written in terms of the linear combinations of the Cartesian solutions.

The solution in the spherical coordinate system should take the form

$$\varphi(r, \theta, \phi) = R_{n\ell m}(r) Y_{\ell m}(\theta, \phi), \tag{6.30}$$

where $Y_{\ell m}(\theta, \phi)$ are the spherical harmonics. The radial function $R_{n\ell m}(r)$ is well defined, but its explicit form is not readily available in the literature. It should take the form

$$R_{n\ell m}(r) = r^m g_{n\ell}(r) e^{-r^2/2}, \tag{6.31}$$

with

$$g_{n\ell}(r) = \sum_k a_{2k} r^{2k}. \tag{6.32}$$

Here

$$\frac{a_{2k+2}}{a_{2k}} = \frac{2(\lambda - \ell - 2k)}{\ell(\ell + 1) - (\ell + 2k + 3)(\ell + 2k + 2)}. \tag{6.33}$$

For large values of k, this ratio becomes $1/k$, which is like the expansion of the exponential $\exp(r^2)$, leading the radial function of equation (6.31) to increase as $\exp(r^2/2)$. Thus the series has to be truncated with

$$\lambda = 2k + \ell. \tag{6.34}$$

The first term a_0 in the series is determined by the normalization condition

$$\int_0^\infty [rR_{n\ell m}(r)]^2 dr = 1. \tag{6.35}$$

The increases in ℓ and n are called the orbital and radial excitations in the literature.

If the system is Lorentz-boosted along the z-direction, according to equation (6.4), the Lorentz-invariant differential equation of equation (6.18) remains invariant. The z and t variables in equation (6.27) become z' and t', respectively, and the wave function becomes modified accordingly. The important point is that the eigenvalue λ remains invariant.

6.5 Further properties of covariant oscillator wave functions

Since the x and y coordinates are not affected, we drop their terms from the differential equation from equation (6.18), and consider the equation

$$\frac{1}{2}\left\{\left[z^2 - \left(\frac{\partial}{\partial z}\right)^2\right] - \left[t^2 - \left(\frac{\partial}{\partial t}\right)^2\right]\right\}\psi(z, t) = \lambda\psi(z, t). \tag{6.36}$$

The solution of this equation should take the form

$$\psi_0^n(z, t) = \left[\frac{1}{\pi 2^n n!}\right]^{1/2} H_n(z) \exp\left[-\frac{1}{2}(z^2 + t^2)\right], \tag{6.37}$$

where $H_n(z)$ is the Hermite polynomial of order n.

The differential equation of equation (6.36) is invariant under the Lorentz boost along the z-direction and is invariant under the replacements of z and t by z' and t', respectively, where

$$\begin{pmatrix} z' \\ t' \end{pmatrix} = \begin{pmatrix} \cosh\eta & -\sinh\eta \\ -\sinh\eta & \cosh\eta \end{pmatrix} \begin{pmatrix} z \\ t \end{pmatrix}. \tag{6.38}$$

This equation is the inverse of the transformation given in equation (6.4). We thus achieve the Lorentz boost of the wave function by writing

$$\begin{aligned} \psi_\eta^n(z, t) &= \psi_0^n(z', t') \\ &= \left[\frac{1}{\pi 2^n n!}\right]^{1/2} H_n(z') \exp\left[-\frac{1}{2}(z'^2 + t'^2)\right]. \end{aligned} \tag{6.39}$$

It is possible to expand this in terms of the functions of z and t variables [23, 28, 35, 36], and the result is

$$\psi_\eta^n(z, t) = \left(\frac{1}{\cosh\eta}\right)^{(n+1)} \sum_k \left[\frac{(n+k)!}{n!k!}\right]^{1/2} (\tanh\eta)^k \chi_{n+k}(z)\chi_k(t), \tag{6.40}$$

where $\chi_n(z)$ is the nth excited state oscillator wave function which takes the familiar form

$$\chi_n(z) = \left[\frac{1}{\sqrt{\pi} 2^n n!}\right]^{1/2} H_n(z) \exp\left(\frac{-z^2}{2}\right). \tag{6.41}$$

If $n = 0$, this formula becomes simplified to [36]

$$\psi_\eta^0(z, t) = \left(\frac{1}{\cosh\eta}\right) \sum_k (\tanh\eta)^k \chi_k(z)\chi_k(t). \tag{6.42}$$

This formula plays an important role also in squeezed states of light [41] and also in continuous-variable entanglement [1, 5, 6, 11, 18, 33, 34, 37]. This subject was discussed systematically in a book entitled *Mathematical Devices in Optical Sciences* by the present authors [3].

According to the Gaussian form of equation (6.8), the ground-state wave function of equation (6.42) takes the form [21]

$$\psi_\eta^0(z, t) = \left(\frac{1}{\pi}\right)^{1/2} \exp\left\{-\frac{1}{4}[e^{-2\eta}(z+t)^2 + e^{2\eta}(z-t)^2]\right\}. \tag{6.43}$$

This is precisely the formula for Dirac's picture of the Lorentz squeeze, as illustrated in figure 6.4.

Relativity and quantum mechanics are the two most important physical theories developed in the 20th century. Thus, the most pressing issue in modern physics is to make quantum mechanics consistent with special relativity and then with general relativity. Since the harmonic oscillator plays the central role in quantum mechanics, the problem is to construct harmonic oscillator wave functions consistent with Lorentz covariance. A comprehensive list of early papers on this problem was given by Kim and Noz in their book [23]. Additionally, there are more recent papers on this subject [2, 16, 30–33, 37].

6.6 Lorentz contraction of harmonic oscillators

Let us now consider the overlap of the wave function $\psi_\eta^n(z, t)$ with $\eta = 0$, as indicated in figure 6.6. We are interested in the integral

$$\int \left(\psi_\eta^{n'}(z, t)\right)^* \psi_0^n(z, t) dz\, dt, \tag{6.44}$$

where $\psi_0^n(z, t)$, given in equation (6.37), can be written as

$$\psi_0^n = \chi_n \chi_0(t). \tag{6.45}$$

Then the evaluation of this integral is straightforward from equation (6.40), and the result is [36]

$$\left(\psi_\eta^{n'}, \psi_0^n\right) = \left(\frac{1}{\cosh \eta}\right)^{(n+1)} \delta_{nn'}. \tag{6.46}$$

This means that the orthogonality relation is preserved between two wave functions in two different frames.

If $n = n'$, the inner product between two wave functions leads to the contraction given on the right-hand side of equation (6.46). In terms of the velocity parameter $\beta = v/c$, where v is the hadronic velocity, we obtain

$$\frac{1}{\cosh \eta} = \sqrt{1 - \beta^2}. \tag{6.47}$$

This expression is more familiar to us, and the right-hand side of equation (6.46) is

$$\left(\sqrt{1 - \beta^2}\right)^{(n+1)}. \tag{6.48}$$

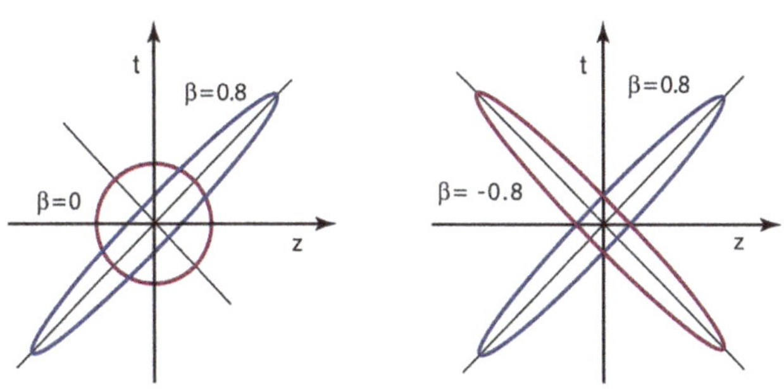

Figure 6.6. Two wave functions that overlap. Sketched in figure 6.4 are the wave functions with $\beta = 0$ and $\beta = 0.8$. The present figure shows that they can overlap. Additionally, the wave functions can move in the opposite directions with $\beta = \pm 0.8$ [21, 23].

For the ground-state wave function with $n = 0$, it is like the Lorentz contraction of a rigid body. For the first excited state, it is like an additional rod. This is not surprising in view of the fact that the excited states are obtained through application of the step-up operator. The nth excited state $|n\rangle$ can be written as

$$\frac{1}{\sqrt{n!}}(a^{\dagger})^{n}|0\rangle. \tag{6.49}$$

The additional contraction factor $\sqrt{1-\beta^2}$ comes from the step-up operator.

If the value of η in one of the wave functions is replaced by nonzero value η', $\cosh\eta$ in equation (6.40) should become $\cosh(\eta - \eta')$. Of particular interest is the case with $\eta' = -\eta$, as shown in figure 6.6. In this case, this is an overlap of two wave functions moving in the opposite directions, and the contraction factor is

$$\left(\frac{1}{\cosh(2\eta)}\right)^{(n+1)}, \tag{6.50}$$

which becomes

$$\left(\frac{1-\beta^2}{1+\beta^2}\right)^{(n+1)}. \tag{6.51}$$

Based on these orthogonality and contraction properties, it appears possible to give a quantum probability interpretation to the covariant harmonic oscillator in the Lorentz-covariant world, as shown in figure 6.7. On the other hand, these wave functions contain the time-separation variable between the constituent particles. What is the meaning of the distribution in terms of this variable, which is thoroughly hidden [25] in the present form of quantum mechanics? In order to clarify this issue, let us examine the concept of Feynman's rest of the Universe.

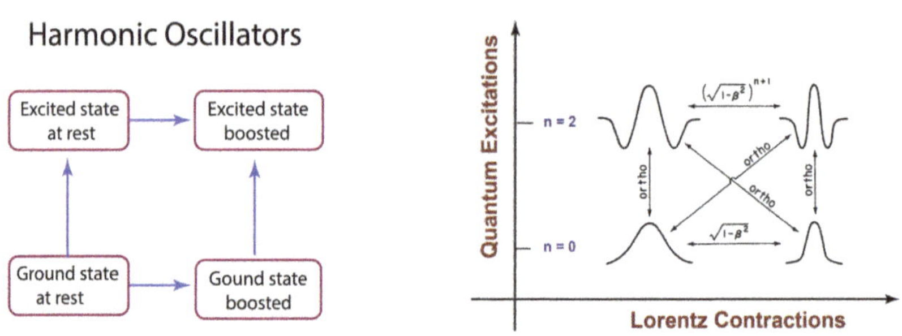

Figure 6.7. The covariant oscillator wave functions have the orthogonal relations displayed here. As illustrated in (a), the harmonic oscillator can be excited, as well as Lorentz boosted. Under Lorentz boosts, the orthogonality relations remain invariant and their inner products possess the Lorentz-contraction property given in (b) [23, 36].

6.7 Feynman's rest of the Universe

In the section above, we noted that in the covariant formulation of the harmonic oscillator wave functions the time-separation variable plays a very important role. Wherever space separation exists, this time-separation variable should exist. The measure of the proton and electron separation in the hydrogen atom is given by the Bohr radius. This radius, therefore, whenever the hydrogen atom moves, picks up this time separation in accordance with Einstein.

The present form of quantum mechanics, however, does not include this time-separation variable. Presently, the best way we can interpret this variable is to treat this time-separation as being in Feynman's rest of the Universe [20]. Feynman states, in his statistical mechanics book [14], that

> *When we solve a quantum-mechanical problem, what we really do is divide the Universe into two parts—the system in which we are interested and the rest of the Universe. We then usually act as if the system in which we are interested comprised the entire Universe. To motivate the use of density matrices, let us see what happens when we include the part of the Universe outside the system.*

The result of failing to include what happens outside the system is that there is an increase of entropy. Computed from the density matrix [38], entropy is, therefore, a measure of our ignorance. When the experimental procedure, to the maximum extent which is possible consistent with quantum mechanics, does not analyze all relevant variables, it is necessary to use the density matrix [12]. An increase in entropy results whenever we do not consider the time-separation variable [26, 29].

The pure-state density matrix for the covariant harmonic oscillator wave functions, which were defined in equation (6.39), is

$$\rho_\eta^n(z, t; z', t') = \psi_\eta^n(z, t)\psi_\eta^n(z', t'). \tag{6.52}$$

This also satisfies the condition that $\rho^2 = \rho$:

$$\rho_\eta^n(z, t; x', t') = \int \rho_\eta^n(z, t; x'', t'')\rho_\eta^n(z'', t''; z', t')dz''dt''. \tag{6.53}$$

However, because presently in quantum mechanics, the possibility to consider the time-separation variables doesn't exist, it is necessary, with respect to the t variable, to take the trace of the density matrix. This results in the density matrix having the form

$$\begin{aligned}\rho_\eta^n(z, z') &= \int \psi_\eta^n(z, t)\psi_\eta^n(z', t)dt \\ &= \left(\frac{1}{\cosh\eta}\right)^{2(n+1)} \sum_k \frac{(n+k)!}{n!k!}(\tanh\eta)^{2k}\, \psi_{n+k}(z)\psi_{n+k}^*(z').\end{aligned} \tag{6.54}$$

The trace of ρ^2 is less than one, as

$$Tr(\rho^2) = \int \rho_\eta^n(z, z')\rho_\eta^n(z', z)dzdz' = \left(\frac{1}{\cosh\eta}\right)^{4(n+1)} \sum_k \left[\frac{(n+k)!}{n!k!}\right]^2 (\tanh\eta)^{4k}, \tag{6.55}$$

even though the trace of the density matrix given in equation (6.54) is one. This result happens because we do not know how, in present formulation of quantum mechanics, to deal with the time-like separation. Thus we see that our knowledge is not complete.

The usual method used to measure this ignorance is generally to calculate the entropy which is defined as

$$S = -Tr(\rho\ln(\rho)). \tag{6.56}$$

This entropy is zero when, pretending we know the distribution along t, we use the pure-state density matrix given in equation (6.52). However, if it is not known how to deal with the distribution along the time-like direction, then the density matrix of equation (6.54) should be used to calculate the entropy. This result is then

$$S = (n+1)\left\{(\cosh\eta)^2\ln(\cosh\eta)^2 - (\sinh\eta)^2\ln(\sinh\eta)^2\right\} - \left(\frac{1}{\cosh\eta}\right)^{2(n+1)} \sum_k \frac{(n+k)!}{n!k!}\ln\left[\frac{(n+k)!}{n!k!}\right](\tanh\eta)^{2k}. \tag{6.57}$$

If we use the velocity parameter β of the hadron, we obtain

$$\tanh\eta = \beta = \frac{v}{c}, \tag{6.58}$$

and

$$S = -(n+1)\left\{\ln(1-\beta^2) + \frac{\beta^2\ln\beta^2}{1-\beta^2}\right\} - (1-\beta^2)^{(n+1)} \sum_k \frac{(n+k)!}{n!k!}\ln\left[\frac{(n+k)!}{n!k!}\right]\beta^{2k}. \tag{6.59}$$

If we return to the wave function given in equation (6.40), figure 6.8 illustrates that the localization property is dictated by the Gaussian factor. This corresponds to the ground-state wave function. Therefore, it is reasonable to expect that the density matrix or the entropy for the nth excited state will behave in a manner which is the same as that for the ground state with $n = 0$.

If we consider that for the ground state with $n = 0$, the density matrix, computed from the Gaussian form of equation (6.42), becomes

$$\rho(z, z') = \left(\frac{1}{\pi\cosh(2\eta)}\right)^{1/2} \exp\left\{-\frac{1}{4}\left[\frac{(z+z')^2}{\cosh(2\eta)} + (z-z')^2\cosh(2\eta)\right]\right\}. \tag{6.60}$$

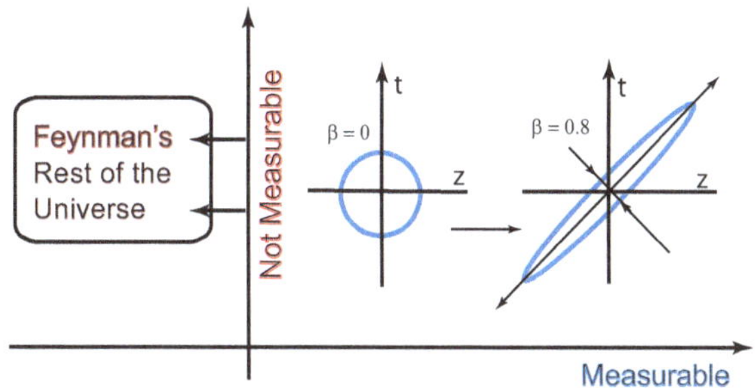

Figure 6.8. Localization property in the zt-plane. The Gaussian form, concentrated within a circular region specified by $(z + t)^2 + (z - t)^2 = 1$, results when the hadron is considered to be at rest. The region becomes deformed to $e^{-2\eta}(z + t)^2 + e^{2\eta}(z - t)^2 = 1$ as the hadronic speed increases. Measurements along the t-direction are not possible to make. Thus, the information available with which we are dealing, is not complete [29].

Then for this ground state, the entropy becomes

$$S = (\cosh\eta)^2 \ln(\cosh\eta)^2 - (\sinh\eta)^2 \ln(\sinh\eta)^2. \tag{6.61}$$

If we use the velocity parameter β, we can write this entropy as

$$S = \frac{1}{1 - \beta^2} \ln\left[\frac{1}{1 - \beta^2}\right] - \beta^2 \ln\beta^2. \tag{6.62}$$

Then $\sqrt{\cosh(2\eta)}$ represents the width of the distribution. In terms of β,

$$\sqrt{\cosh(2\eta)} = \sqrt{\frac{1 + \beta^2}{1 - \beta^2}}, \tag{6.63}$$

which becomes very large as $\beta \to 1$. This is shown in figure 6.9.

Even though, in the Lorentz-covariant world, the time-separation variable exists, we insist on pretending not to know about it. Therefore, it must exist in Feynman's rest of the Universe. It becomes translated into the entropy when we do not measure this time separation.

The Wigner function defined as [24, 39]

$$W(z, q) = \frac{1}{\pi} \int \rho(z + y, z - y) e^{2iqy} dy \tag{6.64}$$

can be used to see the uncertainty in our measurement process. The Wigner function, after integration can be written as

$$W(z, q) = \frac{1}{\pi \cosh(2\eta)} \exp\left\{-\left(\frac{z^2 + q^2}{\cosh(2\eta)}\right)\right\}. \tag{6.65}$$

We use here q instead of p for the momentum conjugate to z. The notation p has been used for the hadronic momentum.

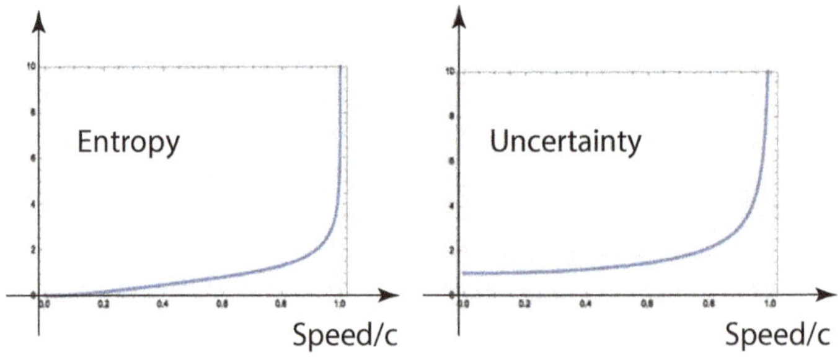

Figure 6.9. Entropy and uncertainty in the Lorentz-covariant system. The horizontal axis measures β in both graphs. The formulas for the entropy and uncertainty are given in equations (6.61) and (6.66), respectively. These quantities arise from our ignorance about the time sepration variable, which is hidden in the present form of quantum mechanics [26].

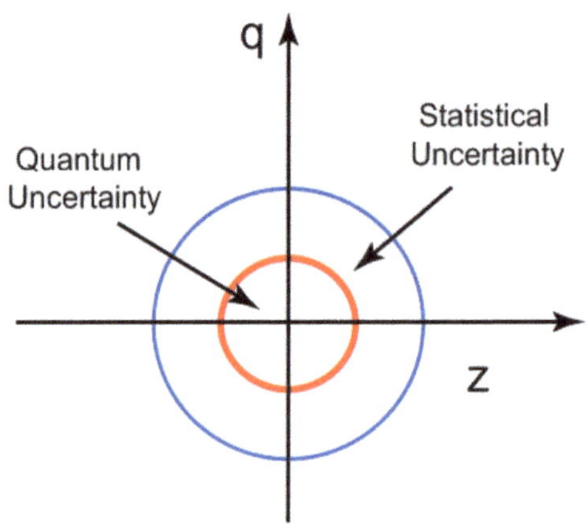

Figure 6.10. The uncertainty that results when time-separation coordinate is hidden. If the hadron is at rest, the small circle represents the minimal uncertainty. If the hadron moves, the result is more added uncertainty. The larger circle, with the radius increased by $\sqrt{\cosh(2\eta)}$, illustrates this [26].

We illustrate in figure 6.10, the Wigner phase-space distribution. The minimal uncertainty of the single oscillator is represented by the smaller inner circle. The total uncertainty, which includes the statistical uncertainty from our failure to observe the time-separation variable, is represented by the large circle. The larger radius is

$$\sqrt{\cosh(2\eta)} = \sqrt{\frac{1+\beta^2}{1-\beta^2}}. \tag{6.66}$$

The behavior of this radius is illustrated in figure 6.9. This radius takes the minimum value of one when $\beta = 0$ and increases rapidly when β becomes close to one.

References

[1] Adesso G and Illuminati F 2007 Entanglement in continuous-variable systems: recent advances and current perspectives *J. Phys. A: Math. Theor.* **40** 7821–80

[2] Bars I 2009 Relativistic harmonic oscillator revisited *Phys. Rev.* D **79**

[3] Başkal S, Kim Y S, and Noz M E 2019 *Mathematical Devices for Optical Sciences* (Bristol, UK: IOP Publishing) (ISBN 978-0-7503-1612-5, OCLC: 1034620988)

[4] Bell J S and Aspect A 2008 *Speakable and Unspeakable in Quantum Mechanics* Collected Papers on Quantum Philosophy 2nd rev. edn reprint edition (Cambridge: Cambridge University Press) (ISBN 978-0-521-52338-7 978-0-521-81862-9, OCLC: 552116191)

[5] Braunstein S L and van Loock P 2005 Quantum information with continuous variables *Rev. Mod. Phys.* **77** 513–77

[6] Chou C-H, Yu T, and Hu B L 2008 Exact master equation and quantum decoherence of two coupled harmonic oscillators in a general environment *Phys. Rev.* E **77** 011112-1–011112-18

[7] Dirac P A M 1927 The quantum theory of dispersion *Proc. R. Soc. A: Math. Phys. Eng. Sci.* **114** 710–28

[8] Dirac P A M 1927 The quantum theory of the emission and absorption of radiation *Proc. R. Soc. A: Math. Phys. Eng. Sci.* **114** 243–65

[9] Dirac P A M 1945 Unitary representations of the Lorentz group *Proc. R. Soc. A: Math. Phys. Eng. Sci.* **183** 284–95

[10] Dirac P A M 1949 Forms of relativistic dynamics *Rev. Mod. Phys.* **21** 392–9

[11] Dodd P J and Halliwell J J 2004 Disentanglement and decoherence by open system dynamics *Phys. Rev.* A **69**

[12] Fano U 1957 Description of states in quantum mechanics by density matrix and operator techniques *Rev. Mod. Phys.* **29** 74–93

[13] Feynman R P 1970 *Relativistic Quark Model* talk presented at the American Physical Society meeting, Washington, DC

[14] Feynman R P 1998 *Statistical Mechanics: A Set of lectures* (*Advanced Book Classics*) (Boulder, CO: Westview Press) (ISBN 978-0-201-36076-9, Originally published 1972; OCLC: ocm60679997)

[15] Feynman R P, Kislinger M, and Ravndal F 1971 Current matrix elements from a relativistic quark model *Phys. Rev.* D **3** 2706–32

[16] Gazeau J P and Graffi S 1997 Quantum harmonic oscillator: a relativistic and statistical point of view *Boll. Unione Mat. Ital.* **11A** 815–39

[17] Gell-Mann M 1964 A schematic model of baryons and mesons *Phys. Lett.* **8** 214–5

[18] Giedke G, Wolf M M, Krüger O, Werner R F, and Cirac J I 2003 Entanglement of formation for symmetric Gaussian states *Phys. Rev. Lett.* **91** 107901-1–4

[19] Han D, Kim Y S, and Noz M E 1981 Physical principles in quantum field theory and in covariant harmonic oscillator formalism *Found. Phys.* **11** 895–905

[20] Han D, Kim Y S, and Noz M E 1999 Illustrative example of Feynman's rest of the Universe *Am. J. Phys.* **67** 61–6

[21] Kim Y S and Noz M E 1973 Covariant harmonic oscillators and the quark model *Phys. Rev.* D **8** 3521–7

[22] Kim Y S and Noz M E 1978 Group theory of covariant harmonic oscillators *Am. J. Phys.* **46** 480–3

[23] Kim Y S and Noz M E 1986 *Theory and Applications of the Poincaré Group* (Dordrecht: Springer) (ISBN 978-94-010-8526-7 978-94-009-4558-6)

[24] Kim Y S and Noz M E 1991 *Phase Space Picture of Quantum Mechanics: Group Theoretical Approach* (*Lecture Notes in Physics* vol 40) (Singapore; Hackensack, NJ: World Scientific Publishing Co.) (ISBN 978-981-02-0360-3 978-981-02-0361-0)

[25] Kim Y S and Noz M E 2003 Feynman's decoherence *Opt. Spectrosc.* **94** 733–40

[26] Kim Y S and Noz M E 2014 Entropy and temperature from entangled space and time *Phys. Sci. Int. J.* **4** 1015–39
[27] Kim Y S, Noz M E, and Oh S H 1979 Representations of the Poincaré group for relativistic extended hadrons *J. Math. Phys.* **20** 1341–4 See also, Physics Auxiliary Publication Service Document No. PAPS JMAPA-20-1336-12
[28] Kim Y S, Noz M E, and Oh S H 1979 A simple method for illustrating the difference between the homogeneous and inhomogeneous Lorentz groups *Am. J. Phys.* **47** 892–7
[29] Kim Y S and Wigner E P 1990 Entropy and Lorentz transformations *Phys. Lett.* A **147** 343–7
[30] Kowalski K and Rembieliński J 2010 Relativistic massless harmonic oscillator *Phys. Rev.* A 81
[31] Navarro D J and Navarro-Salas J 1996 Special-relativistic harmonic oscillator modeled by Klein–Gordon theory in anti-de Sitter space *J. Math. Phys.* **37** 6060–73
[32] Oda M, Ishida M, and Ishida S 1999 Exclusive semi-leptonic decays of heavy mesons in the covariant oscillator quark model *Prog. Theor. Phys.* **101** 1285–311
[33] Pavšič M 2013 Pais-Uhlenbeck oscillator with a benign friction force *Phys. Rev.* D **87** 87
[34] Paz J P and Roncaglia A J 2008 Dynamics of the entanglement between two oscillators in the same environment *Phys. Rev. Lett.* **100** 220401-1–4
[35] Rotbart F C 1981 Complete orthogonality relations for the covariant harmonic oscillator *Phys. Rev.* D **23** 3078–80
[36] Ruiz M J 1974 Orthogonality relation for covariant harmonic-oscillator wave functions *Phys. Rev.* D **10** 4306–7
[37] Simon D S 2011 Supersymmetry and duality in relativistic oscillator systems *Phys. Lett.* A **375** 3751–4
[38] Von Neumann J 1996 *Mathematical Foundations of Quantum MechanicsPrinceton Landmarks in Mathematics and Physics* (Princeton, NJ: Princeton University Press) (ISBN 978-0-691-02893-4 978-0-691-08003-1, Originally published: Princeton University Press, 1955, and in German by Springer, Berlin 1932; OCLC: ocm37904902)
[39] Wigner E P 1932 On the quantum correction for thermodynamic equilibrium *Phys. Rev.* **40** 749–59
[40] Wigner E P 1939 On unitary representations of the inhomogeneous Lorentz group *Ann. Math.* **40** 149–204
[41] Yuen H P 1976 Two-photon coherent states of the radiation field *Phys. Rev.* A **13** 2226–43

IOP Publishing

Physics of the Lorentz Group (Second Edition)

Beyond high-energy physics and optics

Sibel Başkal, Young S Kim, and Marilyn E Noz

Chapter 7

Quarks and partons in the Lorentz-covariant world

Since the time of Einstein and Bohr, there has been an evolution of how we look at quantum bound states, as illustrated in figure 7.1.

The evolution took place in the following three steps:

1. The energy levels of the hydrogen atom played the pivotal role by replacing the electron orbit of the hydrogen atom with a standing wave, leading to bound states in quantum mechanics. However, the hydrogen atom cannot play a role in the Lorentz-covariant world since it cannot be accelerated to a relativistic speed.
2. 1964, the proton became a bound state of the more fundamental constituents called *quarks* [15]. Of course, the proton is different from the hydrogen atom but inherits the same quantum mechanics from the hydrogen atom. Unlike the hydrogen atom, the proton can be accelerated to a speed very close to that of light. Thus, by studying the proton in the quark model, it is possible to study the quantum mechanics of the hydrogen atom or bound states in the Lorentz-covariant world. Figure 7.1 illustrates this transition.
3. In 1969, Feynman observed that the proton, when moving with a velocity close to that of light, appears like a collection of partons with a wide-spread momentum distribution [9, 10]. Partons are like free particles. Quarks and partons are the same particles, but they appear differently to observers in two different reference frames. Therefore, there must be a Lorentz-covariant model for quantum bound states, as illustrated in figure 7.2.

At the time of Einstein and Bohr, both the proton and electron were regarded as point particles. However, the discovery of Hofstadter and McAllister changed our picture of the proton [21]. The proton charge has an internal distribution. Within the framework of quantum electrodynamics, it is possible to calculate the Rutherford formula for electron–proton scattering when both electron and proton are point

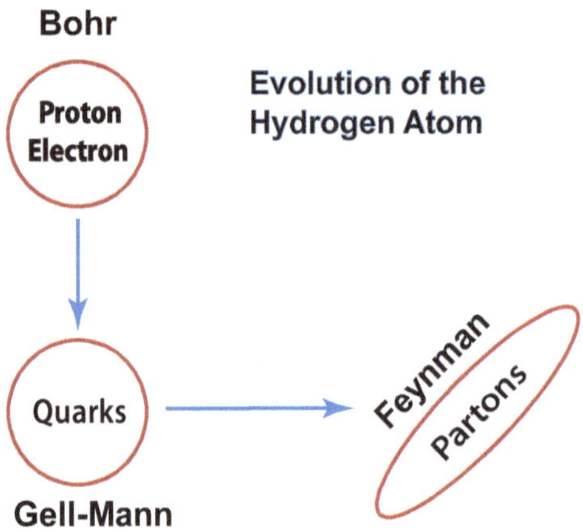

Figure 7.1. Evolution of the hydrogen atom. The electron orbit was replaced by a standing wave, but the hydrogen atom cannot be accelerated. In 1964, the proton as a bound state inherited the quantum mechanics of the hydrogen atom [15]. These days, the proton can move with a speed very close to that of light and exhibits the properties of quantum bound states in the Lorentz-covariant world.

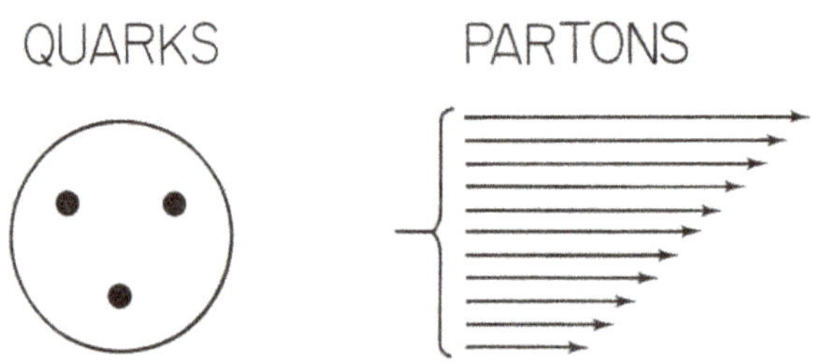

Figure 7.2. The proton appearing in the real world in two distinct ways. It appears as a bound state of three quarks [15] when the proton is at rest. However, the proton appears like a collection of an infinite number of partons [3, 9, 10] when moving with a speed close to that of light. We are then faced with the question of whether quarks and partons might be two different manifestations of the same Lorentz-covariant entity [31].

particles. Because the proton is not a point particle, there is a deviation from the Rutherford formula. We describe this deviation as the *proton form factor*, which depends on the momentum transfer during electron–proton scattering.

Indeed, the study of the proton form factor has been and still is one of the central issues in high-energy physics. The proton form factor decreases as the momentum transfer increases. Its behavior is called the *dipole cut-off*, meaning an inverse-square decrease, and it has been a challenging problem in quantum field theory and other theoretical models [13]. Since the emergence of the quark model in 1964 [15], the hadrons are regarded as quantum bound states of quarks with space–time extensions.

Furthermore, the hadronic mass spectra indicate that the binding force between the quarks is like that of the harmonic oscillator [12]. This leads us to suspect that the quark model with harmonic oscillator wave functions could explain the behavior of the proton form factor. There are indeed many papers written on this subject. We shall return to this problem in section 7.4.

Another problem in high-energy physics is Feynman's parton picture [9, 10]. If the hadron is at rest, we can approach this problem within the framework of bound-state quantum mechanics. If it moves with a speed close to that of light, it appears as a collection of an infinite number of partons, which interact with external signals incoherently. This phenomenon raises the question of whether the Lorentz boost destroys quantum coherence [26].

In section 7.1, we formulate the Lorentz-covariant quark model. We develop in section 7.2 Feynman's parton picture, show the contradictions it contains, and suggest how these contradictions can be resolved. In section 7.3, the proton structure function is introduced. The proton form factor is derived in section 7.4 and shown to be a consequence of Lorentz coherence. In section 7.5, we discuss Lorentz coherence in energy–momentum space. We show in section 7.6 how the temperature of a fast moving hadron increases. This leads to the quarks being heated and appearing as partons.

7.1 Lorentz-covariant quark model

In the quark model, mesons are two-body bound states of one quark and one anti-quark, and baryons are bound states of three quarks. The early successes of the quark model include the ratio of the proton–neutron electromagnetic potential and magnetic moments [2]. In 1971, Feynman *et al* [12] showed the hadronic mass spectra to be similar to that of three-dimensional harmonic oscillators.

The question then is how the mass spectrum for the hydrogen atom, calculated within the framework of non-relativistic quantum mechanics, is valid for the relativistic hadronic mass spectrum. For the hadron, the time-separation variable cannot be ignored. The answer to this question, given in the 1971 paper of Feynman *et al* [12], is not satisfactory. The correct answer is that Wigner's little group for massive particles is like the three-dimensional rotation group, as was detailed in chapters 2 and 6. The role of the time-separation variable is discussed there.

The original question was how the hydrogen atom looks to a moving observer. The question now is how much we can learn about this Bohr–Einstein issue by studying the proton in the quark model based on the three-dimensional harmonic oscillator. For the hydrogen atom, we use the Coulomb potential, while the binding force between quarks is that of the oscillator. The point is that those two different potentials share the same quantum mechanics.

For this purpose, we will need a bound-state wave function, which can be Lorentz-boosted. Here the natural choice is the harmonic oscillator wave function discussed in chapter 6. We can start with the ground-state wave function, which can be Lorentz boosted. Since the harmonic oscillator wave function is separable in the Cartesian coordinate system, we can leave out the transverse components of the wave function and consider only the longitudinal and time-like coordinates. For this purpose, let us rewrite the wave function of equation (6.43) as

$$\psi_\eta(z, t) = \frac{1}{\sqrt{\pi}} \exp\left\{-\frac{1}{4}\left[\left(e^{2\eta}(z + t)^2 + e^{-2\eta}(z - t)^2\right)\right]\right\}, \tag{7.1}$$

which becomes

$$\psi_0(z, t) = \frac{1}{\sqrt{\pi}} \exp\left\{-\frac{1}{2}\left[(z^2 + t^2)\right]\right\}, \tag{7.2}$$

for $\eta = 0$.

7.2 Feynman's parton picture

Let us go back to the problem of the hadron and discuss what happens when the wave function is Lorentz-boosted. For this system, we have considered the Lorentz-squeeze problem in section 6.2. It is a widely accepted view that hadrons are quantum bound states of quarks having a localized probability distribution. As is true for all bound-states, this localization condition is responsible for the existence of discrete mass spectra. The most convincing evidence for this bound-state picture is the hadronic mass spectra [12, 31].

However, this picture of bound states is applicable only to observers in the Lorentz frame in which the hadron is at rest. How would the hadrons appear to observers in other Lorentz frames?

In 1969, Feynman observed that a fast-moving hadron could be regarded as a collection of many *partons* whose properties appear to be quite different from those of the quarks [9, 10, 31]. For example, the number of quarks inside a static proton is three, while the number of partons in a rapidly moving proton appears infinite. The question then is how the proton looking like a bound state of quarks to one observer can appear so differently to an observer in a different Lorentz frame? Feynman made the following systematic observations.

a. The picture is valid only for hadrons moving with a velocity close to that of light.
b. The interaction time between the quarks becomes dilated, and partons behave as free independent particles.
c. The momentum distribution of partons becomes wide-spread as the hadron moves fast.
d. The number of partons seems to be infinite or much larger than that of quarks.

Because the hadron is believed to be a bound state of two or three quarks, each of the above phenomena appears as a paradox, particularly (b) and (c) together. How can a free particle have a wide-spread momentum distribution?

To resolve this paradox, let us construct the energy–momentum wave function corresponding to equation (6.43). If the quarks have the energy–momentum four-vectors p_a and p_b, we can construct two independent energy–momentum four-vector variables [12]

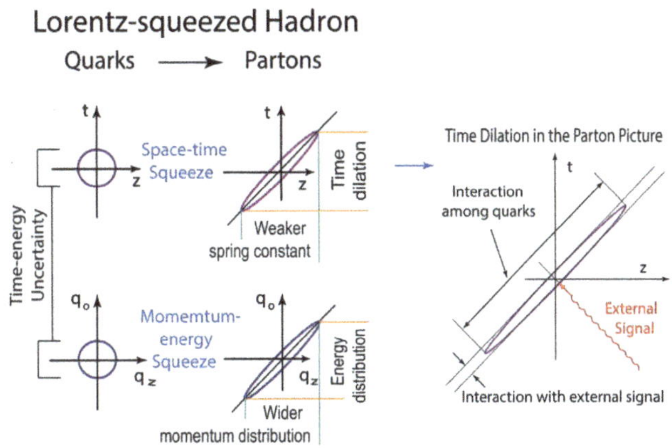

Figure 7.3. Lorentz-squeezed wave functions are pictured in both the space–time and energy–momentum formalism. Both wave functions are concentrated along their respective positive light-cone axes as the hadronic speed approaches that of light. It is this light-cone concentration that leads to Feynman's parton picture [25, 36]. The external signal travels along the negative light-cone axis and thus is moving in the direction opposite to the direction of the hadron. This results in the interaction time of this signal with the bound state being much shorter than the period of oscillation of the quarks inside the hadron. We call this effect Feynman's time dilation [9, 10, 33].

$$P = p_a + p_b \quad \text{and} \quad q = \sqrt{2}(p_a - p_b). \tag{7.3}$$

The energy–momentum four-vector P is the total energy–momentum four-vector and thus the hadronic energy–momentum four-vector, while q measures the energy–momentum four-vector separation between the quarks.

The resulting energy–momentum wave function is

$$\phi_\eta(q_z, q_0) = \left(\frac{1}{\pi}\right)^{1/2} \exp\left\{-\frac{1}{4}\left[e^{-2\eta}(q_z + q_0)^2 + e^{2\eta}(q_z - q_0)^2\right]\right\}. \tag{7.4}$$

For large values of η, we can let $q_0 = q_z$, and the wave function becomes

$$\phi_\eta(q_z) = \left(\frac{1}{\pi}\right)^{1/4} \exp\left\{-\left[e^{-2\eta}(q_z)^2\right]\right\}. \tag{7.5}$$

As we are using here the harmonic oscillator, the mathematical form of the energy–momentum wave function in equation (7.4) is identical to that of the space–time wave function of equation (6.43). The Lorentz squeeze properties of these wave functions are also the same. This aspect of the squeeze has been exhaustively discussed in the literature [25, 36], and it is illustrated again in figure 7.3. The hadronic structure function calculated from this formalism is in reasonable agreement with the experimental data [22].

When the hadron is at rest with $\eta = 0$, both wave functions behave like those for the static bound state of quarks. As η increases, the wave functions become continuously squeezed until they become concentrated along their respective positive light-cone axes. Let us look at the z-axis projection of the space–time wave function. Indeed, the width of the quark distribution increases as the hadronic speed approaches that of light. The position of each quark appears wide-spread to the observer in the laboratory frame, and the quarks appear like free particles.

The energy–momentum wave function is just like the space–time wave function. The longitudinal momentum distribution becomes wide-spread as the hadronic speed approaches the velocity of light. This is in contradiction with our expectation from non-relativistic quantum mechanics that the width of the momentum distribution is inversely proportional to that of the position wave function. Our expectation is that if quarks are free, they must have a sharply defined momenta, not a wide-spread distribution.

However, according to our Lorentz-squeezed space–time and energy–momentum wave functions, the space–time width and the energy–momentum width increase in the same direction as the hadron is boosted. This, of course, is an effect of Lorentz covariance. This indeed leads to the resolution of one of the quark–parton puzzles [25, 31, 36].

Another puzzling problem in the parton picture is that partons appear as incoherent particles, while quarks are coherent when the hadron is at rest. Does this mean that the coherence is destroyed by the Lorentz boost [26, 27]? The answer is NO, and here is the resolution to this puzzle.

When the hadron is boosted, the hadronic matter becomes squeezed and becomes concentrated in the elliptic region along the positive light-cone axis. The length of the major axis becomes expanded by e^{η}, and the minor axis is contracted by $e^{-\eta}$.

This means that the interaction time of the quarks among themselves becomes dilated. Because the wave function becomes wide-spread, the distance between one end of the oscillator well and the other end increases. This effect, first noted by Feynman [9, 10], is universally observed in high-energy hadronic experiments. The period of oscillation increases like e^{η}. On the other hand, since it is moving in the direction opposite to the direction of the hadron, the external signal travels along the negative light-cone axis.

If the hadron contracts along the negative light-cone axis, the interaction time decreases by $e^{-\eta}$. The ratio of the interaction time to the oscillator period becomes $e^{-2\eta}$. The energy of each proton coming out of the large hadron collider is 13 TeV. This leads to the ratio 1.25×10^{-9}. This number is indeed small. The external signal is not able to sense the interaction of the quarks among themselves inside the hadron.

Indeed, the covariant harmonic oscillator formalism provides one Lorentz-covariant entity that produces the quark and parton models as two limiting cases, as is indicated in table 7.1.

7.3 Proton structure function

The quark distribution in energy–momentum space can be measured from the inelastic electron–proton scattering with one-photon exchange [3]. The measured

Table 7.1. Here we add a fourth row to table 2.1 to show the further contents of Einstein's $E = mc^2$. Thus, we are able to view this unified picture of the quark and parton models as further content of Einstein's energy–momentum relation [25, 31].

	Massive Slow	Lorentz Covariance	Massless Fast
Energy-Momentum	$E = p^2/2m$	Einstein's $E = [p^2 + m^2]^{1/2}$	$E = cp$
Spin, Helicity	J_3	Wigner's	Helicity
Gauge Transformation	J_1, J_2	Little Group	Gauge Trans.
Hadron's	Gell-Mann's	Lorentz-covariant	Feynman's
Constituents	Quark Model	Harmonic Oscillator	Parton Picture

distribution is called the proton structure function. We are now interested in how close the Gaussian form of equation (7.4) is to the experimental world.

First of all, in the large-η limit, the proton wave function is within the narrow elliptic region where $q_z = q_0$, and we are left with the wave function depending on only one variable. Thus, this one-variable wave function takes the form

$$\phi_\eta(q_z) = \left(\frac{1}{\pi}\right)^{1/4} \exp\left\{-\left[e^{-2\eta}(q_z)^2\right]\right\}. \tag{7.6}$$

According to equation (7.3), the

$$p_{az} = \left(\frac{P_z}{2} + \frac{q_z}{2\sqrt{2}}\right) \quad \text{and} \quad p_{bz} = \left(\frac{P_z}{2} - \frac{q_z}{2\sqrt{2}}\right). \tag{7.7}$$

If we introduce the parameter

$$x = \frac{p_{az}}{P_z}, \tag{7.8}$$

this is the ratio of the quark momentum to the hadronic momentum. Indeed, this variable is used for measuring the parton distribution in high-energy laboratories.

It is then possible to write the Gaussian form of equation (7.6) in terms of this x variable, and the quark distribution can be written as

$$\rho(x) = \exp\left[-\gamma\left(x - \frac{1}{2}\right)^2\right], \tag{7.9}$$

where the constant γ is to be determined from the level separation in the hadronic mass spectra [12]. The variable x ranges from its minimum value of zero to the maximum value 1. This Gaussian form peaks at $x = 1/2$.

Before attempting to make real contact with the experimental world, we have to face the fact that the proton is a bound state of three quarks. Within the harmonic oscillator regime, the three-body bound system can be separated into a regime of two independent oscillators. This problem was worked out in detail in the 1971 paper of Feynman *et al* [12]. Let us reproduce their calculation.

Let x_a, x_b, x_c represent the space–time coordinates for those quarks. If there is an oscillator force between two quarks, we are led the quadratic form

$$\left[(x_a - x_b)^2 + (x_b - x_c)^2 + (x_c - x_a)^2\right]. \tag{7.10}$$

In order to deal with this expression, Feynman *et al* introduced the following three variables:

$$\begin{aligned} X &= \frac{x_a + x_b + x_c}{3}, \\ r &= \frac{x_a + x_b - 2x_c}{6}, \\ s &= \frac{x_b - x_a}{2}, \end{aligned} \tag{7.11}$$

and

$$\begin{aligned} x_a &= X - 2r, \\ x_b &= X + r - \sqrt{3}\,s, \\ x_c &= X + r + \sqrt{3}\,s. \end{aligned} \tag{7.12}$$

In terms of the r and s variables, the quadratic form becomes

$$18(r^2 + s^2), \tag{7.13}$$

and does not depend on the X variable, which specifies the space–time coordinate of the proton.

As for the energy–momentum four-vectors, let us call them p_a, p_b, and p_c for the quarks a, b, and c, respectively, and introduce the following variables. For the energy–momentum four-vectors, we can introduce the following three variables.

$$\begin{aligned} P &= p_a + p_b + p_c, \\ q &= p_a + p_b - 2p_c, \\ k &= \sqrt{3}(p_b - p_a). \end{aligned} \tag{7.14}$$

Then

$$\begin{aligned} p_a &= \frac{1}{3}P + \frac{1}{6}q - \frac{1}{2\sqrt{3}}k, \\ p_b &= \frac{1}{3}P + \frac{1}{6}q + \frac{1}{2\sqrt{3}}k, \\ p_c &= \frac{1}{3}P - \frac{1}{3}q. \end{aligned} \tag{7.15}$$

In terms of these variables, we are led to consider the quadratic form of

$$18(q^2 + k^2). \tag{7.16}$$

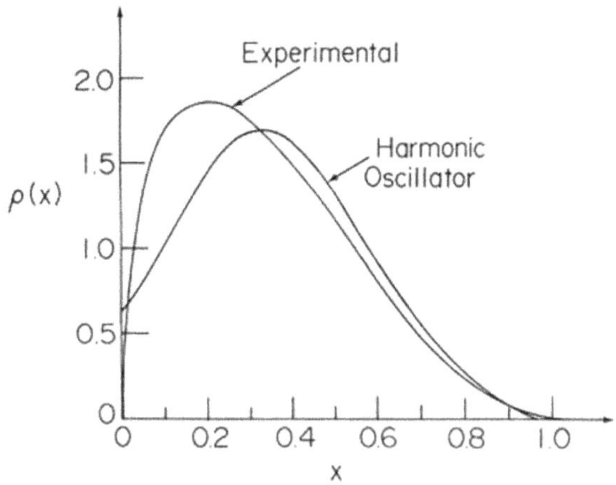

Figure 7.4. The parton distribution function based on the boosted harmonic oscillator is compared here with experimental data. The peak value of the boosted oscillator is at $x = 1/3$. Although this Gaussian form has reasonable agreement with experimental data when x has a large value, the disagreement for small values of x is substantial. This figure is reprinted, with permission, from Paul Hussar's paper [22]. Copyright (1981) by the American Physical Society.

This form does not depend on the variable P, which measures the momentum and energy of the proton.

If the external signal interacts with quark c, its momentum depends only on the q variable, which can be written as

$$q = P - 3p_c. \tag{7.17}$$

We can then define the x variable as

$$x = \frac{p_{cz}}{P_z}. \tag{7.18}$$

Then the quark distribution should take the form

$$\rho(x) = \left(\frac{1}{\pi\gamma}\right)^{1/2} \exp\left[-\gamma\left(x - \frac{1}{3}\right)^2\right]. \tag{7.19}$$

The constant γ is to be determined from the hadronic mass spectra based on the oscillator model [12]. Figure 7.4 shows this Gaussian form and its comparison with what we observe in the real world. The experimental data is measured from inelastic electron–proton scattering [3]. These two curves are somewhat different because the quarks do not interact with the incoming photon as a point particle.

The simplest model is to put all those non-point effects into one additional quark in the oscillator system. This model then leads to the proton as a bound state of four quarks. The fourth quark is responsible for all the unexplained effects [30]. Instead, it is possible is to use the valon model [23, 24], which allows us to screen out all those non-point effects. Using this valon model, Hussar [22] derived the experimental curve to be compared with the Gaussian form, as shown in figure 7.4.

This graph may not be as accurate as we desire. However, the remarkable feature is that the Gaussian form was calculated from the proton at rest. So is the constant γ. It came from the level spacing in the hadronic mass spectra. It is remarkable that these two features manifest themselves for the proton whose speed is very close to that of light.

There are many other models to deal with the problem of providing corrections to the parton distribution. QCD (quantum chromodynamics) is an illustrative example [4]. QCD can provide corrections to the distribution, but it does not produce the distribution from which to start. The covariant harmonic oscillator function provides this starting point.

It is like the case of QED (quantum electrodynamics). QED is quite successful in producing the Lamb shift in the hydrogen energy spectrum but cannot produce the Rydberg energy levels to which the correction is made. The hydrogen energy levels are still obtained from the Schrödinger or Dirac equation with the localization condition on wave functions.

7.4 Proton form factor and Lorentz coherence

Let us now consider the elastic scattering of proton and electron with one photon exchange. If the proton is a point particle, the scattering cross section can be calculated from the one-photon exchange Feynman diagram. The calculation is straightforward if the proton is a point particle. This process is called Rutherford scattering, and the cross section becomes the same as the classical Coulomb scattering if the proton's recoil is negligible.

As the momentum transfer becomes substantial, as indicated in figure 7.5, the cross section deviates from Rutherford scattering, as was observed first by Hofstadter and McAllister in 1955 [21]. Subsequently, it was observed that the cross section decreases as

$$\frac{1}{(\text{momentum transfer})^8}. \tag{7.20}$$

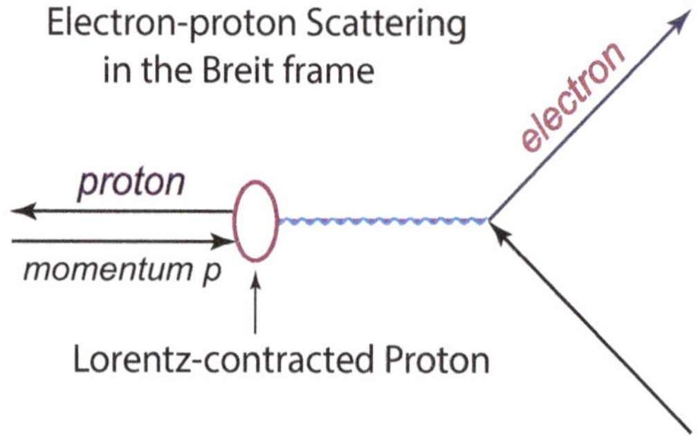

Figure 7.5. Here we illustrate the Breit frame in which the incoming and outgoing protons move with equal magnitude of momentum but travel in opposite directions [31].

This deviation comes from the fact that the proton is not a point particle and that the electric charge inside the proton is distributed with a finite radius. The portion of the scattering amplitude describing this distribution is called the proton form factor. The proton form factor should therefore decrease as

$$\frac{1}{(\text{momentum transfer})^4}. \tag{7.21}$$

This behavior of decrease is known as the dipole cut-off in the literature. This dipole cut-off and possible deviations from it constitute one of the major branches of high-energy physics. There have been, in the past, some far-reaching theoretical models to deal with this problem [13].

In this section, we are interested in approaching this problem using the harmonic oscillator formalism developed in chapter 6. We shall show that the dipole cut-off is a consequence of the coherence between the contraction of the proton wave function and the decrease in the wavelength of the incoming signal.

While the formalism of chapter 6 is primarily based on the papers written by Dirac and Wigner, it is interesting to note that the same harmonic oscillator functions can be derived from those authors who attempted to understand the proton form factor. These authors were not aware of the works of Dirac and Wigner. Let us briefly review what they did.

In 1953, Yukawa was interested in constructing harmonic oscillator wave functions that can be Lorentz-transformed [45]. His primary interest was in the mass spectrum produced by his Lorentz-invariant differential equation. However, at that time, his mass spectrum did not appear to have anything to do with the physical world.

After witnessing a non-zero charge radius of the proton observed by Hofstadter and McAllister [20, 21], Markov in 1956 considered using Yukawa's oscillator formalism for calculating the proton form factor [40].

However, the constituent particles of the oscillator wave functions were not defined at that time. Shortly after the emergence of the quark model in 1964 [15], Ginzburg and Man'ko [16] considered relativistic harmonic oscillators for bound-states of quarks.

Even though they did not mention Yukawa's 1953 paper, Fujimura, Kobayashi, and Namiki used the quark model based on Yukawa's relativistic oscillator wave function to calculate the proton form factor, and obtained the dipole cut-off [14]. In the same year, Licht and Pagnamenta derived the same result using Lorentz-contracted oscillator wave functions. They used the Breit coordinate system to bypass the time-separation variable appearing in the covariant formalism [29, 38].

In 1971, Feynman, Kislinger, and Ravndal noted that the observed hadronic mass spectra could be explained in terms of the degeneracies of three-dimensional harmonic oscillators [12], confirming the earlier suggestion made by Yukawa in 1953. They quoted the paper by Fujimura *et al* [14], but they did not mention Yukawa's 1953 paper [45]. This is the reason why Feynman *et al* could not write down normalizable wave functions.

Let us go back to the formalism developed in chapter 6. When considering the scattering of one electron and one proton by exchanging one photon, it is possible to choose the Lorentz frame in which the incoming and outgoing protons are moving in opposite directions with the same speed. Let us assume that the proton is moving along the z-direction as indicated in figure 7.5, and let p be the magnitude of the momentum. Then the initial and final energy–momentum four-vectors are

$$(p, E) \quad \text{and} \quad (-p, E), \tag{7.22}$$

respectively, where $E = \sqrt{1 + p^2}$. The momentum transfer in this Breit frame is

$$(p, E) - (-p, E) = (2p, 0), \tag{7.23}$$

with zero energy component.

The proton form factor then becomes

$$F(p) = \int e^{2ipz}\left(\psi_\eta(z, t)\right)^* \psi_{-\eta}(z, t)\, dz\, dt. \tag{7.24}$$

If we use the ground-state harmonic oscillator wave function, this integral becomes

$$\frac{1}{\pi}\int e^{2ipz} \exp\left\{-\cosh(2\eta)(z^2 + t^2)\right\} dz\, dt. \tag{7.25}$$

The physics of $\cosh(2\eta)$ in this expression was explained in equation (6.50).

In the Fourier integral of equation (7.25), the exponential function does not depend on the t variable. Thus, after the t integration, equation (7.25) becomes

$$F(p) = \frac{1}{\sqrt{\pi \cosh(2\eta)}}\int e^{2ipz} \exp\left\{-z^2 \cosh(2\eta)\right\} dz. \tag{7.26}$$

If we complete this integral, the proton form factor becomes

$$F(p) = \frac{1}{\cosh(2\eta)} \exp\left\{\frac{-p^2}{\cosh(2\eta)}\right\}. \tag{7.27}$$

If we use the expression of $\cosh(2\eta)$ given in equation (6.50), this proton form factor becomes

$$F(p) = \frac{1}{1 + 2p^2} \exp\left(\frac{-p^2}{1 + 2p^2}\right), \tag{7.28}$$

which decreases as $1/p^2$ for large values of p.

To illustrate the effect of the role of this Lorentz contraction in more detail, let us perform the integral of equation (7.26) without the contraction factor $\cosh(2\eta)$. This means that the wave function $\psi_\eta(z, t)$ in equation (7.24) is replaced by the Gaussian form $\psi_0(z, t)$ of equation (7.2). With this non-squeezed wave function, the Fourier integral becomes

$$G(p) = \int e^{2ipz}(\psi_0(z, t))^* \psi_0(z, t)\, dz\, dt. \tag{7.29}$$

The result of this integral is

$$G(p) = \frac{1}{\sqrt{\pi}} \exp(-p^2). \tag{7.30}$$

This leads to a Gaussian cut-off of the proton form factor. This Gaussian cut-off does not happen in the real world, and the calculation of $G(p)$ is for an illustrative purpose only.

Let us go back to the Fourier integrals of equations (7.24) and (7.29). The only difference is the $\cosh(2\eta)$ factor in equation (7.24). This factor is in the normalization constant and comes from the integration over the t variable, which does not affect the Fourier integral.

However, it causes the Gaussian width to shrink by $1/\sqrt{2}p$ for large values of p. The wavelength of the sinusoidal factor is inversely proportional to the momentum $2p$. Thus, both the Gaussian width and the wavelength of the incoming signal shrink at the same rate of $1/p$ as p becomes large. Without this coherence, the cut-off is Gaussian, as noted in equation (7.30). The effect of this Lorentz coherence is illustrated in figure 7.6.

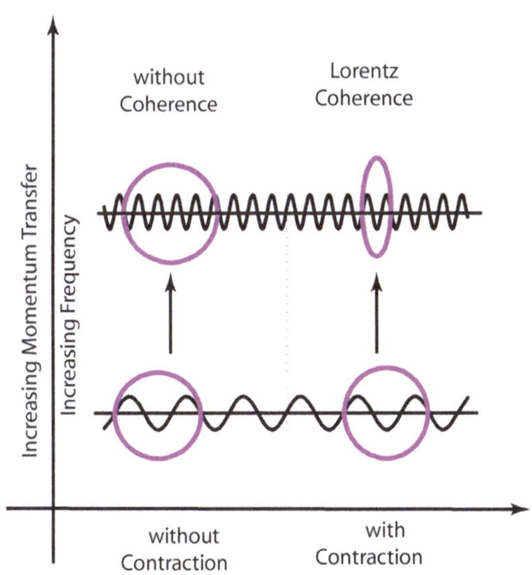

Figure 7.6. This figure shows the coherence between the proton size and the wavelength. By referring back to figure 7.5, it is evident that the proton sees the incoming photon. As the momentum transfer increases, the wavelength of this photon becomes smaller. Should the proton size remain unchanged, the result would be a rapid oscillation cut-off for the form factor in the Fourier integral that leads to a Gaussian cut-off. On the other hand, when the proton size decreases coherently with the wavelength, there are no oscillation effects, which leads to a polynomial decrease of the form factor [31, 34].

There is still a gap between $F(p)$ of equation (7.28) and the real world. Before comparing this expression with the experimental data, we have to realize that there are three quarks inside the proton with two oscillator modes.

One of the modes goes through the Lorentz coherence process discussed in this section. The other mode goes through the contraction process given in equation (6.50). The net effect is

$$F_3(p) = \left(\frac{1}{1 + 2p^2}\right)^2 \exp\left(\frac{-p^2}{1 + 2p^2}\right). \tag{7.31}$$

This will lead to the desired dipole cut-off of $(1/p^2)^2$.

In addition, the effect of the quark spin should be addressed. There are also reports of deviations from the exact dipole cut-off. There have been attempts to study the proton form factors based on the four-dimensional rotation group with an imaginary time coordinate. There are also many papers based on the lattice QCD. A brief list of the references to these efforts is given in [34].

The purpose of this section was limited to studying in detail the role of Lorentz coherence in keeping the proton form factor from the steep Gaussian cut-off in the momentum transfer variable. The coherence problem is one of the primary issues of the current trend in physics.

7.5 Coherence in energy–momentum space

We are now interested in how Lorentz coherence manifests itself in energy–momentum space. We start with the Lorentz-squeezed wave function in energy–momentum space, which can be written as

$$\phi_\eta(q_z, q_0) = \frac{1}{2\pi} \int e^{-i(q_z z - q_0 t)} \psi_\eta(z, t) dt\, dz. \tag{7.32}$$

This is a Fourier transformation of the Lorentz-squeezed wave function of equation (7.1), where q_z and q_0 are Fourier conjugate variables to z and t, respectively. The result of this integral is

$$\phi_\eta(q_z, q_0) = \frac{1}{\sqrt{\pi}} \exp\left\{-\frac{1}{4}\left[e^{-2\eta}(q_z + q_0)^2 + e^{2\eta}(q_z - q_0)^2\right]\right\}. \tag{7.33}$$

In terms of this energy–momentum wave function, the proton form factor of equation (7.24) can be written as

$$F(p) = \int \phi^*_{-\eta}(q_0, q_z - p)\phi_\eta(q_0, q_z + p) dq_0\, dq_z. \tag{7.34}$$

The evaluation of this integral leads to the proton form factor $F(p)$ given in equation (7.28).

To see the effect of the Lorentz coherence, let us look at two wave functions in figure 7.7. The integral is carried out over the $q_z\, q_o$-plane. As the momentum p increases, the two wave functions become separated. Without the Lorentz squeeze,

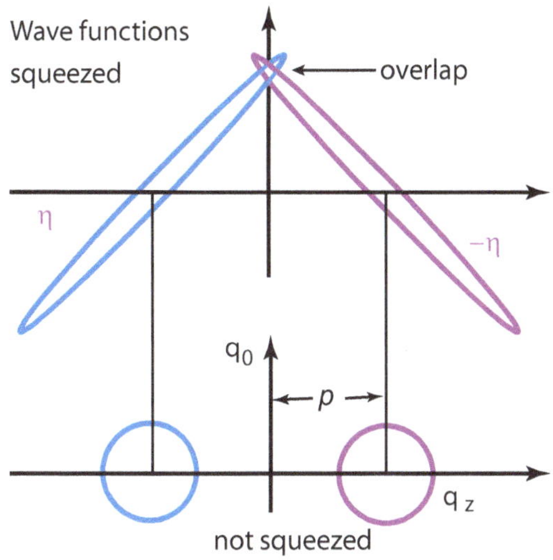

Figure 7.7. Momentum-energy space is shown in terms of Lorentz coherence with both squeezed and non-squeezed wave functions. The two wave functions in equation (7.34), as p increases, become separated. If there were no squeeze, then there would be no overlap, leading to a Gaussian cut-off. The squeezed wave functions, however, maintain an overlap, that leads to a slower polynomial cut-off [31].

the wave functions do not overlap, leading to a sharp Gaussian cut-off as in the case of $G(p)$ of equation (7.30).

On the other hand, the squeezed wave functions have an overlap, as shown in figure 7.7. This overlap becomes smaller as p increases, leading to a slower polynomial cut-off [31, 34].

The discovery of the non-zero size of the proton opened a new era of physics [20]. The proton is no longer a point particle. One way to measure its internal structure is to study the proton–electron scattering amplitude with one photon exchange and its dependence on the momentum transfer. The deviation from the case with the point-particle proton is called the proton form factor.

On the experimental front, the dipole cut-off has been firmly established. Yes, there are also experimental results indicating deviations from this dipole behavior [1, 41]. However, in the present section, no attempts have been made to review all the papers written on the corrections. From the theoretical point of view, those deviations are corrections from the basic dipole behavior.

While the study of the proton form factor is still a major subject in physics, it is gratifying to note that the proton's dipole cut-off comes from the coherence between the Lorentz contraction of the proton's longitudinal size and the decrease in the wavelength of the incoming signal.

7.6 Hadronic temperature and boiling quarks

Harmonic oscillator wave functions are used for all branches of physics. The single-variable ground-state harmonic oscillator can be excited in the following three different ways:

1. Energy level excitations, with the energy eigenvalues $\hbar\omega(n + 1/2)$.

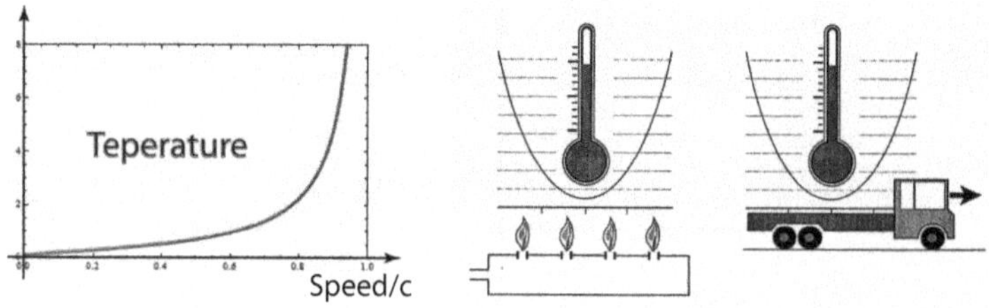

Figure 7.8. The hadronic temperature is plotted against the speed, β. As the hadronic speed increases, the quarks inside the hadron become excited and the result is in a rise in temperature. Those quarks start to boil and become partons when the temperature is sufficiently high, [35].

2. Coherent state excitations resulting in

$$|\alpha\rangle = e^{-\alpha\alpha^*/2}\sum_n \frac{\alpha^n}{\sqrt{n!}}|n\rangle. \tag{7.35}$$

3. Thermal excitations resulting in the density matrix of the form

$$\rho_T(z, z') = (1 - e^{-\hbar\omega/kT})\sum_k e^{-k\hbar\omega/kT}\phi_k(z)\phi_k^*(z'), \tag{7.36}$$

where $\hbar\omega$ and k are the oscillator energy separation and Boltzmann's constant, respectively. This form of the density matrix is well known [5, 17, 28, 37].

We are now interested in thermal excitation. If the temperature is measured in units of $\hbar\omega/k$, the density matrix of equation (7.36) can be written as

$$\rho_T(z, z') = (1 - e^{-1/T})\sum_k e^{-1/T}\phi_k(z)\phi_k^*(z'). \tag{7.37}$$

If we compare this expression with the density matrix of equation (6.60), we are led to

$$\tanh^2\eta = \exp(-1/T), \tag{7.38}$$

and to

$$T = \frac{-1}{\ln(\tanh^2\eta)}. \tag{7.39}$$

The temperature can be calculated as a function of $\tanh(\eta)$, and this calculation is plotted in figure 7.8.

Earlier in equation (6.58), we noted that $\tanh(\eta)$ is proportional to the velocity of the hadron and $\tanh(\eta) = v/c$. Thus, the oscillator becomes thermally excited as it moves, as is illustrated in figure 7.8.

Let us look at the velocity dependence of the temperature again. It is almost proportional to the velocity from $\tanh(\eta) = 0$ to 0.7, and again from $\tanh(\eta) = 0.9$ to 1 with different slopes.

While the physical motivation for this section was based on Feynman's time separation variable [12] and his rest of the Universe [11], we should note that many authors have discussed field theoretic approaches to derive the density matrix of equation (7.36). Among them are two-mode squeezed states of light [19, 32, 44, 47] and thermo-field-dynamics [8, 39, 42, 43].

The mathematics of two-mode squeezed states of light is the same as that for the covariant harmonic oscillator formalism discussed in chapter 6 [6, 19, 32, 46]. Instead of the z and t coordinates, there are two measurable photons. If we choose not to observe one of them [7, 48], it belongs to Feynman's rest of the Universe [18].

Another remarkable feature of two-mode squeezed states of light is that its formalism is identical to thermo-field-dynamics. The temperature is related to the squeeze parameter in the two-mode case. Therefore, it is possible to define the temperature of a Lorentz-squeezed hadron within the framework of the covariant harmonic oscillator model. We shall do this in chapter 9.

References

[1] Alkofer R, Höll A, Kloker M, Krassnigg A, and Roberts C D 2005 On nucleon electromagnetic form factors *Few-Body Syst.* **37** 1–31

[2] Bég M A B, Lee B W, and Pais A 1964 SU(6) and electromagnetic interactions *Phys. Rev. Lett.* **13** 514–7

[3] Bjorken J D and Paschos E A 1969 Inelastic electron–proton and γ-proton scattering and the structure of the nucleon *Phys. Rev.* **185** 1975–82

[4] Buras A J 1980 Asymptotic freedom in deep inelastic processes in the leading order and beyond *Rev. Mod. Phys.* **52** 199–276

[5] Davies R W and Davies K T R 1975 On the Wigner distribution function for an oscillator *Ann. Phys.* **89** 261–73

[6] Dirac P A M 1963 A remarkable representation of the 3 + 2 de Sitter group *J. Math. Phys.* **4** 901–9

[7] Ekert A K and Knight P L 1989 Correlations and squeezing of two-mode oscillations *Am. J. Phys.* **57** 692–7

[8] Fetter A L and Walecka J D 2003 *Quantum Theory of Many-Particle Systems* (Mineola, NY: Dover Publications) (ISBN 978-0-486-42827-7, Originally published: New York, NY: McGraw-Hill, 1971)

[9] Feynman R P 1969 The behavior of Hadron collisions at extreme energies *Proc. of the 3rd Int. Conf. on High Energy Collisions (New York, 5–6 September)*R P Yang *et al* (Stony Brook, NY: Gordon and Breach) pp 237–49

[10] Feynman R P 1969 Very high-energy collisions of hadrons *Phys. Rev. Lett.* **23** 1415–7

[11] Feynman R P 1998 *Statistical Mechanics: A Set of lecturesAdvanced Book Classics* (Boulder, CO: Westview Press) (ISBN 978-0-201-36076-9, Originally published 1972; OCLC: ocm60679997)

[12] Feynman R P, Kislinger M, and Ravndal F 1971 Current matrix elements from a relativistic quark model *Phys. Rev.* D **3** 2706–32

[13] Frazer W R and Fulco J R 1960 Effect of a Pion–Pion scattering resonance on nucleon structure. II *Phys. Rev.* **117** 1609–14

[14] Fujimura K, Kobayashi T, and Namiki M 1970 Nucleon electromagnetic form factors at high momentum transfers in an extended particle model based on the quark model *Prog. Theor. Phys.* **43** 73–9

[15] Gell-Mann M 1964 A schematic model of baryons and mesons *Phys. Lett.* **8** 214–5

[16] Ginzburg V L and Man'ko V I 1965 Relativistic oscillator models of elementary particles *Nucl. Phys.* **74** 577–88

[17] Han D, Kim Y S, and Noz M E 1990 Lorentz-squeezed hadrons and hadronic temperature *Phys. Lett.* A **144** 111–5

[18] Han D, Kim Y S, and Noz M E 1999 Illustrative example of Feynman's rest of the Universe *Am. J. Phys.* **67** 61–6

[19] Han D, Kim Y S, Noz M E, and Yeh L 1993 Symmetries of two-mode squeezed states *J. Math. Phys.* **34** 5493–508

[20] Hofstadter R 1956 Electron scattering and nuclear structure *Rev. Mod. Phys.* **28** 214–54

[21] Hofstadter R and McAllister R W 1955 Electron scattering from the proton *Phys. Rev.* **98** 217–8

[22] Hussar P E 1981 Valons and harmonic oscillators *Phys. Rev.* D **23** 2781–3

[23] Hwa R C 1980 Evidence for valence-quark clusters in nucleon structure functions *Phys. Rev.* D **22** 759–64

[24] Hwa R C and Zahir M S 1981 Parton and valon distributions in the nucleon *Phys. Rev.* D **23** 2539–53

[25] Kim Y S 1989 Observable gauge transformations in the parton picture *Phys. Rev. Lett.* **63** 348–51

[26] Kim Y S 1998 Does Lorentz boost destroy coherence? *Fortschr. Phys.* **46** 713–23

[27] Kim Y S 2004 Einstein, Wigner, and Feynman: from $E = mc^2$ to Feynman's decoherence via Wigner's little groups *Acta Phys. Hungarica A: Heavy Ion Phys.* **19** 317–28

[28] Kim Y S and Li M 1989 Squeezed states and thermally excited states in the Wigner phase-space picture of quantum mechanics *Phys. Lett.* A **139** 445–8

[29] Kim Y S and Noz M E 1973 Covariant harmonic oscillators and the quark model *Phys. Rev.* D **8** 3521–7

[30] Kim Y S and Noz M E 1978 Quarks, partons, and Lorentz-deformed hadrons *Prog. Theor. Phys.* **60** 801–16

[31] Kim Y S and Noz M E 1986 *Theory and Applications of the Poincaré Group* (Dordrecht: Springer) (ISBN 978-94-010-8526-7 978-94-009-4558-6)

[32] Kim Y S and Noz M E 1991 *Phase Space Picture of Quantum Mechanics: Group Theoretical Approach* (*Lecture Notes in Physics* vol 40) (Singapore; Hackensack, NJ: World Scientific Publishing Co.) (ISBN 978-981-02-0360-3 978-981-02-0361-0)

[33] Kim Y S and Noz M E 2005 Coupled oscillators, entangled oscillators, and Lorentz-covariant harmonic oscillators *J. Opt. B: Quant. Semiclass. Opt.* **7** S458–67

[34] Kim Y S and Noz M E 2011 Lorentz harmonics, squeeze harmonics, and their physical applications *Symmetry* **3** 16–36

[35] Kim Y S and Noz M E 2014 Entropy and temperature from entangled space and time *Phys. Sci. Int. J.* **4** 1015–39

[36] Kim Y S and Noz M E 1977 Covariant harmonic oscillators and the parton picture *Phys. Rev.* D **15** 335–8

[37] Landau L D and Lifshitz E M 2008 *Statistical Physics, Part 1* (*Course of Theoretical Physics*) vol 5 3rd edn (Amsterdam: Elsevier) (ISBN 978-0-7506-3372-7, First published in 1958; OCLC: 837367500)
[38] Licht A L and Pagnamenta A 1970 Wave functions and form factors for relativistic composite particles. I *Phys. Rev.* D **2** 1150–6
[39] Mann A and Revzen M 1989 Thermal coherent states *Phys. Lett.* A **134** 273–5
[40] Markov M 1956 On dynamically· deformable form factors in the theory of elementary particles *Il Nuovo Cimento* **3** 760–72
[41] Matevosyan H H, Thomas A W, and Miller G A 2005 Study of lattice QCD form factors using the extended Gari–Krümpelmann model *Phys. Rev.* C **72** 065204-1–5
[42] Ojima I 1981 Gauge fields at finite temperatures—'Thermo field dynamics' and the KMS condition and their extension to gauge theories *Ann. Phys.* **137** 1–32
[43] Umezawa H, Matsumoto H, and Tachiki M 1982 *Thermo Field Dynamics and Condensed States* (Amsterdam; New York: North-Holland Pub. Co.) (ISBN 978-0-444-86361-4)
[44] Yuen H P 1976 Two-photon coherent states of the radiation field *Phys. Rev.* A **13** 2226–43
[45] Yukawa H 1953 Structure and mass spectrum of elementary particles. I. General considerations *Phys. Rev.* **91** 415–6
[46] Yurke B 1986 Quantizing the damped harmonic oscillator *Am. J. Phys.* **54** 1133–9
[47] Yurke B, McCall S L, and Klauder J R 1986 SU(2) and SU(1,1) interferometers *Phys. Rev.* A **33** 4033–54
[48] Yurke B and Potasek M 1987 Obtainment of thermal noise from a pure quantum state *Phys. Rev.* A **36** 3464–6

Sibel Başkal, Young S Kim, and Marilyn E Noz

Chapter 8

Wigner functions and their symmetries

In both classical and quantum mechanics, the position variable and the momentum variable are treated equally. In classical mechanics, these two variables constitute a two-dimensional phase space. In quantum mechanics, they are q-numbers. If the coordinate is x, the momentum takes a differential operator. The question then is whether it is possible in quantum mechanics to treat x and p as c-numbers and define a phase space.

The Wigner phase-space distribution function allows us to define a phase space in quantum mechanics [13]. Let us start with the Schrödinger wave function. The Wigner function is defined as [13]

$$W(x, p) = \frac{1}{\pi} \int \psi(x + x')\psi^*(x - x') \exp(2ipx')\, dx'. \tag{8.1}$$

This Wigner phase-space distribution function is normalized as

$$\int W(x, p)\, dxdp = 1. \tag{8.2}$$

Furthermore, the density distribution $|\psi(x)|^2$ is recovered from the Wigner function as

$$|\psi(x)|^2 = \psi^*(x)\psi(x) = \int W(x, p)\, dp. \tag{8.3}$$

The momentum wave function for $\phi(p)$ is

$$\phi(p) = \frac{1}{\sqrt{2\pi}} \int \psi(x) \exp(ipx)\, dx. \tag{8.4}$$

The momentum distribution can be derived from the Wigner function as

$$|\phi(p)|^2 = \phi^*(p)\phi(p) = \int W(x, p)\, dx. \tag{8.5}$$

doi:10.1088/978-0-7503-3607-9ch8

We can check these features using the normalized Gaussian wave function

$$\psi(x) = \left(\frac{1}{2\pi}\right)^{1/4} \exp\left(\frac{-x^2}{2}\right). \tag{8.6}$$

The Wigner function becomes

$$W(x, p) = \left(\frac{1}{\pi}\right) \exp\left(-[x^2 + p^2]\right). \tag{8.7}$$

If there are two sets of variables, namely (x_1, p_1) and (x_2, p_2), the Wigner function is defined as

$$\begin{aligned} &W(x_1, x_2 \colon p_1, p_2) \\ &= \frac{1}{\pi} \int \psi\left(x_1 + x_1', x_2 + x_2'\right) \psi^*\left(x_1 - x_1', x_2 - x_2'\right) \\ &\times \exp\left(2i\left[p_1 x_1' + p_2 x_2'\right]\right) dx_1' dx_2'. \end{aligned} \tag{8.8}$$

In the case of two ground-state oscillators, the Schrödinger wave function is

$$\psi(x_1, x_2) = \left(\frac{1}{\pi}\right) \exp\left(-\frac{1}{2}\left[x_1^2 + x_2^2\right]\right), \tag{8.9}$$

and the Wigner function becomes the form

$$W(x_1, x_2 \colon p_1, p_2) = \left(\frac{1}{\pi}\right)^2 \exp\left(-\left[x_1^2 + x_2^2 + p_1^2 + p_2^2\right]\right). \tag{8.10}$$

Here, the Wigner function is defined in the four-dimensional phase space. As we shall see in this chapter, the symmetries in this four-dimensional phase space are much richer and produce interesting properties of the Lorentz group. If we integrate out the x_2 and p_2 variables:

$$W(x_1, p_1) = \int W(x_1, x_2 \colon p_1, p_2)\, dx_2 dp_2, \tag{8.11}$$

the result is the Wigner function for the single pair of the variables.

In section 8.1, we shall first study the symmetries contained in the Wigner function for a single pair of the variables. Heisenberg's uncertainty relation is defined in terms of a finite area in the two-dimensional phase space. The rotation and squeeze are the area-preserving linear transformations. They are called area-preserving linear transformations in classical mechanics. We shall see the two-dimensional phase space from a single pair of variables, produces the Lorentz group applicable to the three-dimensional space consisting of one time and two space coordinates.

In section 8.2, it is noted that the two pairs of position and momentum variables lead to the four-dimensional phase space. In section 8.3, canonical transformations with ten generators are considered. This group is called $Sp(4)$, the four-dimensional symplectic group. As Dirac noted in 1963 [2], the Lie algebra of this group produces the five-dimensional de Sitter group $O(3, 2)$, which is the Lorentz group applicable to three space and two time dimensions.

The generators of this $Sp(4)$ group are ten four-by-four traceless matrices with imaginary elements, while there are fifteen possible four-by-four matrices. Indeed, the group generated by all those fifteen matrices is called $SL(4, r)$. In section 8.4, we study this group with fifteen generators. In section 8.5, it is shown to be possible to write those fifteen matrices in terms of Dirac's gamma matrices in the Majorana representation where all elements are imaginary numbers.

Finally, in section 8.6, it is shown that the algebra of this $SL(4, r)$ can generate the Lorentz group applicable to three space dimensions and three time dimensions. The generators take the form of six-by-six matrices. They are tabulated in this section.

8.1 Symmetries and the uncertainty principle in the Wigner phase space

In classical mechanics, the motion can be described in terms of a point tracing a line in phase space. In quantum mechanics, the uncertainty relation does not allow a point. The minimum uncertainty is defined by the Gaussian function given in equation (8.7). We are quite familiar with the case of the invariance of the uncertainty relation for the squeeze transformation:

$$\begin{pmatrix} x' \\ p' \end{pmatrix} = \begin{pmatrix} \exp(\eta/2) & 0 \\ 0 & \exp(-\eta/2) \end{pmatrix} \begin{pmatrix} x \\ p \end{pmatrix}. \tag{8.12}$$

This is an area-preserving transformation. The phase-space distribution of the Gaussian form of equation (8.7) can be described by a circle in the phase space. The squeeze transformation of equation (8.12) does not change the area and thus does not change the uncertainty relation.

Classical mechanics allows rotations in phase space, and this is also possible for the Wigner function, with the coordinate rotation:

$$\begin{pmatrix} x' \\ p' \end{pmatrix} = \begin{pmatrix} \cos(\alpha/2) & -\sin(\alpha/2) \\ \sin(\alpha/2) & \cos(\alpha/2) \end{pmatrix} \begin{pmatrix} x \\ p \end{pmatrix}. \tag{8.13}$$

This rotation will not change the area. Let us call the squeeze matrix of equation (8.12) $S_1(\eta)$ and the rotation matrix of equation (8.13) $R(\alpha)$. Then the squeeze along the $\alpha/2$-direction is

$$\begin{aligned} S(\alpha, \eta) &= R(\alpha)S_1(\eta)R^{-1}(\alpha) \\ &= \begin{pmatrix} \cosh(\eta/2) + [\sinh(\eta/2)]\cos(\alpha) & [\sinh(\eta/2)]\sin(\alpha) \\ [\sinh(\eta/2)]\sin(\alpha) & \cosh(\eta/2) - [\sinh(\eta/2)]\cos(\alpha) \end{pmatrix}. \end{aligned} \tag{8.14}$$

This expression becomes a unit matrix when $\eta = 0$. When $\alpha = 0$, it becomes the squeeze along the z-axis

$$S_1(\eta) = \begin{pmatrix} \exp(\eta/2) & 0 \\ 0 & \exp(-\eta/2) \end{pmatrix}. \tag{8.15}$$

When $\alpha = 90^o$, this two-by-two matrix becomes

$$S_2(\eta) = \begin{pmatrix} \cosh(\eta/2) & \sinh(\eta/2) \\ \sinh(\eta/2) & \cosh(\eta/2) \end{pmatrix}. \tag{8.16}$$

This is the squeeze made along the 45° line, as shown in figure 8.1.

We note that the rotation matrix $R(\alpha)$ and the squeeze matrices $S_1(\eta)$ and $S_2(\eta)$ are generated, respectively, by

$$J_2 = \frac{1}{2}\begin{pmatrix} 0 & -i \\ i & 0 \end{pmatrix}, \quad K_3 = \frac{1}{2}\begin{pmatrix} i & 0 \\ 0 & -i \end{pmatrix}, \quad \text{and} \quad K_1 = \frac{1}{2}\begin{pmatrix} 0 & i \\ i & 0 \end{pmatrix}, \tag{8.17}$$

with

$$R = \exp(-i\alpha J_2), \quad S_1 = \exp(-i\eta K_3), \quad \text{and} \quad S_2 = \exp(-i\eta K_1). \tag{8.18}$$

These generators satisfy the closed set of commutation relations:

$$[K_1, K_3] = iJ_2, \quad [K_1, J_2] = iK_3, \quad \text{and} \quad [K_3, J_2] = -iK_1. \tag{8.19}$$

These commutation relations form the Lie algebra for the $Sp(2)$ group. By looking at the generators, K_1, J_2, and K_3, we notice that they have all imaginary elements and that the transformation variables $S_1(\eta)$, $R(\alpha)$, and $S_2(\eta)$ have all real elements. This is the definition of the $Sp(2)$ group. The operators applicable to the Wigner function $W(x, p)$ take the form

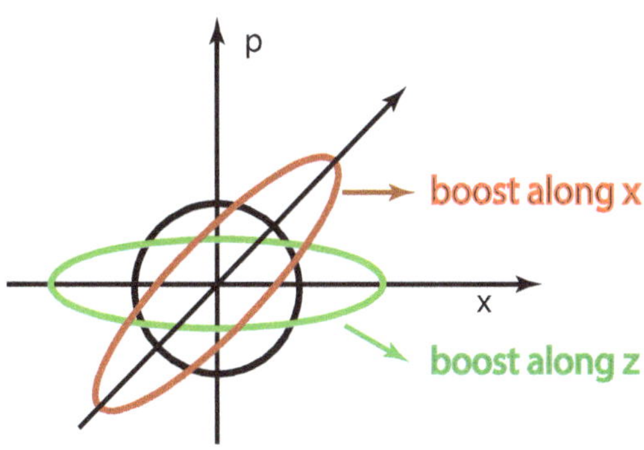

Figure 8.1. Symmetries of the Wigner function in the two-dimensional phase space. The circle in this figure corresponds to the Gaussian distribution given in equation (8.7). This function can be squeezed along the x-direction by the Lorentz boost along the z-axis. It can also be squeezed along the p-direction by the Lorentz boost along the x-axis. This squeeze can be rotated. The squeeze does not change the area, and the minimum uncertainty can be defined in terms of the area even though the distribution in both x and p becomes extended. The area of the ellipse does not change when the ellipse is rotated around the y-axis.

$$K_3 = -i\left(x\frac{\partial}{\partial x} - p\frac{\partial}{\partial p}\right), \quad K_1 = -i\left(x\frac{\partial}{\partial p} + p\frac{\partial}{\partial x}\right), \quad \text{and } J_2 = -i\left(x\frac{\partial}{\partial p} - p\frac{\partial}{\partial x}\right). \tag{8.20}$$

These operators satisfy the *Sp*(2) Lie algebra given in equation (8.19).

Let us go back to the rotation matrix, $R(\alpha)$, of equation (8.13). This rotation matrix satisfies the condition

$$R(\alpha)J\tilde{R}(\alpha) = J \quad \text{with} \quad J = \begin{pmatrix} 0 & 1 \\ -1 & 0 \end{pmatrix}, \tag{8.21}$$

where $\tilde{R}$ is the transpose of the R matrix. The squeeze matrix $S(\alpha, \eta)$ of equation (8.14) also satisfies this condition. If all the matrices of a group satisfy this condition, we call this group the *two-dimensional symplectic group* or *Sp*(2) [1, 3, 4, 10].

The Lie algebra of equation (8.19) is constructed from the two-by-two matrices. However, it is possible to construct the same Lie algebra using a different set of generators in the form of three-by-three matrices. Let us consider

$$K_1 = \begin{pmatrix} 0 & i & 0 \\ i & 0 & 0 \\ 0 & 0 & 0 \end{pmatrix}, \quad K_2 = \begin{pmatrix} 0 & 0 & i \\ 0 & 0 & 0 \\ i & 0 & 0 \end{pmatrix}, \quad \text{and} \quad J_3 = \begin{pmatrix} 0 & 0 & 0 \\ 0 & 0 & -i \\ 0 & i & 0 \end{pmatrix}. \tag{8.22}$$

These matrices satisfy the Lie algebra of the group *Sp*(2) given in equation (8.19), and lead to the transformation matrices of the form

$$\exp(-i\eta K_1) = \begin{pmatrix} \cosh\eta & \sinh\eta & 0 \\ \sinh\eta & \cosh\eta & 0 \\ 0 & 0 & 1 \end{pmatrix}, \quad \exp(-i\eta K_2) = \begin{pmatrix} \cosh\eta & 0 & \sinh\eta \\ 0 & 1 & 0 \\ \sinh\eta & 0 & \cosh\eta \end{pmatrix},$$
$$\exp(-i\alpha J_3) = \begin{pmatrix} 1 & 0 & 0 \\ 0 & \cos\alpha & -\sin\alpha \\ 0 & \sin\alpha & \cos\alpha \end{pmatrix}. \tag{8.23}$$

It is thus clear that, when applied to the column matrix of (t, x, y), they perform Lorentz transformations. The first two matrices, respectively, lead to the Lorentz boosts along the x- and y-directions:

$$\begin{aligned} x &\to x\cosh\eta + t\sinh\eta, \quad & t &\to t\cosh\eta + x\sinh\eta, \\ y &\to y\cosh\eta + t\sinh\eta, \quad & t &\to t\cosh\eta + y\sinh\eta. \end{aligned} \tag{8.24}$$

and the third matrix leads to the rotation:

$$x \to x\cos\alpha - y\sin\alpha, \qquad y \to y\cos\alpha + x\sin\alpha. \tag{8.25}$$

Indeed, the three-by-three matrices given in equation (8.22) are the generators of the Lorentz group applicable to two space-like directions and one time-like variable. We use the notation $O(2, 1)$ for this Lorentz group.

If two groups share the same Lie algebra, they are *like* each other (locally isomorphic). Hence it is apparent that the Lorentz group $O(2, 1)$ is like the symplectic group *Sp*(2). However, the generators J_3 and K_2 from table 1.1 do not have all imaginary elements. The Lorentz group $O(2, 1)$ and symplectic group $Sp(2)$ are not the same.

The generators of this $O(2, 1)$ can, of course, be written in terms of four-by-four matrices applicable to the space of (t, z, x, y). In that case, we can expand the generators of equation (8.22) into four-by-four matrices with zero elements on both the second row and column as can be seen from equations (1.3) and (1.9) and table 1.1.

8.2 Four-dimensional phase space

Let us now consider the Wigner function for two pairs of variables. The Wigner function is defined in equation (8.8), and its form for the ground-state oscillator is given in equation (8.10). The symmetry of this function is thus defined in the four-dimensional phase space of

$$(x_1, p_1, x_2, p_2). \tag{8.26}$$

Let us start with the Gaussian form of equation (8.10). This is the Wigner function for two independent harmonic oscillators. Since each oscillator has its $Sp(2)$ symmetry, we can construct its generators as

$$A_1 = \frac{1}{2}\begin{pmatrix} \sigma_2 & 0 \\ 0 & 0 \end{pmatrix}, \quad B_1 = \frac{i}{2}\begin{pmatrix} \sigma_3 & 0 \\ 0 & 0 \end{pmatrix}, \quad \text{and} \quad C_1 = \frac{i}{2}\begin{pmatrix} \sigma_1 & 0 \\ 0 & 0 \end{pmatrix}, \tag{8.27}$$

for the first oscillator. These generators satisfy the Lie algebra for $Sp(2)$:

$$[A_1, B_1] = iC_1, \quad [B_1, C_1] = -iA_1, \quad \text{and} \quad [C_1, A_1] = iB_1. \tag{8.28}$$

For the second oscillator, the generators become

$$A_2 = \frac{1}{2}\begin{pmatrix} 0 & 0 \\ 0 & \sigma_2 \end{pmatrix}, \quad B_2 = \frac{i}{2}\begin{pmatrix} 0 & 0 \\ 0 & \sigma_3 \end{pmatrix}, \quad \text{and} \quad C_2 = \frac{i}{2}\begin{pmatrix} 0 & 0 \\ 0 & \sigma_1 \end{pmatrix}. \tag{8.29}$$

These generators also satisfy the Lie algebra of equation (8.28). We are interested here in constructing the symmetry group for the coupled harmonic oscillators by soldering together the two $Sp(2)$ groups generated by A_1, B_1, C_1 and A_2, B_2, C_2, respectively.

It will be more convenient to use the linear combinations:

$$A_+ = A_1 + A_2, \quad B_+ = B_1 + B_2, \quad \text{and} \quad C_+ = C_1 + C_2, \tag{8.30}$$

and

$$A_- = A_1 - A_2, \quad B_- = B_1 - B_2, \quad \text{and} \quad C_- = C_1 - C_2. \tag{8.31}$$

These matrices take the form

$$A_+ = \frac{1}{2}\begin{pmatrix} \sigma_2 & 0 \\ 0 & \sigma_2 \end{pmatrix}, \quad B_+ = \frac{i}{2}\begin{pmatrix} \sigma_3 & 0 \\ 0 & \sigma_3 \end{pmatrix}, \quad \text{and} \quad C_+ = \frac{i}{2}\begin{pmatrix} \sigma_1 & 0 \\ 0 & \sigma_1 \end{pmatrix}, \tag{8.32}$$

and

$$A_- = \frac{1}{2}\begin{pmatrix} \sigma_2 & 0 \\ 0 & -\sigma_2 \end{pmatrix}, \quad B_- = \frac{i}{2}\begin{pmatrix} \sigma_3 & 0 \\ 0 & -\sigma_3 \end{pmatrix}, \quad \text{and} \quad C_- = \frac{i}{2}\begin{pmatrix} \sigma_1 & 0 \\ 0 & -\sigma_1 \end{pmatrix}. \tag{8.33}$$

The sets (A_+, B_+, C_+) and (A_+, B_-, C_-) satisfy the Lie algebra of equation (8.28). The same is true for (A_-, B_+, C_-) and (A_-, B_-, C_+).

It is possible to write these four-by-four matrices as

$$A_+ = \frac{1}{2}\begin{pmatrix} I & 0 \\ 0 & I \end{pmatrix}\sigma_2, \quad B_+ = \frac{i}{2}\begin{pmatrix} I & 0 \\ 0 & I \end{pmatrix}\sigma_3, \quad \text{and} \quad C_+ = \frac{i}{2}\begin{pmatrix} I & 0 \\ 0 & I \end{pmatrix}\sigma_1, \tag{8.34}$$

and

$$A_- = \frac{1}{2}\begin{pmatrix} I & 0 \\ 0 & -I \end{pmatrix}\sigma_2, \quad B_- = \frac{i}{2}\begin{pmatrix} I & 0 \\ 0 & -I \end{pmatrix}\sigma_3, \quad \text{and} \quad C_- = \frac{i}{2}\begin{pmatrix} I & 0 \\ 0 & -I \end{pmatrix}\sigma_1. \tag{8.35}$$

Here I is the two-by-two unit matrix.

Next, let us couple the oscillators through a rotation generated by

$$A_0 = \frac{i}{2}\begin{pmatrix} 0 & -I \\ I & 0 \end{pmatrix}I. \tag{8.36}$$

In view of the fact that the first two coordinate variables are for the phase space of the first oscillator, and the third and fourth are for that of the second oscillator, this matrix generates parallel rotations in the (x_1, x_2) and (p_1, p_2) coordinates. As the coordinates (x_1, x_2) are coupled through a two-by-two matrix, the coordinates (p_1, p_2) are coupled through the same two-by-two matrix.

Then, A_0 commutes with A_+, B_+, C_+, and the following commutation relations generate new operators A_3, B_3, and C_3:

$$[A_0, A_-] = iA_3, \qquad [A_0, B_-] = iB_3, \qquad \text{and} \quad [A_0, C_-] = iC_3, \tag{8.37}$$

with

$$A_3 = \frac{1}{2}\begin{pmatrix} 0 & I \\ I & 0 \end{pmatrix}\sigma_2, \quad B_3 = \frac{i}{2}\begin{pmatrix} 0 & I \\ I & 0 \end{pmatrix}\sigma_3, \quad \text{and} \quad C_3 = \frac{i}{2}\begin{pmatrix} 0 & I \\ I & 0 \end{pmatrix}\sigma_1. \tag{8.38}$$

In this section, we started with the generators of the symmetry groups for two independent oscillators. They are A_1, B_1, C_1, and A_2, B_2, C_2. We then introduced A_0, which generates the coupling of two oscillators. This process produced three additional generators A_3, B_3, C_3. Remarkably, C_3, B_3, and A_+ form another set of generators for another $Sp(2)$ group. They satisfy the commutation relations

$$[B_3, C_3] = -iA_+, \quad [C_3, A_+] = iB_3, \quad \text{and} \quad [A_+, B_3] = iC_3. \tag{8.39}$$

The same can be said about the sets A_+, B_1, C_1, and A_+, B_2, C_2. These $Sp(2)$-like groups are associated with the coupling of the two pairs of conjugate variables.

We observed in this section that this four-dimensional phase space contains three $Sp(2)$-like groups with nine generators plus A_0 of equation (8.36). There are thus ten generators.

8.3 Canonical transformations

Table 8.1 lists the ten four-by-four generators, which are the same as those ten given in section 8.2. They are matched to a set of ten generators that we will define and use in this section.

For a dynamical system consisting of two pairs of canonical variables x_1, p_1 and x_2, p_2, we have introduced the four-dimensional coordinate system. As in the case of equation (8.21), the transformation is canonical if the transformation matrix M satisfies the condition [5, 7]

$$MJ\tilde{M} = J, \tag{8.40}$$

where

$$J = \begin{pmatrix} 0 & 1 & 0 & 0 \\ -1 & 0 & 0 & 0 \\ 0 & 0 & 0 & 1 \\ 0 & 0 & -1 & 0 \end{pmatrix}. \tag{8.41}$$

It is possible to construct a set of generators for these canonical transformation matrices. We can write the matrix, from equation (8.41), in a two-by-two block form as

$$J = \begin{pmatrix} i\sigma_2 & 0 \\ 0 & i\sigma_2 \end{pmatrix} \quad \text{or} \quad J = \begin{pmatrix} i & 0 \\ 0 & i \end{pmatrix} \sigma_2. \tag{8.42}$$

First, the following four rotation-like generators produce canonical transformations.

$$\begin{gathered} J_1 = -\frac{1}{2}\begin{pmatrix} 0 & I \\ I & 0 \end{pmatrix}\sigma_2, \quad J_2 = \frac{i}{2}\begin{pmatrix} 0 & -I \\ I & 0 \end{pmatrix} I, \quad J_3 = \frac{1}{2}\begin{pmatrix} -I & 0 \\ 0 & I \end{pmatrix}\sigma_2, \\ S_3 = \frac{1}{2}\begin{pmatrix} I & 0 \\ 0 & I \end{pmatrix}\sigma_2. \end{gathered} \tag{8.43}$$

In addition, there are six squeeze-like generators. They are

Table 8.1. The four-by-four A, B, and C generators of section 8.2 and the J, K, Q, and S generators in this section. These two sets share the same set of four-by-four matrices.

$A_+ = S_3$	$A_- = -J_3$	$A_3 = -J_1$	$A_0 = J_2$
$B_+ = K_2$	$B_- = -Q_1$	$B_3 = Q_3$	
$C_+ = Q_2$	$C_- = K_1$	$C_3 = -K_3$	

$$K_1 = \frac{i}{2}\begin{pmatrix} I & 0 \\ 0 & -I \end{pmatrix}\sigma_1, \quad K_2 = \frac{i}{2}\begin{pmatrix} I & 0 \\ 0 & I \end{pmatrix}\sigma_3, \quad K_3 = -\frac{i}{2}\begin{pmatrix} 0 & I \\ I & 0 \end{pmatrix}\sigma_1,$$
$$Q_1 = \frac{i}{2}\begin{pmatrix} -I & 0 \\ 0 & I \end{pmatrix}\sigma_3, \quad Q_2 = \frac{i}{2}\begin{pmatrix} I & 0 \\ 0 & I \end{pmatrix}\sigma_1, \quad Q_3 = \frac{i}{2}\begin{pmatrix} 0 & I \\ I & 0 \end{pmatrix}\sigma_3. \tag{8.44}$$

These ten generators, listed in table 8.2, satisfy the following Lie algebra and thus form the four-dimensional symplectic group $Sp(4)$.

$$[J_i, J_j] = i\varepsilon_{ijk}J_k, \quad [J_i, K_j] = i\varepsilon_{ijk}K_k, \quad [K_i, K_j] = -i\varepsilon_{ijk}J_k,$$
$$[J_i, Q_j] = i\varepsilon_{ijk}Q_k, \quad [Q_i, Q_j] = -i\varepsilon_{ijk}J_k, \tag{8.45}$$

$$[K_i, Q_j] = -i\delta_{ij}S_3, \quad [J_i, S_3] = 0, \quad [K_i, S_3] = -iQ_i,$$
$$[Q_i, S_3] = iK_i. \tag{8.46}$$

Let us look at the first row of equation (8.45). It contains the Lie algebra of the three-dimensional rotation group. Thus, the group $Sp(4)$ contains the three-dimensional rotation group as a subgroup. If we next consider the entire first row of equation (8.45), we see that the J_i and the K_i operators constitute the Lie algebra

Table 8.2. Generators of the $Sp(4)$ group applicable to the Wigner function $W(x_1, p_1, x_2, p_2)$. There are four rotation-like and six squeeze-like generators.

Generator	Matrix	Applicable to the Wigner function
J_1,	$-\frac{1}{2}\begin{pmatrix} 0 & I \\ I & 0 \end{pmatrix}\sigma_2$	$+\frac{i}{2}\left\{\left(x_1\frac{\partial}{\partial p_2} - p_2\frac{\partial}{\partial x_1}\right) + \left(x_2\frac{\partial}{\partial p_1} - p_1\frac{\partial}{\partial x_2}\right)\right\}$
J_2	$\frac{i}{2}\begin{pmatrix} 0 & -I \\ I & 0 \end{pmatrix}I$	$+\frac{i}{2}\left\{\left(x_1\frac{\partial}{\partial x_2} - x_2\frac{\partial}{\partial x_1}\right) + \left(p_1\frac{\partial}{\partial p_2} - p_2\frac{\partial}{\partial p_1}\right)\right\}$
J_3	$\frac{1}{2}\begin{pmatrix} -I & 0 \\ 0 & I \end{pmatrix}\sigma_2$	$-\frac{i}{2}\left\{\left(x_1\frac{\partial}{\partial p_1} - p_1\frac{\partial}{\partial x_1}\right) - \left(x_2\frac{\partial}{\partial p_2} - p_2\frac{\partial}{\partial x_2}\right)\right\}$
S_3	$\frac{1}{2}\begin{pmatrix} I & 0 \\ 0 & I \end{pmatrix}\sigma_2$	$-\frac{i}{2}\left\{\left(x_1\frac{\partial}{\partial p_1} - p_1\frac{\partial}{\partial x_1}\right) + \left(x_2\frac{\partial}{\partial p_2} - p_2\frac{\partial}{\partial x_2}\right)\right\}$
K_1,	$\frac{i}{2}\begin{pmatrix} I & 0 \\ 0 & -I \end{pmatrix}\sigma_1$	$-\frac{i}{2}\left\{\left(x_1\frac{\partial}{\partial p_1} + p_1\frac{\partial}{\partial x_1}\right) - \left(x_2\frac{\partial}{\partial p_2} + p_2\frac{\partial}{\partial x_2}\right)\right\}$
K_2	$\frac{i}{2}\begin{pmatrix} I & 0 \\ 0 & I \end{pmatrix}\sigma_3$	$+\frac{i}{2}\left\{\left(x_1\frac{\partial}{\partial x_1} + x_2\frac{\partial}{\partial x_2}\right) - \left(p_1\frac{\partial}{\partial p_1} + p_2\frac{\partial}{\partial p_2}\right)\right\}$
K_3	$-\frac{i}{2}\begin{pmatrix} 0 & I \\ I & 0 \end{pmatrix}\sigma_1$	$-\frac{i}{2}\left\{\left(x_1\frac{\partial}{\partial p_2} + p_2\frac{\partial}{\partial x_1}\right) + \left(x_2\frac{\partial}{\partial p_1} + p_1\frac{\partial}{\partial x_2}\right)\right\}$
Q_1	$\frac{i}{2}\begin{pmatrix} -I & 0 \\ 0 & I \end{pmatrix}\sigma_3$	$\frac{i}{2}\left\{\left(x_1\frac{\partial}{\partial x_1} - p_1\frac{\partial}{\partial p_1}\right) - \left(x_2\frac{\partial}{\partial x_2} - p_2\frac{\partial}{\partial p_2}\right)\right\}$
Q_2	$\frac{i}{2}\begin{pmatrix} I & 0 \\ 0 & I \end{pmatrix}\sigma_1$	$-\frac{i}{2}\left\{\left(x_1\frac{\partial}{\partial p_1} + p_1\frac{\partial}{\partial x_1}\right) + \left(x_2\frac{\partial}{\partial p_2} + p_2\frac{\partial}{\partial x_2}\right)\right\}$
Q_3	$\frac{i}{2}\begin{pmatrix} 0 & I \\ I & 0 \end{pmatrix}\sigma_3$	$-\frac{i}{2}\left\{\left(x_2\frac{\partial}{\partial x_1} + x_1\frac{\partial}{\partial x_2}\right) - \left(p_2\frac{\partial}{\partial p_1} + p_1\frac{\partial}{\partial p_2}\right)\right\}$

Table 8.3. Ten Generators of the $O(3, 2)$ de Sitter group in the five-dimensional space (s, t, z, x, y). The J matrices generate rotations in the three-dimensional $(z, x, y,)$ space. The K matrices generate Lorentz boosts with respect to the time variable t. The Q matrices are for the boosts with respect to the time-like coordinate s. The generator S_3 is for rotations in the (s, t)-plane.

$$J_1 = \begin{pmatrix} 0 & 0 & 0 & 0 & 0 \\ 0 & 0 & 0 & 0 & 0 \\ 0 & 0 & 0 & 0 & i \\ 0 & 0 & 0 & 0 & 0 \\ 0 & 0 & -i & 0 & 0 \end{pmatrix} \quad J_2 = \begin{pmatrix} 0 & 0 & 0 & 0 & 0 \\ 0 & 0 & 0 & 0 & 0 \\ 0 & 0 & 0 & -i & 0 \\ 0 & 0 & i & 0 & 0 \\ 0 & 0 & 0 & 0 & 0 \end{pmatrix} \quad J_3 = \begin{pmatrix} 0 & 0 & 0 & 0 & 0 \\ 0 & 0 & 0 & 0 & 0 \\ 0 & 0 & 0 & 0 & 0 \\ 0 & 0 & 0 & 0 & -i \\ 0 & 0 & 0 & i & 0 \end{pmatrix}$$

$$K_1 = \begin{pmatrix} 0 & 0 & 0 & 0 & 0 \\ 0 & 0 & 0 & i & 0 \\ 0 & 0 & 0 & 0 & 0 \\ 0 & i & 0 & 0 & 0 \\ 0 & 0 & 0 & 0 & 0 \end{pmatrix} \quad K_2 = \begin{pmatrix} 0 & 0 & 0 & 0 & 0 \\ 0 & 0 & 0 & 0 & i \\ 0 & 0 & 0 & 0 & 0 \\ 0 & 0 & 0 & 0 & 0 \\ 0 & i & 0 & 0 & 0 \end{pmatrix} \quad K_3 = \begin{pmatrix} 0 & 0 & 0 & 0 & 0 \\ 0 & 0 & i & 0 & 0 \\ 0 & i & 0 & 0 & 0 \\ 0 & 0 & 0 & 0 & 0 \\ 0 & 0 & 0 & 0 & 0 \end{pmatrix}$$

$$Q_1 = \begin{pmatrix} 0 & 0 & 0 & i & 0 \\ 0 & 0 & 0 & 0 & 0 \\ 0 & 0 & 0 & 0 & 0 \\ i & 0 & 0 & 0 & 0 \\ 0 & 0 & 0 & 0 & 0 \end{pmatrix} \quad Q_2 = \begin{pmatrix} 0 & 0 & 0 & 0 & i \\ 0 & 0 & 0 & 0 & 0 \\ 0 & 0 & 0 & 0 & 0 \\ 0 & 0 & 0 & 0 & 0 \\ i & 0 & 0 & 0 & 0 \end{pmatrix} \quad Q_3 = \begin{pmatrix} 0 & 0 & i & 0 & 0 \\ 0 & 0 & 0 & 0 & 0 \\ i & 0 & 0 & 0 & 0 \\ 0 & 0 & 0 & 0 & 0 \\ 0 & 0 & 0 & 0 & 0 \end{pmatrix}$$

$$S_3 = \begin{pmatrix} 0 & -i & 0 & 0 & 0 \\ i & 0 & 0 & 0 & 0 \\ 0 & 0 & 0 & 0 & 0 \\ 0 & 0 & 0 & 0 & 0 \\ 0 & 0 & 0 & 0 & 0 \end{pmatrix}$$

for the Lorentz group applicable to the Minkowski space with three space and one time coordinates. The same can be said for the J_i operators given in the first row and the Q_i operators given in the second row. There are two Lorentz groups as subgroups of $Sp(4)$.

As Dirac noted in 1963 [2], this $Sp(4)$ group shares the same Lie algebra as the generators of the de Sitter group $O(3, 2)$, which is the Lorentz group applicable to the five-dimensional space consisting of three space and two time coordinates. The generators are thus five-by-five matrices. These are listed explicitly in table 8.3. This group plays important roles in later chapters on squeezed states of light, polarization optics, and the synthesis of quantum mechanics and special relativity.

8.4 $SL(4, r)$ symmetry

In the four-by-four regime, if the transformation matrices are all real, there are sixteen generators with imaginary elements. If the determinants of all these matrices are to be one, there are fifteen generators. The group of all these matrices is called the four-dimensional special linear group with real elements, or $SL(4, r)$.

In section 8.3, we discussed the group $Sp(4)$ whose transformation matrices satisfy the condition given in equation (8.40). It was noted there that this $Sp(4)$ group is generated by ten four-by-four matrices all of which have imaginary elements. The question then is what happens if the remaining five matrices are added to the picture.

Let us consider a generator of the form

$$G_3 = \frac{i}{2}\begin{pmatrix} I & 0 \\ 0 & -I \end{pmatrix} I. \tag{8.47}$$

This matrix generates the four-by-four transformation matrix of the form

$$M = \begin{pmatrix} I\exp(\eta) & 0 \\ 0 & I\exp(-\eta) \end{pmatrix}. \tag{8.48}$$

This matrix does not satisfy the condition of equation (8.40), and it does not perform a canonical transformation.

The matrix G_3 is not among the ten matrices given in equation (8.45), which generate canonical transformations. Thus, we are led to consider its commutators with all those canonical generators.

The G_3 matrix commutes with S_3, J_3, K_1, K_2, Q_1, and Q_2. However, its commutators with the rest of the matrices produce four additional generators:

$$[G_3, J_1] = iG_2, \quad [G_3, J_2] = -iG_1, \quad [G_3, K_3] = iS_2, \quad \text{and} \quad [G_3, Q_3] = -iS_1, \tag{8.49}$$

with

$$\begin{aligned} G_1 &= \frac{i}{2}\begin{pmatrix} 0 & I \\ I & 0 \end{pmatrix} I & G_2 &= \frac{1}{2}\begin{pmatrix} 0 & -I \\ I & 0 \end{pmatrix}\sigma_2, \\ S_1 &= -\frac{i}{2}\begin{pmatrix} 0 & -I \\ I & 0 \end{pmatrix}\sigma_3, & S_2 &= \frac{i}{2}\begin{pmatrix} 0 & -I \\ I & 0 \end{pmatrix}\sigma_1. \end{aligned} \tag{8.50}$$

We list these generators explicitly in table 8.4.

If we take into account the above five generators in addition to the ten generators of $Sp(4)$, there are fifteen generators. These generators satisfy the following set of commutation relations.

$$\begin{aligned} [J_i, J_j] &= i\varepsilon_{ijk}J_k, \quad [S_i, S_j] = i\varepsilon_{ijk}S_k, \quad [J_i, S_j] = 0, \quad [J_i, G_j] = i\varepsilon_{ijk}G_k, \\ [J_i, K_j] &= i\varepsilon_{ijk}K_k, \quad [J_i, Q_j] = i\varepsilon_{ijk}Q_k, \\ [K_i, K_j] &= [Q_i, Q_j] = [G_i, G_j] = -i\varepsilon_{ijk}J_k, \end{aligned} \tag{8.51}$$

and

$$\begin{aligned} &[K_i, Q_j] = -i\delta_{ij}S_3, \quad [Q_i, G_j] = -i\delta_{ij}S_1, \\ &[G_i, K_j] = -i\delta_{ij}S_2, \quad [K_i, S_3] = -iQ_i, \\ &[Q_i, S_3] = iK_i, \quad [K_i, S_1] = 0, \\ &[G_i, S_3] = 0, \quad [Q_i, S_2] = 0, \quad [Q_i, S_1] = -iG_i, \\ &[G_i, S_1] = iQ_i, \quad [K_i, S_2] = iG_i, \quad [G_i, S_2] = -iK_i. \end{aligned} \tag{8.52}$$

These commutation relations form the Lie algebra for the $O(3, 3)$ Lorentz group with three space and three time coordinates.

Table 8.4. Generators not leading to canonical transformations. There are six generators in this table. However, S_3 produces a canonical transformation and is listed in table 8.3. There are thus five new generators. The S_3 generator is here to complete the set (S_1, S_2, S_3) for a three-dimensional rotation group.

Generator	Matrix	Applicable to the Wigner Function
S_1,	$-\frac{i}{2}\begin{pmatrix} 0 & -I \\ I & 0 \end{pmatrix}\sigma_3$	$-\frac{i}{2}\left\{\left(x_1\frac{\partial}{\partial x_2} - x_2\frac{\partial}{\partial x_1}\right) - \left(p_1\frac{\partial}{\partial p_2} - p_2\frac{\partial}{\partial p_1}\right)\right\}$
S_2	$\frac{i}{2}\begin{pmatrix} 0 & -I \\ I & 0 \end{pmatrix}\sigma_1$	$-\frac{i}{2}\left\{\left(x_1\frac{\partial}{\partial p_2} - p_2\frac{\partial}{\partial x_1}\right) + \left(x_2\frac{\partial}{\partial p_1} - p_1\frac{\partial}{\partial x_2}\right)\right\}$
S_3	$\frac{1}{2}\begin{pmatrix} I & 0 \\ 0 & I \end{pmatrix}\sigma_2$	$-\frac{i}{2}\left\{\left(x_1\frac{\partial}{\partial p_1} - p_1\frac{\partial}{\partial x_1}\right) + \left(x_2\frac{\partial}{\partial p_2} - p_2\frac{\partial}{\partial x_2}\right)\right\}$
G_1	$\frac{i}{2}\begin{pmatrix} 0 & I \\ I & 0 \end{pmatrix}I$	$-\frac{i}{2}\left\{\left(x_1\frac{\partial}{\partial x_2} + x_2\frac{\partial}{\partial x_1}\right) + \left(p_1\frac{\partial}{\partial p_2} + p_2\frac{\partial}{\partial p_1}\right)\right\}$
G_2	$\frac{1}{2}\begin{pmatrix} 0 & -I \\ I & 0 \end{pmatrix}\sigma_2$	$\frac{i}{2}\left\{\left(x_1\frac{\partial}{\partial p_2} + p_2\frac{\partial}{\partial x_1}\right) - \left(x_2\frac{\partial}{\partial p_1} + p_1\frac{\partial}{\partial x_2}\right)\right\}$
G_3	$\frac{i}{2}\begin{pmatrix} I & 0 \\ 0 & -I \end{pmatrix}I$	$-\frac{i}{2}\left\{\left(x_1\frac{\partial}{\partial x_1} + p_1\frac{\partial}{\partial p_1}\right) + \left(x_2\frac{\partial}{\partial p_2} + p_2\frac{\partial}{\partial x_2}\right)\right\}$

8.5 Dirac matrices for $O(3, 3)$

In the regime of four-by-four matrices, if all the elements are imaginary, there are sixteen independent matrices. If we require the traces of those matrices to be zero, there are only fifteen matrices. The Dirac matrices are also four-by-four and, if chosen in the Majorana representation [8, 9, 11, 12], all the elements are imaginary. The Dirac matrices are for the internal space–time symmetries of spinning particles.

There are fifteen four-by-four matrices in section 8.4. It is possible to construct fifteen matrices from Dirac's four γ matrices. It is thus of interest to see whether the fifteen matrices from section 8.3 and section 8.4 can be constructed from the Dirac matrices [11].

The fifteen matrices, which start from Dirac's four γ matrices in the Majorana representation [12], are

$$\gamma_1 = i\begin{pmatrix} I & 0 \\ 0 & I \end{pmatrix}\sigma_3, \quad \gamma_2 = \begin{pmatrix} 0 & -I \\ I & 0 \end{pmatrix}\sigma_2,$$
$$\gamma_3 = -i\begin{pmatrix} I & 0 \\ 0 & I \end{pmatrix}\sigma_1, \quad \gamma_0 = \begin{pmatrix} 0 & I \\ I & 0 \end{pmatrix}\sigma_2. \tag{8.53}$$

These are Lorentz-transformed like the four-vector (x, y, z, t). With these four matrices, we can construct one pseudo-scalar matrix

$$\gamma_5 = i\gamma_0\gamma_i\gamma_2\gamma_3 = \begin{pmatrix} I & 0 \\ 0 & -I \end{pmatrix}\sigma_2, \tag{8.54}$$

and a one pseudo-vector $i\gamma_5\gamma_\mu$ consisting of

Table 8.5. Fifteen generators of $O(3, 3)$ written in the language of Dirac's γ matrices.

$J_1 = -\frac{1}{2}\gamma_0$	$J_2 = -\frac{i}{2}\gamma_5\gamma_0$	$J_3 = -\frac{1}{2}\gamma_5$
$S_1 = \frac{i}{2}\gamma_2\gamma_3$	$S_2 = \frac{i}{2}\gamma_1\gamma_2$	$S_3 = -\frac{i}{2}\gamma_1\gamma_3$
$K_1 = \frac{i}{2}\gamma_1\gamma_5$	$K_2 = \frac{1}{2}\gamma_1$	$K_3 = \frac{i}{2}\gamma_0\gamma_1$
$Q_1 = \frac{i}{2}\gamma_5\gamma_3,$	$Q_2 = -\frac{1}{2}\gamma_3$	$Q_3 = -\frac{i}{2}\gamma_0\gamma_3$
$G_1 = \frac{i}{2}\gamma_2\gamma_5$	$G_2 = \frac{1}{2}\gamma_2$	$G_3 = \frac{i}{2}\gamma_0\gamma_2$

$$
\begin{aligned}
i\gamma_5\gamma_1 &= i\begin{pmatrix} -I & 0 \\ 0 & I \end{pmatrix}\sigma_1, \qquad & i\gamma_5\gamma_2 &= -i\begin{pmatrix} 0 & I \\ I & 0 \end{pmatrix} I, \\
i\gamma_5\gamma_3 &= -i\begin{pmatrix} I & 0 \\ 0 & -I \end{pmatrix}\sigma_3, \qquad & i\gamma_5\gamma_0 &= i\begin{pmatrix} 0 & I \\ -I & 0 \end{pmatrix} I.
\end{aligned}
\tag{8.55}
$$

In addition, we can construct the tensor of the form

$$
T_{\mu\nu} = \frac{i}{2}(\gamma_\mu\gamma_\nu - \gamma_\nu\gamma_\mu), \tag{8.56}
$$

resulting in the following six matrices.

$$
\begin{aligned}
i\gamma_0\gamma_1 &= -i\begin{pmatrix} 0 & I \\ I & 0 \end{pmatrix}\sigma_1, \quad & i\gamma_0\gamma_2 &= i\begin{pmatrix} I & 0 \\ 0 & -I \end{pmatrix} I, \quad & i\gamma_0\gamma_3 &= -i\begin{pmatrix} 0 & I \\ I & 0 \end{pmatrix}\sigma_3, \\
i\gamma_1\gamma_2 &= i\begin{pmatrix} 0 & -I \\ I & 0 \end{pmatrix}\sigma_1, \quad & i\gamma_2\gamma_3 &= i\begin{pmatrix} 0 & I \\ -I & 0 \end{pmatrix}\sigma_3, \quad & i\gamma_3\gamma_1 &= \begin{pmatrix} I & 0 \\ 0 & I \end{pmatrix}\sigma_2.
\end{aligned}
\tag{8.57}
$$

Thus, it is possible to write these J, S, K, Q, and G matrices in terms of Dirac's γ matrices, as given in table 8.5.

8.6 $O(3, 3)$ symmetry

In the regime of $SL(4, r)$, there are two sets of rotation generators and three sets of squeeze generators [6] as given in equations (8.43), (8.44), (8.47), and (8.50). Their Lie algebra is given in equations (8.51) and (8.52).

From this Lie algebra, it is possible to construct fifteen six-by-six matrices which generate the Lorentz group applicable to the six-dimensional space consisting of three space and three time coordinate. There are two rotation subgroups with six generators. Thus, the rotation matrix could take the form

$$
\begin{pmatrix} \text{3-by-3 rotation} & 0 \\ 0 & \text{3-by-3 rotation} \end{pmatrix}\begin{pmatrix} \text{3-dimensional space} \\ \text{3-dimensional time} \end{pmatrix}. \tag{8.58}
$$

We now introduce a six-dimensional space with (z, x, y) as the space coordinates and (u, s, t) as the three-dimensional time coordinates. These variables can be ordered as (u, s, t, z, x, y) [6].

In this six-by-six representation, the Lorentz boost along x with respect to the time variable t becomes

$$\begin{pmatrix} 1 & 0 & 0 & 0 & 0 & 0 \\ 0 & 1 & 0 & 0 & 0 & 0 \\ 0 & 0 & \cosh\eta & 0 & \sinh\eta & 0 \\ 0 & 0 & 0 & 1 & 0 & 0 \\ 0 & 0 & \sinh\eta & 0 & \cosh\eta & 0 \\ 0 & 0 & 0 & 0 & 0 & 1 \end{pmatrix} \begin{pmatrix} u \\ s \\ t \\ z \\ x \\ y \end{pmatrix} = \begin{pmatrix} u \\ s \\ x\sinh\eta + t\cosh\eta \\ z \\ x\cosh\eta + t\sinh\eta \\ y \end{pmatrix}. \tag{8.59}$$

The generator for this boost takes the form

$$K_1 = \begin{pmatrix} 0 & 0 & 0 & 0 & 0 & 0 \\ 0 & 0 & 0 & 0 & 0 & 0 \\ 0 & 0 & 0 & 0 & i & 0 \\ 0 & 0 & 0 & 0 & 0 & 0 \\ 0 & 0 & i & 0 & 0 & 0 \\ 0 & 0 & 0 & 0 & 0 & 0 \end{pmatrix}. \tag{8.60}$$

Likewise, there are six-by-six matrices for the two sets of rotation generators and three sets of boost generators. They are tabulated in tables 8.6 and 8.7.

We are not able to see direct physical applications of this extensive space–time symmetry group. However, it is interesting to note that this symmetry is derivable from the four-dimensional phase space. The Lorentz group $O(3, 1)$ and the de Sitter group $O(3, 2)$ are subgroups of this $O(3, 3)$ group.

Let us borrow the language of harmonic oscillators. Each oscillator contains the symmetry of the $Sp(2)$ group with three generators. Their generators are given in the first two rows of table 8.8 with the $Sp(2)$ generators, namely σ_2, $i\sigma_3$, and $i\sigma_1$. Nine others are from the way they are coupled mathematically. We shall study the physical coupling of the oscillators in detail in chapter 9 on coupled harmonic oscillators and squeezed states of light.

Table 8.6. Rotation generators of $O(3, 3)$. The J matrices generate rotations in the three-dimensional space of (z, x, y), and the S matrices generate rotations in the (u, s, t) space.

$$J_1 = \begin{pmatrix} 0 & 0 & 0 & 0 & 0 & 0 \\ 0 & 0 & 0 & 0 & 0 & 0 \\ 0 & 0 & 0 & 0 & 0 & 0 \\ 0 & 0 & 0 & 0 & 0 & i \\ 0 & 0 & 0 & 0 & 0 & 0 \\ 0 & 0 & 0 & -i & 0 & 0 \end{pmatrix} \quad J_2 = \begin{pmatrix} 0 & 0 & 0 & 0 & 0 & 0 \\ 0 & 0 & 0 & 0 & 0 & 0 \\ 0 & 0 & 0 & 0 & 0 & 0 \\ 0 & 0 & 0 & 0 & -i & 0 \\ 0 & 0 & 0 & i & 0 & 0 \\ 0 & 0 & 0 & 0 & 0 & 0 \end{pmatrix} \quad J_3 = \begin{pmatrix} 0 & 0 & 0 & 0 & 0 & 0 \\ 0 & 0 & 0 & 0 & 0 & 0 \\ 0 & 0 & 0 & 0 & 0 & 0 \\ 0 & 0 & 0 & 0 & 0 & 0 \\ 0 & 0 & 0 & 0 & 0 & -i \\ 0 & 0 & 0 & 0 & i & 0 \end{pmatrix}$$

$$S_1 = \begin{pmatrix} 0 & i & 0 & 0 & 0 & 0 \\ -i & 0 & 0 & 0 & 0 & 0 \\ 0 & 0 & 0 & 0 & 0 & 0 \\ 0 & 0 & 0 & 0 & 0 & 0 \\ 0 & 0 & 0 & 0 & 0 & 0 \\ 0 & 0 & 0 & 0 & 0 & 0 \end{pmatrix} \quad S_2 = \begin{pmatrix} 0 & 0 & -i & 0 & 0 & 0 \\ 0 & 0 & 0 & 0 & 0 & 0 \\ i & 0 & 0 & 0 & 0 & 0 \\ 0 & 0 & 0 & 0 & 0 & 0 \\ 0 & 0 & 0 & 0 & 0 & 0 \\ 0 & 0 & 0 & 0 & 0 & 0 \end{pmatrix} \quad S_3 = \begin{pmatrix} 0 & 0 & 0 & 0 & 0 & 0 \\ 0 & 0 & i & 0 & 0 & 0 \\ 0 & -i & 0 & 0 & 0 & 0 \\ 0 & 0 & 0 & 0 & 0 & 0 \\ 0 & 0 & 0 & 0 & 0 & 0 \\ 0 & 0 & 0 & 0 & 0 & 0 \end{pmatrix}$$

Table 8.7. Boost generators of $O(3, 3)$. The K matrices generate the Lorentz boosts with respect to the time variable t, while Q and G matrices lead to the boosts with respect to the time variables s and u.

$K_1 = \begin{pmatrix} 0&0&0&0&0&0 \\ 0&0&0&0&0&0 \\ 0&0&0&0&i&0 \\ 0&0&0&0&0&0 \\ 0&0&i&0&0&0 \\ 0&0&0&0&0&0 \end{pmatrix}$	$K_2 = \begin{pmatrix} 0&0&0&0&0&0 \\ 0&0&0&0&0&0 \\ 0&0&0&0&0&i \\ 0&0&0&0&0&0 \\ 0&0&0&0&0&0 \\ 0&0&i&0&0&0 \end{pmatrix}$	$K_3 = \begin{pmatrix} 0&0&0&0&0&0 \\ 0&0&0&0&0&0 \\ 0&0&0&i&0&0 \\ 0&0&i&0&0&0 \\ 0&0&0&0&0&0 \\ 0&0&0&0&0&0 \end{pmatrix}$
$Q_1 = \begin{pmatrix} 0&0&0&0&0&0 \\ 0&0&0&0&i&0 \\ 0&0&0&0&0&0 \\ 0&0&0&0&0&0 \\ 0&i&0&0&0&0 \\ 0&0&0&0&0&0 \end{pmatrix}$	$Q_2 = \begin{pmatrix} 0&0&0&0&0&0 \\ 0&0&0&0&0&i \\ 0&0&0&0&0&0 \\ 0&0&0&0&0&0 \\ 0&0&0&0&0&0 \\ 0&i&0&0&0&0 \end{pmatrix}$	$Q_3 = \begin{pmatrix} 0&0&0&0&0&0 \\ 0&0&0&i&0&0 \\ 0&0&0&0&0&0 \\ 0&i&0&0&0&0 \\ 0&0&0&0&0&0 \\ 0&0&0&0&0&0 \end{pmatrix}$
$G_1 = \begin{pmatrix} 0&0&0&0&i&0 \\ 0&0&0&0&0&0 \\ 0&0&0&0&0&0 \\ 0&0&0&0&0&0 \\ i&0&0&0&0&0 \\ 0&0&0&0&0&0 \end{pmatrix}$	$G_2 = \begin{pmatrix} 0&0&0&0&0&i \\ 0&0&0&0&0&0 \\ 0&0&0&0&0&0 \\ 0&0&0&0&0&0 \\ 0&0&0&0&0&0 \\ i&0&0&0&0&0 \end{pmatrix}$	$G_3 = \begin{pmatrix} 0&0&0&i&0&0 \\ 0&0&0&0&0&0 \\ 0&0&0&0&0&0 \\ i&0&0&0&0&0 \\ 0&0&0&0&0&0 \\ 0&0&0&0&0&0 \end{pmatrix}$

Table 8.8. Fifteen generators of the $O(3, 3)$. The K matrices generate the Lorentz boosts with respect to the time variable t, while Q and G matrices lead to the boosts with respect to the time variables s and u.

$\frac{i}{2}I$	$\frac{1}{2}\sigma_2$	$\frac{i}{2}\sigma_3$	$\frac{i}{2}\sigma_1$
	$\frac{1}{2}\begin{pmatrix} I & 0 \\ 0 & I \end{pmatrix}\sigma_2$ S_3	$\frac{i}{2}\begin{pmatrix} I & 0 \\ 0 & I \end{pmatrix}\sigma_3$ K_2	$\frac{i}{2}\begin{pmatrix} I & 0 \\ 0 & I \end{pmatrix}\sigma_1$ Q_2
$\frac{i}{2}\begin{pmatrix} I & 0 \\ 0 & -I \end{pmatrix}I$ G_3	$\frac{1}{2}\begin{pmatrix} -I & 0 \\ 0 & I \end{pmatrix}\sigma_2$ J_3	$\frac{i}{2}\begin{pmatrix} -I & 0 \\ 0 & I \end{pmatrix}\sigma_3$ Q_1	$\frac{i}{2}\begin{pmatrix} I & 0 \\ 0 & -I \end{pmatrix}\sigma_1$ K_1
$\frac{i}{2}\begin{pmatrix} 0 & -I \\ I & 0 \end{pmatrix}I$ J_2	$\frac{1}{2}\begin{pmatrix} 0 & -I \\ I & 0 \end{pmatrix}\sigma_2$ G_2	$-\frac{i}{2}\begin{pmatrix} 0 & -I \\ I & 0 \end{pmatrix}\sigma_3$ S_1	$\frac{i}{2}\begin{pmatrix} 0 & -I \\ I & 0 \end{pmatrix}\sigma_1$ S_2
$\frac{i}{2}\begin{pmatrix} 0 & I \\ I & 0 \end{pmatrix}I$ G_1	$-\frac{1}{2}\begin{pmatrix} 0 & I \\ I & 0 \end{pmatrix}\sigma_2$ J_1	$\frac{i}{2}\begin{pmatrix} 0 & I \\ I & 0 \end{pmatrix}\sigma_3$ Q_3	$-\frac{i}{2}\begin{pmatrix} 0 & I \\ I & 0 \end{pmatrix}\sigma_1$ K_3

References

[1] Arnold V I 1997 *Mathematical Methods of Classical Mechanics* (*Graduate Texts in Mathematics* vol 60) 2nd edn (New York: Springer) (ISBN 978-0-387-96890-2), This is an English translation by K. Vogtmann and A. Weinstein of the Russian original edition: Maremaricheskie Merody Klassicheskoi Mekhaniki (Nauka, Moscow, 1974)

[2] Dirac P A M 1963 A remarkable representation of the 3 + 2 de Sitter group *J. Math. Phys.* **4** 901–9

[3] Guillemin V and Sternberg S 2001 *Symplectic Techniques in Physics* reprinted edition (Cambridge: Cambridge University Press) (ISBN 978-0-521-38990-7, Originally published 1984; OCLC: 248721606)

[4] Han D, Kim Y S, and Noz M E 1988 Linear canonical transformations of coherent and squeezed states in the Wigner phase space *Phys. Rev.* A **37** 807–14

[5] Han D, Kim Y S, and Noz M E 1990 Linear canonical transformations of coherent and squeezed states in the Wigner phase space. III. Two-mode states *Phys. Rev.* A **41** 6233–44

[6] Han D, Kim Y S, and Noz M E 1995 *O* (3,3)–like symmetries of coupled harmonic oscillators *J. Math. Phys.* **36** 3940–54

[7] Han D, Kim Y S, Noz M E, and Yeh L 1993 Symmetries of two-mode squeezed states *J. Math. Phys.* **34** 5493–508

[8] Itzykson C and Zuber J B 2005 *Quantum Field Theory (Dover Books on Physics)* (Mineola, NY: Dover Publications) (ISBN 978-0-486-44568-7). With new preface and list of errata. Originally published: McGraw-Hill, New York, 1980, in series: International Series in Pure and Applied Physics

[9] Kim Y S and Noz M E 2012 Dirac matrices and Feynman's rest of the Universe *Symmetry* **4** 626–43

[10] Kim Y S and Wigner E P 1990 Canonical transformation in quantum mechanics *Am. J. Phys.* **58** 439–48

[11] Lee D-G 1995 The Dirac gamma matrices as 'relics' of a hidden symmetry?: as fundamental representations of the algebra sp(4,R) *J. Math. Phys.* **36** 524–30

[12] Majorana E 1932 Teoria relativistica di particelle con momento intrinseco arbitrario relativistic particles with arbitrary intrinsic angular momentum *Il Nuovo Cimento* **9** 335–44

[13] Wigner E P 1932 On the quantum correction for thermodynamic equilibrium *Phys. Rev.* **40** 749–59

IOP Publishing

Physics of the Lorentz Group (Second Edition)
Beyond high-energy physics and optics
Sibel Başkal, Young S Kim, and Marilyn E Noz

Chapter 9

Coupled harmonic oscillators and squeezed states of light

The physics of two coupled harmonic oscillators [11] provides the mathematical basis for many soluble models in physics, including the Lee model in quantum field theory [18, 19], the Bogoliubov transformation in superconductivity [4, 9, 20], relativistic models of elementary particles [14, 21], and squeezed states of light [2, 5, 6, 23]. The problem of finding a coordinate transformation which will decouple the oscillators, leads us to study the symmetry of two-mode squeezed states using two coupled harmonic oscillators. Although the two-mode coherent photon state, commonly called the squeezed state of light, has been the central issue in optical sciences since 1976 [23], it has a much broader mathematical basis, namely, it is largely based on the group of Lorentz transformations.

It was Dirac, who in 1963 [6], carefully studied the symmetries generated by two harmonic oscillators and constructed the symmetry group consisting of ten generators. Indeed, this study of Dirac led to the study of the symmetries in the four-dimensional phase space, as we studied in chapter 8 on the Wigner functions. Also in his 1963 paper, the two-oscillator system leads to the Lorentz group applicable to three spatial and two time dimensions commonly called the $O(3, 2)$ de Sitter group. In chapter 10, we shall discuss in detail how the two-oscillator system will lead to a single set of equations leading to both Heisenberg's quantum mechanics and Einstein's special relativity.

We shall study here how the coupled oscillators play roles in physical applications, especially in the symmetry problems in two-photon systems. In section 9.1, we introduce the Hamiltonians for two single oscillators, and the coupling of the two oscillators. We then rescale the variables for the study of symmetry problems. In section 9.2, it is noted that the Lorentz-covariant oscillators share many symmetry features with those of the coupled oscillators.

It is well known that the algebra of harmonic oscillators is applicable to quantum optics where photons are created and annihilated. In section 9.3, the two-photon

doi:10.1088/978-0-7503-3607-9ch9

coherent states are discussed within the framework of the Lorentz group. These two-photon states are commonly called *squeezed states*. In section 9.4, further physical implications of squeezed states are discussed.

In section 9.5, we study the $O(3, 2)$ symmetry of Dirac's two-oscillator system. For the two-oscillator system, there are fifteen generators in the four-dimensional phase space. Ten of them generate canonical transformations. Five of them lead to non-canonical transformations. In section 9.6, the role of those non-canonical transformations is discussed.

For the system of two photons or two oscillators, the problem arises when one of them is not observed. This will lead to the increase in entropy and increase in temperature. This aspect is studied in connection with the non-canonical transformations in section 9.7.

9.1 Coupled oscillators

Let us start with the Hamiltonian for two harmonic oscillators whose Hamiltonians are

$$H_1 = \frac{1}{2}\left(\frac{p_1^2}{m_1} + k_1 x_1^2\right), \qquad H_2 = \frac{1}{2}\left(\frac{p_2^2}{m_2} + k_2 x_2^2\right). \tag{9.1}$$

If they are coupled, the Hamiltonian becomes

$$H = \frac{1}{2}\left(\frac{p_1^2}{m_1} + \frac{p_2^2}{m_2} + k_1 x_1^2 + k_2 x_2^2\right) + K(x_1 - x_2)^2. \tag{9.2}$$

Let us make the following scale changes.

$$x_1 \to \frac{x_1}{\sqrt{k_1}}, \qquad x_2 \to \frac{x_2}{\sqrt{k_2}}, \quad \text{and} \quad p_1 \to p_1\sqrt{m_1}, \qquad p_2 \to p_2\sqrt{m_2}. \tag{9.3}$$

The Hamiltonian then can be written as

$$H = \frac{1}{2}\left(p_1^2 + p_2^2 + x_1^2 + x_2^2\right) + K\left(\frac{x_1}{\sqrt{k_1}} - \frac{x_2}{\sqrt{k_2}}\right)^2. \tag{9.4}$$

Next, we can rotate in the space of (x_1, x_2) and (p_1, p_2) in the following way.

$$\begin{aligned} x_1 &\to x_1 \cos\theta - x_2 \sin\theta, \qquad & x_2 &\to x_1 \sin\theta + x_2 \cos\theta, \\ p_1 &\to p_1 \cos\theta - p_2 \sin\theta, \qquad & p_2 &\to p_1 \sin\theta + p_2 \cos\theta. \end{aligned} \tag{9.5}$$

After this rotation, the Hamiltonian becomes

$$\begin{aligned} H = &\frac{1}{2}\left(p_1^2 + p_2^2 + x_1^2 + x_2^2\right) \\ &+ K\left\{\left(\frac{x_1\cos\theta - x_2\sin\theta}{\sqrt{k_1}}\right) - \left(\frac{x_1\sin\theta + x_2\cos\theta}{\sqrt{k_2}}\right)\right\}^2, \end{aligned} \tag{9.6}$$

which becomes

$$H = \frac{1}{2}\left(p_1^2 + p_2^2 + x_1^2 + x_2^2\right) + K\{A(x_1 - x_2) + B(x_1 + x_2)\}^2, \tag{9.7}$$

with

$$\begin{aligned} A &= \frac{1}{2}\left\{\left(\frac{1}{\sqrt{k_1}} + \frac{1}{\sqrt{k_2}}\right)\cos\theta + \left(\frac{1}{\sqrt{k_1}} - \frac{1}{\sqrt{k_2}}\right)\sin\theta\right\}, \\ B &= \frac{1}{2}\left\{\left(\frac{1}{\sqrt{k_1}} - \frac{1}{\sqrt{k_2}}\right)\cos\theta - \left(\frac{1}{\sqrt{k_1}} + \frac{1}{\sqrt{k_2}}\right)\sin\theta\right\}. \end{aligned} \tag{9.8}$$

If we demand $B = 0$,

$$\tan\theta = \frac{\sqrt{k_2} - \sqrt{k_1}}{\sqrt{k_2} + \sqrt{k_1}}. \tag{9.9}$$

This leads to

$$A = \sqrt{\frac{1}{2}\left(\frac{1}{k_1} + \frac{1}{k_2}\right)}. \tag{9.10}$$

The Hamiltonian becomes

$$H = \frac{1}{2}\left(p_1^2 + p_2^2\right) + \frac{1}{2}\left\{x_1^2 + x_2^2 + \left[\frac{K(k_1 + k_2)}{k_1 k_2}\right](x_1 - x_2)^2\right\}. \tag{9.11}$$

Again, we make the coordinate transformation:

$$x_+ = \frac{x_1 + x_2}{\sqrt{2}}, \quad x_- = \frac{x_1 - x_2}{\sqrt{2}}, \quad \text{and} \quad p_+ = \frac{p_1 + p_2}{\sqrt{2}}, \quad p_- = \frac{p_1 - p_2}{\sqrt{2}}. \tag{9.12}$$

The Hamiltonian then becomes

$$H = \frac{1}{2}\left[\left(p_+^2 + p_-^2\right) + \left(x_+^2 + e^{4\eta}x_-^2\right)\right], \tag{9.13}$$

where

$$e^{4\eta} = 1 + \frac{K(k_1 + k_2)}{k_1 k_2}. \tag{9.14}$$

This parameterization is convenient when we examine whether the coupling is a canonical transformation because if $\eta = 0$, there here is no coupling with $K = 0$.

If we multiply this Hamiltonian by $e^{-\eta}$, this Hamiltonian is renormalized to

$$H = \frac{1}{2}\left[\left(x_+^2 + p_+^2\right)e^{-\eta} + \left(x_-^2 e^{2\eta} + p_-^2 e^{-2\eta}\right)e^{\eta}\right]. \tag{9.15}$$

Let us now make the following transformations:

$$x_+ \to e^{\eta/2} x_+, \quad p_+ \to e^{\eta/2} p_+, \quad \text{and} \quad x_- \to e^{-\eta/2} x_-, \quad p_- \to e^{-\eta/2} p_-, \tag{9.16}$$

leading to the Hamiltonian of the form

$$H = \frac{1}{2}\left(x_-^2 e^{2\eta} + p_-^2 e^{-2\eta}\right) + \frac{1}{2}\left(x_+^2 + p_+^2\right). \tag{9.17}$$

Now, only within the system of x_+ and p_+, we make the scale change of

$$x_+ \to e^{-\eta} x_+, \quad \text{and} \quad p_+ \to e^{\eta} p_+, \tag{9.18}$$

with the resulting Hamiltonian of the form

$$H = \frac{1}{2}\left(e^{-2\eta} x_+^2 + e^{2\eta} p_+^2\right) + \frac{1}{2}\left(e^{2\eta} x_-^2 + e^{-2\eta} p_-^2\right). \tag{9.19}$$

The ground-state wave function for this Hamiltonian is

$$\psi_\eta(x_1, x_2) = \frac{1}{\sqrt{\pi}} \exp\left\{-\frac{1}{4}\left[(x_1 + x_2)^2 e^{-2\eta} + (x_1 - x_2)^2 e^{2\eta}\right]\right\}. \tag{9.20}$$

The momentum wave function becomes

$$\phi_\eta(p_1, p_2) = \frac{1}{2\pi} \int \exp\left(i[p_1 x_1 + p_2 x_2]\right) \psi(x_1, x_2) dx_1 dx_2, \tag{9.21}$$

leading to

$$\phi_\eta(p_1, p_2) = \frac{1}{\sqrt{\pi}} \exp\left\{-\frac{1}{4}\left[(p_1 + p_2)^2 e^{2\eta} + (p_1 - p_2)^2 e^{-2\eta}\right]\right\}. \tag{9.22}$$

The Wigner function for this system of two coupled oscillators takes the form

$$W_{cp}(x_1, p_1; x_2, p_2) = \left(\frac{1}{\pi}\right)^2 \exp\left\{-\left(x_+^2 e^{-2\eta} + x_-^2 e^{2\eta} + p_+^2 e^{2\eta} + p_-^2 e^{-2\eta}\right)\right\}. \tag{9.23}$$

9.2 Lorentz-covariant oscillators

In chapter 6 on Lorentz-covariant oscillators, the oscillator differential equation is separable in the space and time variables. Among those, the transverse components remain unchanged when the system is boosted along the z direction. Thus, only the z and t variables play a role when the system is Lorentz-boosted. Here, we use x_1 and x_2 for the z and t variables, respectively. Then, the Hamiltonian for the harmonic oscillator function can be written as:

$$H_- = \frac{1}{2}\left(x_1^2 + p_1^2\right) - \frac{1}{2}\left(x_2^2 + p_2^2\right), \tag{9.24}$$

which can be written as

$$H_{-} = \frac{1}{2}\left(x_1^2 - x_2^2\right) + \frac{1}{2}\left(p_1^2 - p_2^2\right). \tag{9.25}$$

This form is invariant under the Lorentz transformation of the form

$$\begin{aligned} x_1 &\to x_1 \cosh\eta + x_2 \sinh\eta, & x_2 &\to x_1 \sinh\eta + x_2 \cosh\eta, \\ p_1 &\to p_1 \cosh\eta + p_2 \sinh\eta, & p_2 &\to p_1 \sinh\eta + p_2 \cosh\eta. \end{aligned} \tag{9.26}$$

In terms of $x_\pm$ and $p_\pm$, this transformation can be written as

$$x_\pm \to e^{\pm\eta} x_\pm \quad \text{and} \quad p_\pm \to e^{\pm\eta} p_\pm. \tag{9.27}$$

The wave function takes the form

$$f_\eta(x_1, x_2) = \frac{1}{\sqrt{\pi}} \exp\left\{-\frac{1}{4}\left[(x_1 + x_2)^2 e^{-2\eta} + (x_1 - x_2)^2 e^{2\eta}\right]\right\}, \tag{9.28}$$

which is the same as the one for the coupled oscillators given in equation (9.20). However, in the case of the covariant harmonic oscillators, the momentum wave function is

$$g_\eta(p_1, p_2) = \frac{1}{2\pi} \int \exp\left(i[p_1 x_1 - p_2 x_2]\right) \psi(x_1, x_2) dx_1 dx_2, \tag{9.29}$$

leading to

$$g_\eta(p_1, p_2) = \frac{1}{\sqrt{\pi}} \exp\left\{-\frac{1}{4}\left[(p_1 + p_2)^2 e^{-2\eta} + (p_1 - p_2)^2 e^{2\eta}\right]\right\}. \tag{9.30}$$

The Wigner function takes the form

$$W_{cov}(x_1, p_1; x_2, p_2) = \left(\frac{1}{\pi}\right)^2 \exp\left\{-\left(x_+^2 e^{-2\eta} + x_-^2 e^{2\eta} + p_+^2 e^{-2\eta} + p_-^2 e^{2\eta}\right)\right\}. \tag{9.31}$$

The difference between the coupled and covariant harmonic oscillators are illustrated in figure 9.1

Let us check whether they are canonical transformations. In order to check, we need the vector (x_1, p_1, x_2, p_2), instead of (x_+, x_-, p_+, p_-). The transformation to the former from the latter can be written as

$$\begin{pmatrix} x_+ \\ p_+ \\ x_- \\ p_- \end{pmatrix} = \frac{1}{\sqrt{2}} \begin{pmatrix} 1 & 0 & 1 & 0 \\ 0 & 1 & 0 & 1 \\ 1 & 0 & -1 & 0 \\ 0 & 1 & 0 & -1 \end{pmatrix} \begin{pmatrix} x_1 \\ p_1 \\ x_2 \\ p_2 \end{pmatrix}. \tag{9.32}$$

This matrix can be written as

$$\frac{1}{\sqrt{2}} \begin{pmatrix} I & I \\ I & -I \end{pmatrix}, \tag{9.33}$$

and its transpose and its inverse take the same form.

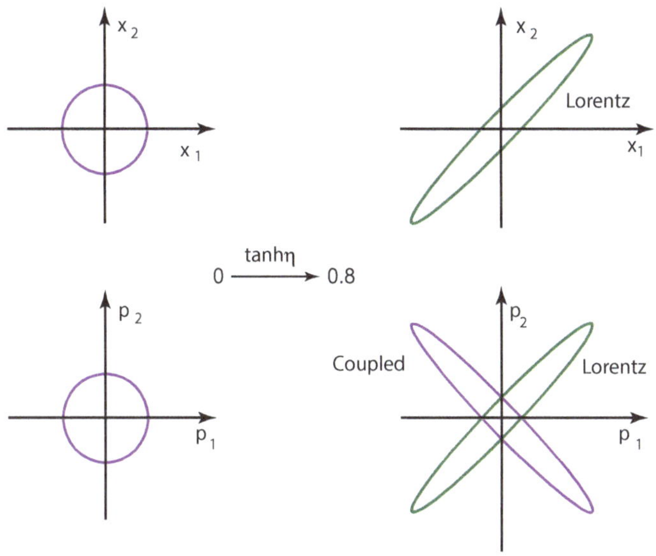

Figure 9.1. Covariant oscillator and coupled oscillator. The wave functions in the space coordinates transform in the same way, but in momentum space the wave functions are transformed differently.

In order that the given four-by-four matrix M be canonical, it should satisfy the condition

$$M\ J\ \tilde{M} = J, \tag{9.34}$$

where

$$J = i\begin{pmatrix} \sigma_2 & 0 \\ 0 & \sigma_2 \end{pmatrix}. \tag{9.35}$$

This condition was given earlier in chapter 8 on Wigner functions.

This J matrix commutes with the two-by-two matrix of equation (9.33), as well as with its transverse and its inverse. Thus, the condition of equation (9.34) originally for the vector $(x_1,\ p_1,\ x_2,\ p_2)$ is applicable to the vector $(x_+,\ p_+,\ x_-,\ p_-)$.

The transformation matrix for the coupled oscillators becomes

$$\begin{pmatrix} e^{\eta} & 0 & 0 & 0 \\ 0 & e^{-\eta} & 0 & 0 \\ 0 & 0 & e^{-\eta} & 0 \\ 0 & 0 & 0 & e^{\eta} \end{pmatrix}\begin{pmatrix} x_+ \\ p_+ \\ x_- \\ p_- \end{pmatrix}. \tag{9.36}$$

This is a canonical transformation, whose generator is given in table 9.1 as

$$i\begin{pmatrix} I & 0 \\ 0 & -I \end{pmatrix}\sigma_3. \tag{9.37}$$

For the coupled oscillators, the exponent of the Fourier integral is

$$p_1 \cdot x_1 + p_2 \cdot x_2 = p_+ \cdot x_+ + p_- \cdot x_-. \tag{9.38}$$

Table 9.1. Canonical and non-canonical transformations. Canonical transformations are generated by the nine matrices on the first three rows and three columns, and one matrix on the fourth row and column. The non-canonical transformations are generated by the first three matrices on the fourth column and the first two matrices on the fourth row. Similar tables are given in chapter 8. Here, for simplicity, the 1/2 factor was dropped from the matrices.

$i\begin{pmatrix} I & 0 \\ 0 & I \end{pmatrix}\sigma_3$ K_2	$i\begin{pmatrix} I & 0 \\ 0 & -I \end{pmatrix}\sigma_3$ Q_1	$i\begin{pmatrix} 0 & I \\ I & 0 \end{pmatrix}\sigma_3$ Q_3	$i\begin{pmatrix} 0 & -I \\ I & 0 \end{pmatrix}\sigma_3$ non-can
$i\begin{pmatrix} I & 0 \\ 0 & I \end{pmatrix}\sigma_1$ Q_2	$i\begin{pmatrix} I & 0 \\ 0 & -I \end{pmatrix}\sigma_1$ K_1	$i\begin{pmatrix} 0 & I \\ I & 0 \end{pmatrix}\sigma_1$ K_3	$i\begin{pmatrix} 0 & -I \\ I & 0 \end{pmatrix}\sigma_1$ non-can
$\begin{pmatrix} I & 0 \\ 0 & I \end{pmatrix}\sigma_2$ S_3	$\begin{pmatrix} I & 0 \\ 0 & -I \end{pmatrix}\sigma_2$ J_3	$\begin{pmatrix} 0 & I \\ I & 0 \end{pmatrix}\sigma_2$ J_1	$\begin{pmatrix} 0 & -I \\ I & 0 \end{pmatrix}\sigma_2$ non-can
	$i\begin{pmatrix} I & 0 \\ 0 & -I \end{pmatrix}I$ non-can	$i\begin{pmatrix} 0 & I \\ I & 0 \end{pmatrix}I$ non-can	$i\begin{pmatrix} 0 & -I \\ I & 0 \end{pmatrix}I$ J_2

On the other hand, for the covariant oscillators, the exponent becomes

$$p_1 \cdot x_1 - p_2 \cdot x_2 = p_+ \cdot x_- - p_- \cdot x_+. \tag{9.39}$$

Let us write the Lorentz boost in the covariant oscillator system as

$$\begin{pmatrix} e^{\eta} & 0 & 0 & 0 \\ 0 & e^{-\eta} & 0 & 0 \\ 0 & 0 & e^{-\eta} & 0 \\ 0 & 0 & 0 & e^{\eta} \end{pmatrix} \begin{pmatrix} x_+ \\ p_- \\ x_- \\ p_+ \end{pmatrix}, \tag{9.40}$$

as noted in chapter 6.

For both cases, the transformation in xy space is the same, but they are different in momentum space, as illustrated in figure 9.1 and 9.2.

9.3 Squeezed states of light

The mathematics of harmonic oscillators is used for the number of particles in a given quantum state. The nth excited state means there are n particles in the same quantum state. This aspect is well known in quantum field theory. In this section, we are interested in the states of two photons. We are interested in how these two photons can be in a coherent state. This two-photon coherent state is called the *squeezed state* or *two-mode squeezed state.*

Let us write differential equations for the two oscillators of equation (9.1) as

$$H_1 = \frac{1}{2}\left(p_1^2 + x_1^2\right), \qquad H_2 = \frac{1}{2}\left(p_2^2 + x_2^2\right). \tag{9.41}$$

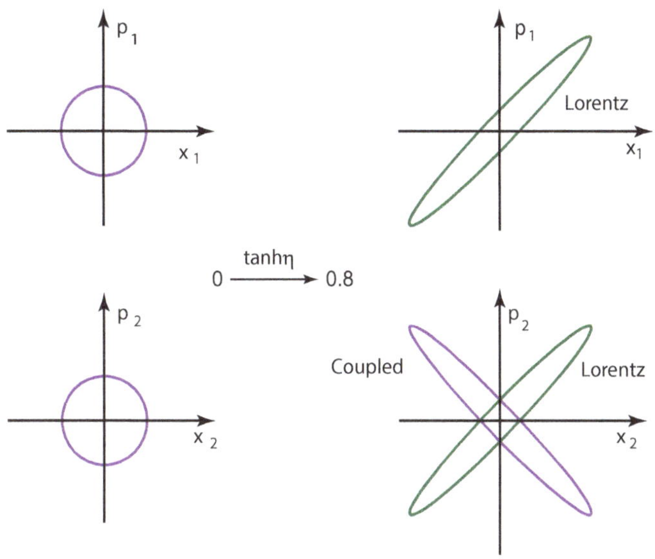

Figure 9.2. Phase-space pictures of the first and second oscillators. They are transformed in two-different directions.

We then write the corresponding solutions of the Schrödinger equation as

$$\chi_n(x) = \left(\frac{1}{2^n n!\sqrt{\pi}}\right)^{1/2} H_n(x)\exp(-x/2), \tag{9.42}$$

for each x_i. This is for the oscillator in the nth excited state with the Hermite polynomial $H_n(x)$. This form is well known.

For each photon, we can then use the operators a and $a^\dagger$ defined as,

$$a = \frac{1}{\sqrt{2}}\left(x + \frac{\partial}{\partial x}\right) \quad \text{and} \quad a^\dagger = \frac{1}{\sqrt{2}}\left(x - \frac{\partial}{\partial x}\right). \tag{9.43}$$

When these operators are applied to the wave function, we obtain

$$a\,\chi_n(x) = \sqrt{n}\,\chi_{n-1}(x) \quad \text{and} \quad a^\dagger\,\chi_n = \sqrt{n+1}\,\chi_{n+1}(x). \tag{9.44}$$

Thus $\chi_n(x)$ can be used for the state of n photons, while a and $a^\dagger$ can serve as the annihilation and creation operators, respectively. Thus, we can use the notation $\chi_n(x) = |n\rangle$, and write

$$a^\dagger|0\rangle = |1\rangle, \quad (a^\dagger)^2|0\rangle = \sqrt{2}\,|2\rangle, \quad \text{and} \quad (a^\dagger)^n|0\rangle = \sqrt{n!}\,|n\rangle, \tag{9.45}$$

leading to

$$|n\rangle = \frac{1}{\sqrt{n!}}(a^\dagger)^n|0\rangle. \tag{9.46}$$

The single-photon coherent state takes the form

$$|\alpha\rangle = e^{-\alpha\alpha^*/2} \sum_n \frac{\alpha^n}{\sqrt{n!}} |n\rangle, \tag{9.47}$$

which can be expanded as

$$|\alpha\rangle = e^{-\alpha\alpha^*/2} \sum_n \frac{\alpha^n}{n!} (a^\dagger)^n |0\rangle = \{e^{-\alpha\alpha^*/2}\} \exp\{\alpha a^\dagger\} |0\rangle. \tag{9.48}$$

This aspect of the single-photon coherent state is well known. Here we are dealing with one kind of photon, namely with a given momentum and polarization. The state $|n\rangle$ means there are n photons of the same kind, and $|0\rangle$ is for the zero-photon vacuum state corresponding to the ground state oscillator wave function.

Let us next consider a state of two kinds of photons, and write $|n_1, n_2\rangle$ as the state of n_1 photons of the first kind, and n_2 photons of the second kind [23]. We are then led to the exponential form

$$|\beta\rangle = (1 - \beta^2)^{1/2} \exp\left\{\beta a_1^\dagger a_2^\dagger\right\} |0, 0\rangle, \tag{9.49}$$

where β is a real number smaller than one. The Taylor expansion of this formula leads to

$$|\beta\rangle = (1 - \beta^2)^{1/2} \sum_k \beta^k |k, k\rangle, \tag{9.50}$$

which is the two-photon coherent state [23].

This two-photon state is normalized with $\langle\beta|\beta\rangle = 1$, but the operator $i(a_1^\dagger a_2^\dagger)$ is not Hermitian, and the operator

$$\exp\left(\beta a_1^\dagger a_2^\dagger\right) \tag{9.51}$$

is not unitary.

In order to obtain the unitary operator which will change $|0\rangle$ to $|\beta\rangle$, let us rewrite the wave function for the Lorentz-covariant oscillators given in equation (9.28) as:

$$f_\eta(x_1, x_2) = \frac{1}{\sqrt{\pi}} \exp\left\{-\frac{1}{4}\left[(x_1 + x_2)^2 e^{-2\eta} + (x_1 - x_2)^2 e^{2\eta}\right]\right\}, \tag{9.52}$$

where the η is the boost parameter. This form is the same as the wave function for the coupled oscillators given in equation (9.20) with η as the coupling parameter. This wave function can be expanded as

$$\psi_\eta(x_1, x_2) = f_\eta(x_1, x_2) = \frac{1}{\cosh\eta} \sum_k (\tanh\eta)^k \chi_k(x_1)\chi_k(x_2). \tag{9.53}$$

This form of expansion was discussed in detail for the orthogonality of the Lorentz-covariant oscillator wave functions in chapter 6. The operator which transforms the ground state wave function to equation (9.53) is generated by

$$i\left(x_1 \frac{\partial}{\partial x_2} + x_2 \frac{\partial}{\partial x_1}\right). \tag{9.54}$$

In terms of the step-up and step-down operators, this generator can be written as

$$Q_3 = \frac{i}{2}\left(a_1^\dagger a_2^\dagger - a_1 a_2\right). \tag{9.55}$$

This is a Hermitian operator, and we now write the two-photon coherent state as

$$|n, n\rangle = \exp\left(i\eta Q_3\right)|0, 0\rangle. \tag{9.56}$$

The expression of equation (9.53) is the same as that of equation (9.50) for the two-photon state, if

$$\beta = \tanh \eta. \tag{9.57}$$

Indeed, Q_3 is the Hermitian generator of the two-photon states. In view of the graphic illustration given in figure 9.1, it is appropriate to call this two-photon state a *squeezed state*.

In the vacuum state with $\beta = 0$ and $\eta = 0$, the total number of photons is zero. If the system becomes squeezed, the number of each photon becomes

$$(1 - \beta^2)\sum_k k\beta^{2k} = \frac{\beta}{2}(1 - \beta^2)\frac{\partial}{\partial\beta}\sum_k \beta^{2k}. \tag{9.58}$$

Thus, the total number of each photon becomes

$$\frac{\beta^2}{1 - \beta^2} = \sinh^2 \eta. \tag{9.59}$$

The total energy then becomes

$$\hbar(\omega_1 + \omega_2)\sinh^2 \eta, \tag{9.60}$$

where ω_1 and ω_2 are the frequencies of the photons. When $\eta = 0$, the system is in a vacuum state, and the number of photons is zero. When the system is squeezed, the number increases rapidly with the increasing value of η.

9.4 Further notes on squeezed states

Early in this chapter, we discussed three different physical systems sharing the same form wave function, namely the exponential form of equations (9.20), (9.28), and (9.52). Let us rewrite this wave function:

$$\psi_\eta(x_1, x_2) = \frac{1}{\sqrt{\pi}}\exp\left\{-\frac{1}{4}\left[(x_1 + x_2)^2 e^{-2\eta} + (x_1 - x_2)^2 e^{2\eta}\right]\right\}. \tag{9.61}$$

These systems share the same wave function even though the squeezed state of photons and the coupled oscillators are based on the addition of the two Hamiltonians while the covariant oscillators are based on the difference between the oscillators. We are thus led to examine symmetries of two-photon states within the framework of the Lorentz group. As illustrated in figure 9.3, it is possible to

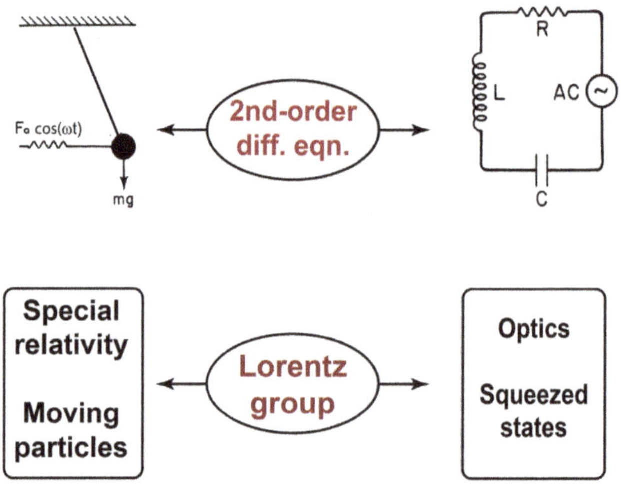

Figure 9.3. Lorentz group for special relativity and optical sciences. It is well known that the same differential equation can explain the resonance effect in both mechanics and electronic circuits. Likewise the same set of formulas can be used for Einstein's special theory of relativity and for optical sciences. The application of the Lorentz group to classical optics will be discussed in later chapters of this book.

transmit the physical interpretations from the squeezed state or the coupled oscillators to the two-photon system, or vice versa.

For the squeezed state, both photons are observable. The case is the same for the two coupled oscillators. For the covariant oscillators, it is possible to establish the following correspondence:

$$\begin{aligned} \text{first photon} \quad &\rightarrow \quad \text{longitudinal separation,} \\ \text{second photon} \quad &\rightarrow \quad \text{time–like separation.} \end{aligned}$$

In the case of squeezed states, both the first and second photons are observable, but the time-separation variable is not observable for the covariant oscillators while the longitudinal variable is observable. This unobservable variable is in Feynman's rest of the Universe, as illustrated in figure 9.4.

The rest of the Universe is discussed in chapter 6 on Lorentz-covariant harmonic oscillators, but the coupled oscillators provides a more transparent illustration of Feynman's rest of the Universe [12]. The best way to illustrate Feynman's rest of the Universe is to use two coupled oscillators [12], as illustrated in figure 9.5. The mathematics of the two-photon problem is the same as that of the coupled oscillators.

In his book on statistical mechanics [10], Feynman recommends the use of density matrices. This problem for the two-photon case was addressed by Yurke and Potasek [25] and by Ekert and Knight [7]. They used the density matrix formalism and integrated out the second-photon states. The density matrix for this system is

$$\rho_\eta(x_1, x_2; x_1', x_2') = \psi_\eta(x_1, x_2)\psi_\eta^*(x_1', x_2'). \tag{9.62}$$

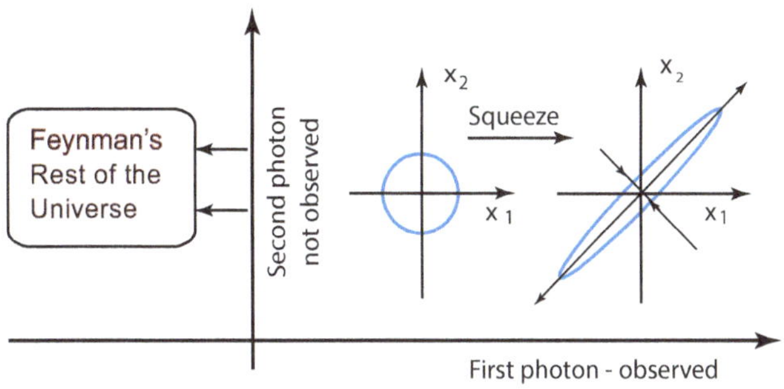

Figure 9.4. Feynman's rest of the Universe in the entangled system of two photons. If one of the photons is not observed, it is in the rest of the Universe. The result is the entropy increase in the system. The same figure is given with $x_1 = z$ and $x_2 = t$ in the earlier chapter on Lorentz-covariant harmonic oscillators.

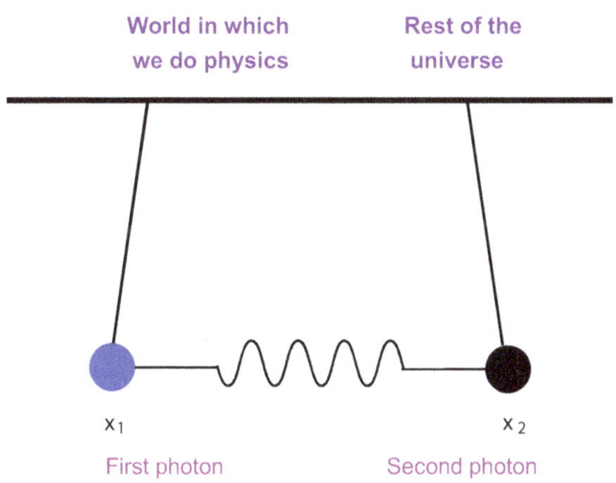

Figure 9.5. Two coupled oscillators illustrating Feynman's rest of the Universe. If one of them is not observed, it belongs to the rest of the Universe [12]. For the two-mode squeezed state, the second photon belongs to the rest of the Universe.

If the second photon is not observed, this density matrix becomes

$$\rho_\eta(x_1, x_1') = \int \rho_\eta(x_1, x_2; x_1', x_2)\, dx_2. \tag{9.63}$$

Let us go to the series expansion of $\psi_\eta(x_1, x_2)$, given in equation (9.53), and perform this integral. Then the result is

$$\rho_\eta(x, x') = \frac{1}{\cosh^2 \eta} \sum_n (\tanh \eta)^{2n} \chi_n(x) \chi_n(x'). \tag{9.64}$$

As was noted in chapter 6, this formula is identical to that of the single oscillator in thermal excitation:

$$\rho_T(x, x') = (1 - e^{-1/T}) \sum_n \exp(-n/T) \chi_n(x) \chi_n^*(x'). \tag{9.65}$$

Here the temperature T is measured in the unit of Boltzmann's constant. In terms of the β parameter, the temperature is

$$T = \frac{-1}{2\ln(\tanh\eta)} \quad \text{or} \quad \eta = \frac{1}{2}\ln\left(\frac{1+\exp(-1/T)}{1-\exp(-1/T)}\right). \tag{9.66}$$

Since $\tanh\eta$ is always smaller than one, T is always positive. It vanishes when $\eta = 0$, and increases while η becomes large.

The entropy of the system is

$$S(\eta) = -Tr\left\{\rho_\eta \ln(\rho_\eta)\right\}. \tag{9.67}$$

The result of this calculation is

$$S(\eta) = \ln(\cosh^2\eta) - (\sinh^2\eta)\ln(\tanh^2\eta). \tag{9.68}$$

This entropy is zero when $\eta = 0$. For large values of η, $S(\eta) \to 2\eta$, consistent with our expectation. This aspect of entropy will be discussed in more detail in section 9.7.

It is interesting indeed that the same set of formulas is applicable to three different physical cases, namely the two coupled oscillators, two-mode squeezed state, and the Lorentz-covariant harmonic oscillator.

The two-photon problem raises symmetry problems beyond the coupling through the η parameter. According to Yurke *et al* [24], it is possible to consider interferometers applicable to two-mode states. First of all, they picked the following three Hermitian operators

$$\begin{aligned} J_1 &= \frac{1}{2}\left(a_1^\dagger a_2 + a_2^\dagger a_1\right), \\ J_2 &= \frac{1}{2i}\left(a_1^\dagger a_2 - a_2^\dagger a_1\right), \\ J_3 &= \frac{1}{2}\left(a_1^\dagger a_1 - a_2^\dagger a_2\right), \end{aligned} \tag{9.69}$$

in addition to the number operator

$$N = a_1^\dagger a_1 + a_2^\dagger a_2. \tag{9.70}$$

Since these J_i operators satisfy the commutation relations

$$[J_i, J_j] = i\varepsilon_{ijk}J_j, \tag{9.71}$$

it is possible to study the symmetry property of the three-dimensional rotation group.

In addition, they wrote down another set of operators which could lead to effects on the two-mode states. They are

$$\begin{aligned} Q_3 &= \frac{i}{2}\left(a_1^\dagger a_2^\dagger - a_1 a_2\right), \\ K_3 &= \frac{1}{2}\left(a_1^\dagger a_2^\dagger + a_1 a_2\right), \\ S_3 &= \frac{1}{2}\left(a_1^\dagger a_1 + a_2 a_2^\dagger\right), \end{aligned} \tag{9.72}$$

with the closed set of commutation relations:

$$[K_3, Q_3] = -iS_3, \qquad [Q_3, S_3] = iK_3, \qquad [S_3, K_3] = iQ_3. \tag{9.73}$$

These relations are like those for the $SU(1, 1)$ group or the Lorentz group $O(2, 1)$, applicable to two space dimensions and one time dimension.

Among these three operators, the role of Q_3 was discussed in section 9.3 as the generator of the two-mode squeezed state.

The question is whether it is possible to construct one set of closed commutation relations with the six operators given in equations (9.69) and (9.72). The answer to this question was given by Dirac in 1963 [6].

9.5 $O(3, 2)$ symmetry from Dirac's coupled oscillators

In his 1963 paper [6], Dirac started with the ground-state wave function for two-harmonic oscillators. He then considered unitary transformations generated by the following ten Hermitian operators:

$$\begin{aligned} J_1 &= \frac{1}{2}\left(a_1^\dagger a_2 + a_2^\dagger a_1\right), \qquad J_2 = \frac{1}{2i}\left(a_1^\dagger a_2 - a_2^\dagger a_1\right), \\ J_3 &= \frac{1}{2}\left(a_1^\dagger a_1 - a_2^\dagger a_2\right), \qquad S_3 = \frac{1}{2}\left(a_1^\dagger a_1 + a_2 a_2^\dagger\right), \\ K_1 &= -\frac{1}{4}\left(a_1^\dagger a_1^\dagger + a_1 a_1 - a_2^\dagger a_2^\dagger - a_2 a_2\right), \\ K_2 &= \frac{i}{4}\left(a_1^\dagger a_1^\dagger - a_1 a_1 + a_2^\dagger a_2^\dagger - a_2 a_2\right), \\ K_3 &= \frac{1}{2}\left(a_1^\dagger a_2^\dagger + a_1 a_2\right), \\ Q_1 &= -\frac{i}{4}\left(a_1^\dagger a_1^\dagger - a_1 a_1 - a_2^\dagger a_2^\dagger + a_2 a_2\right), \\ Q_2 &= -\frac{1}{4}\left(a_1^\dagger a_1^\dagger + a_1 a_1 + a_2^\dagger a_2^\dagger + a_2 a_2\right), \\ Q_3 &= \frac{i}{2}\left(a_1^\dagger a_2^\dagger - a_1 a_2\right), \end{aligned} \tag{9.74}$$

where $a^\dagger$ and a are the step-up and step-down (creation and annihilation) operators applicable to the harmonic oscillator wave functions. These operators satisfy the following set of commutation relations:

$$\begin{aligned} &[J_i, J_j] = i\varepsilon_{ijk}J_k, \quad [J_i, K_j] = i\varepsilon_{ijk}K_k, \quad [J_i, Q_j] = i\varepsilon_{ijk}Q_k, \\ &[K_i, K_j] = [Q_i, Q_j] = -i\varepsilon_{ijk}J_k, \qquad [J_i, S_3] = 0, \\ &[K_i, Q_j] = -i\delta_{ij}S_3, \quad [K_i, S_3] = -iQ_i, \quad [Q_i, S_3] = iK_i. \end{aligned} \tag{9.75}$$

In his paper of 1963 [6], Dirac determined that the commutation relations of equation (9.75) constitute also the Lie algebra for the $O(3, 2)$ de Sitter group with ten generators. This de Sitter group is the Lorentz group applicable to three space coordinates and two time coordinates. Let us use the notation (s, t, z, x, y), with (z, x, y) as space coordinates and (s, t) as two time coordinates. The rotation and squeeze generators are explicitly given in five-by-five matrix form also in chapter 8 on Wigner functions. The harmonic oscillator serves as the language of quantum mechanics. The Lorentz group is a subgroup of the de Sitter group. Indeed, Dirac's observation of 1963 encourages us to look for one mathematical base for both quantum mechanics and special relativity. This issue will be discussed in chapter 10.

Also in his 1963 paper [6], Dirac noted that this set of commutation relations can serve as the Lie algebra for the four-dimensional symplectic group $Sp(4)$. The four dimensions consist of two pairs of position and momentum variables. The generators of this group in the form of four-by-four matrices are discussed in chapter 8. They are also given in table 9.1.

There are fifteen matrices in table 9.1, and ten of them serve as the generators of Dirac's oscillators and canonical transformations. The remaining five matrices lead to non-canonical transformations.

9.6 Canonical and non-canonical transformations from the coupled oscillators

As we noted in chapter 8, among the fifteen Dirac matrices, ten of them can be used for canonical transformations in classical mechanics, and thus in quantum mechanics. As we noted here, they play a special role in quantum optics [13, 14, 23, 24].

The remaining five of them generate transformations which allow the change in area of each oscillator. In quantum mechanics, the area can be increased, but it has a lower limit called Planck's constant. In classical mechanics, this constraint does not exist. The mathematical formalism given in this chapter allows us to study this aspect of the system of coupled oscillators.

There are fifteen matrices in table 9.1, ten of them generate canonical transformations. The remaining five of them lead to non-canonical transformations. They are

$$G_3 = i\begin{pmatrix} I & 0 \\ 0 & -I \end{pmatrix} I, \quad G_1 = i\begin{pmatrix} 0 & I \\ I & 0 \end{pmatrix} I,$$
$$S_1 = i\begin{pmatrix} 0 & -I \\ I & 0 \end{pmatrix} \sigma_3, \quad S_2 = i\begin{pmatrix} 0 & -I \\ I & 0 \end{pmatrix} \sigma_1, \tag{9.76}$$
$$G_2 = \begin{pmatrix} 0 & -I \\ I & 0 \end{pmatrix} \sigma_2,$$

where the notations S_i and G_i, in addition to J_i, K_1 and Q_i, are introduced in chapter 8 in connection with the fifteen generators for the Lorentz group applicable to three space and three time dimensions.

Let us go to table 9.1 and look at the second column. There are three generators for canonical transformations, and one generator for the non-canonical transformation.

Let us choose the generator Q_1, which leads to the four-by-four matrix for the canonical transformation applicable to the column vector (x_1, p_1, x_2, p_2):

$$\exp(-i\eta Q_1) = \begin{pmatrix} e^{\eta} & 0 & 0 & 0 \\ 0 & e^{-\eta} & 0 & 0 \\ 0 & 0 & e^{-\eta} & 0 \\ 0 & 0 & 0 & e^{\eta} \end{pmatrix}. \tag{9.77}$$

This matrix squeezes the first phase space, and squeezes the second phase space in the opposite direction, as illustrated in figure 9.6. The area of each phase space is preserved.

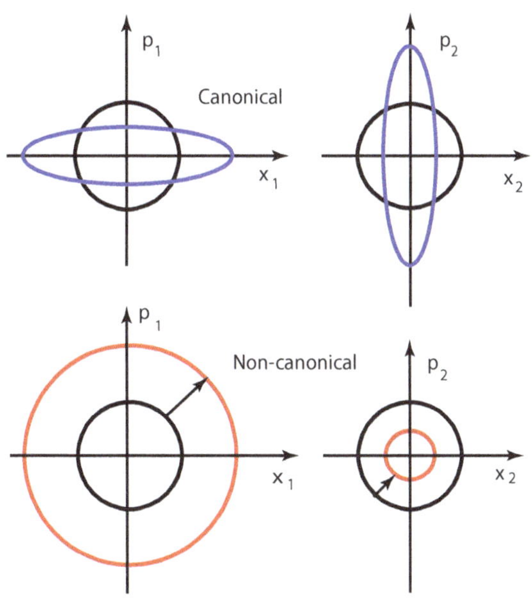

Figure 9.6. Illustration of both canonical and non-canonical transformations. The canonical transformation preserves the area of each phase space. The non-canonical transformation has the affect that one of the areas expands while the other shrinks. However, the product of these two areas remains invariant. The statistical uncertainty is reflected in this increase or decrease. Thus, even though the decrease in uncertainty may be below the quantum limit, it does not violate the uncertainty principle [1].

The non-canonical generator leads to the transformation matrix of the form

$$\exp(-i\eta G_3) = \begin{pmatrix} e^{\eta} & 0 & 0 & 0 \\ 0 & e^{\eta} & 0 & 0 \\ 0 & 0 & e^{-\eta} & 0 \\ 0 & 0 & 0 & e^{-\eta} \end{pmatrix}. \tag{9.78}$$

For positive values of η, this matrix expands the phase space of the first oscillator while shrinking that of the second oscillator. This expansion-shrink effect is illustrated in figure 9.6.

The Wigner function serves useful purposes for these transformations in phase space. We can start with the Wigner function

$$W_0(x_1, p_1; x_2, p_2) = \left(\frac{1}{\pi}\right)^2 \exp\left\{-\left(x_1^2 + p_1^2 + x_2^2 + p_2^2\right)\right\}. \tag{9.79}$$

Under the canonical transformation, the Wigner function becomes

$$W_c(\eta) = \left(\frac{1}{\pi}\right)^2 \exp\left\{-\left(e^{-2\eta}x_1^2 + e^{+2\eta}p_1^2 + e^{2\eta}x_2^2 + e^{-2\eta}p_2^2\right)\right\}, \tag{9.80}$$

while it becomes

$$W_n(\eta) = \left(\frac{1}{\pi}\right)^2 \exp\left\{-\left(e^{-2\eta}x_1^2 + e^{-2\eta}p_1^2 + e^{2\eta}x_2^2 + e^{2\eta}p_2^2\right)\right\}, \tag{9.81}$$

under the non-canonical transformation. These transformed Wigner functions are illustrated in figure 9.6.

9.7 Entropy and the expanding Wigner phase space

In section 9.4, the concept of Feynman's rest of Universe was introduced. In the language of the Wigner functions, we can integrate out the phase space of the second oscillator. As was noted earlier, the area of one of the phase spaces can become smaller than Planck's constant. Since this area is not observable in quantum mechanics it can be integrated out. Let us go to the non-canonical Wigner function of equation (9.81).

$$W_\eta(x_1, p_1) = \int W_n(x_1, p_1; x_2, p_2)\, dx_2 dp_2. \tag{9.82}$$

The result of this integration is

$$W_\eta(x_1, p_1) = \left(\frac{e^{2\eta}}{\pi}\right) \exp\left\{-e^{-2\eta}\left(x_1^2 + p_1^2\right)\right\}. \tag{9.83}$$

As η becomes large, the radius of the Wigner distribution becomes $\exp(2\eta)$.

Let us go back to the density matrix. The standard way to measure this ignorance is to calculate the entropy defined as [3, 8, 15–17, 22]:

$$S = -Tr(\rho \ln(\rho)). \tag{9.84}$$

As given in equation (9.68), the result of this integration is

$$S(\eta) = \ln(\cosh^2 \eta) - (\sinh^2 \eta)\ln(\tanh^2 \eta). \tag{9.85}$$

The value of this entropy is zero when $\eta = 0$, and becomes $S = 2\eta$ for large values of η. It is interesting to note that the ln of the radius of the Wigner distribution is 2η.

References

[1] Başkal S, Kim Y S, and Noz M E 2019 *Mathematical Devices for Optical Sciences* (Bristol, UK: IOP Publishing) (ISBN 978-0-7503-1612-5, OCLC: 1034620988)

[2] Bishop R F and Vourdas A 1988 General two-mode squeezed states *Z. Physik B: Condensed Matter* **71** 527–9

[3] Blum K 2012 *Density Matrix Theory and Applications* 3rd edn (Heidelberg; New York: Springer) Number 64 in Springer Series on Atomic, Optical and Plasma Physics (ISBN 978-3-642-20560-6, Originally published 1981)

[4] Bogoliubov N N 1958 On a new method in the theory of superconductivity *Nuovo Cimcnto* **7** 794–804

[5] Caves C M 1981 Quantum-mechanical noise in an interferometer *Phys. Rev.* D **23** 1693–708

[6] Dirac P A M 1963 A remarkable representation of the 3 + 2 de Sitter group *J. Math. Phys.* **4** 901–9

[7] Ekert A K and Knight P L 1989 Correlations and squeezing of two-mode oscillations *Am. J. Phys.* **57** 692–7

[8] Fano U 1957 Description of states in quantum mechanics by density matrix and operator techniques *Rev. Mod. Phys.* **29** 74–93

[9] Fetter A L and Walecka J D 2003 *Quantum Theory of Many-Particle Systems* (Mineola, NY: Dover Publications) (ISBN 978-0-486-42827-7, Originally published: New York, NY: McGraw-Hill, 1971)

[10] Feynman R P 1998 *Statistical Mechanics: A Set of lectures* (Boulder, CO: Westview Press) Advanced Book Classics (ISBN 978-0-201-36076-9, Originally published 1972; OCLC: ocm60679997)

[11] Goldstein H 1980 *Classical Mechanics* 2nd edn (Reading, MA: Addison-Wesley Pub. Co.) Addison-Wesley Series in Physics (ISBN 978-0-201-02918-5)

[12] Han D, Kim Y S, and Noz M E 1999 Illustrative example of Feynman's rest of the universe *Am. J. Phys.* **67** 61–6

[13] Han D, Kim Y S, Noz M E, and Yeh L 1993 Symmetries of two-mode squeezed states *J. Math. Phys.* **34** 5493–508

[14] Kim Y S and Noz M E 1991 *Phase Space Picture of Quantum Mechanics: Group Theoretical Approach* (*Lecture Notes in Physics* vol 40) 2nd edn (Singapore; Hackensack, NJ: World Scientific Publishing Co.) (ISBN 978-981-02-0360-3 978-981-02-0361-0)

[15] Kim Y S and Noz M E 2014 Entropy and temperature from entangled space and time *Phys. Sci. Int. J.* **4** 1015–39

[16] Kim Y S and Wigner E P 1990 Entropy and Lorentz transformations *Phys. Lett.* A **147** 343–7

[17] Landau L D and Lifshitz E M 2008 *Statistical Physics, Part 1. Course of Theoretical Physics* (vol 5) 3rd edn (Amsterdam: Elsevier) (ISBN 978-0-7506-3372-7, First published in 1958; OCLC: 837367500)

[18] Lee T D 1954 Some special examples in renormalizable field theory *Phys. Rev.* **95** 1329–34

[19] Schweber S S 2005 *An Introduction to Relativistic Quantum Field Theory (Dover Books on Physics)* (New York: Dover Publications, Inc.) (ISBN 978-0-486-44228-0, Originally published 1961, New York, NY, USA: Harper & Row, Publishers)

[20] Tinkham M 2004 *Introduction to Superconductivity* (*Dover Books on Physics*) 2nd edn (Mineola, NY: Dover Publications) (ISBN 978-0-486-43503-9, Original Publication Krieger, Malabar; Florida; 1975; OCLC: 728146785)

[21] Van Dam H, Ng Y J, and Biedenharn L C 1985 A comment on fermionic tachyons and Poincaré representations *Phys. Lett.* B **158** 227–30

[22] Von Neumann J 1996 *Mathematical Foundations of Quantum Mechanics. Princeton Landmarks in Mathematics and Physics* (Princeton, NJ: Princeton University Press) (ISBN 978-0-691-02893-4 978-0-691-08003-1, Originally published: Princeton University Press, 1955, and in German by Springer, Berlin 1932; OCLC: ocm37904902)

[23] Yuen H P 1976 Two-photon coherent states of the radiation field *Phys. Rev.* A **13** 2226–43

[24] Yurke B, McCall S L, and Klauder J R 1986 SU(2) and SU(1,1) interferometers *Phys. Rev.* A **33** 4033–54

[25] Yurke B and Potasek M 1987 Obtainment of thermal noise from a pure quantum state *Phys. Rev.* A **36** 3464–6

IOP Publishing

Physics of the Lorentz Group (Second Edition)
Beyond high-energy physics and optics
Sibel Başkal, Young S Kim, and Marilyn E Noz

Chapter 10

Special relativity from quantum mechanics?

Heisenberg's uncertainty relation can be written in terms of the step-up and step-down operators in the harmonic oscillator representation and the Gaussian wave function plays the fundamental role in interpreting the uncertainty principle. It was noted in chapter 8 that the single-variable Heisenberg commutation relation can be represented as the minimum area in phase space in the Wigner phase-space picture of quantum mechanics. In this two-dimensional phase space, the uncertainty principle allows area-preserving linear transformations among which are rotations and squeezes. The group of these transformations is called the two-dimensional symplectic group or the group $Sp(2)$. This group is known to be locally isomorphic to the group $O(2, 1)$ or the Lorentz group applicable to two space dimensions with one time variable. Thus, the single-variable uncertainty relation already contains the basic elements of the Lorentz group.

As we noted in chapters 8 and 9, the system of two oscillators leads to the four-dimensional phase space with the four-dimensional symplectic group or $Sp(4)$. As Dirac noted first [4], this group is isomorphic to the $O(3, 2)$ de Sitter group, the Lorentz group applicable to three space and two time variables. We are quite familiar with the Lorentz group applicable to one time coordinate. Thus, the problem is how to deal with the second time variable. While the Lorentz group is generated by three rotation and three boost generators, this de Sitter group has three additional boost generators and one generator which leads to rotations between the two time variables. The question is how to convert these four extra variables to those applicable to Einstein's Lorentzian system.

In section 10.1, we show that Heisenberg's uncertainty relations lead to the Galilean system of space–time symmetry consisting of three translations and three rotations. This symmetry could be called the inhomogeneous rotation group or the Galilean group.

Likewise, the uncertainty relations in Einstein's Lorentzian system can be stated in terms of the Lie algebra of the inhomogeneous Lorentz group or the Poincaré group.

Its Lie algebra consists of three rotation generators, three boost generators, plus four space–time translation generators. This Poincaré group is known as the inhomogeneous Lorentz group [4], and the present form of quantum field theory is a representation of this inhomogeneous Lorentz group. An introduction to this group was given in section 1.4.

In section 10.2, it is noted that the best way to study the symmetry of the Heisenberg commutation relation is to use the Wigner function as the Gaussian function for the oscillator state. For the single oscillator, the uncertainty is defined in terms of the area in the phase space. Thus, the area-preserving transformations are of special interest.

Since this single oscillator contains the essential aspect of the Lorentz group, the system of two oscillators should give richer contents of the Lorentz group. The system of two oscillators was discussed in detail in chapter 9. In section 10.3, we consider again this two-oscillators system in preparation for converting it into the symmetry of the inhomogeneous Lorentz group. In section 10.4, we carry out the group contraction procedure to convert the four generators involving the second time variable into four translation variables. These four translation generators correspond to the energy–momentum variables satisfying

$$E^2 = p_x^2 + p_y^2 + p_z^2 + m^2. \tag{10.1}$$

This energy–momentum relation is known as Einstein's $E = mc^2$.

10.1 Definition of the problem

Heisenberg's uncertainty relation starts with the Poisson brackets in classical mechanics. The momentum variable p_i becomes a differential operator

$$p_i \rightarrow P_i = -i\frac{\partial}{\partial x_i}, \tag{10.2}$$

which leads to the commutation relations

$$\left[x_i, P_j\right] = i\,\delta_{ij}. \tag{10.3}$$

It is then possible to define

$$J_i = \varepsilon_{ijk} x_j P_k. \tag{10.4}$$

These three operators, then, satisfy the following closed set of commutation relations:

$$\left[J_i, J_j\right] = i\varepsilon_{ijk} J_k. \tag{10.5}$$

In quantum mechanics, P_i and J_i generate the translation along and the rotation around the ith direction, respectively. They satisfy the commutation relations

$$\left[J_i, P_j\right] = i\varepsilon_{ijk} P_k. \tag{10.6}$$

This set of commutation relations together with those given earlier in equations (10.3) and (10.5), which constitute a closed set for both P_i and J_i, form the Lie algebra of the Galilean group. Thus, the basic symmetry group for the Schrödinger or non-relativistic quantum mechanics is contained in this group. In other words, the Lie algebra is the statement of Heisenberg's uncertainty principle in the Galilean world.

Next, let us go to the Lorentzian world, where the time variable comes in as the 0th dimension. We call this time variable x_0. Then, in this notation, the Minkowski space–time four-vector becomes

$$x_\mu = (t, z, x, y) = (x_0, x_3, x_1, x_2). \tag{10.7}$$

Under Lorentz transformations, the Lorentz scalar is defined as

$$x_\mu^2 = t^2 - z^2 - x^2 - y^2 = x_0^2 - x_3^2 - x_1^2 - x_2^2, \tag{10.8}$$

which remains constant under Lorentz transformations.

The energy–momentum four-vector becomes extended to

$$p_\mu = (p_0, p_3, p_1, p_2) \tag{10.9}$$

with

$$p_\mu^2 = p_0^2 - p_3^2 - p_1^2 - p_2^2 = m^2 \quad \text{or} \quad p_0^2 = m^2 + p_3^2 + p_1^2 + p_2^2. \tag{10.10}$$

Here p_0 is the energy variable. The mass of the particle remains invariant under Lorentz transformations.

In addition to the operators applicable to the Galilean world, we need the time translation generator

$$P_0 = i\frac{\partial}{\partial t}, \tag{10.11}$$

with

$$[t, P_0] = -i. \tag{10.12}$$

In addition, we need the operators which will generate Lorentz boosts, and the generator of boosts along the ith direction is

$$K_i = i\left(x_i\frac{\partial}{\partial t} + t\frac{\partial}{\partial x_i}\right). \tag{10.13}$$

These boost generators are subject to the commutation relations

$$[K_i, K_j] = -i\varepsilon_{ijk}J_k. \tag{10.14}$$

Thus, these three K_i generators alone cannot constitute a Lie algebra (closed set of commutation relations). With P_i, they satisfy the relations

$$[P_i, K_j] = i\delta_{ij}P_i \quad \text{and} \quad [P_0, K_j] = -iP_i. \tag{10.15}$$

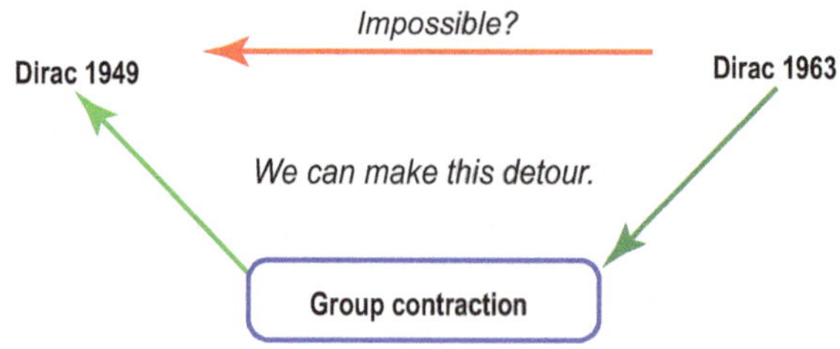

Figure 10.1. In his 1949 paper [3] Dirac said that the task of constructing a Lorentz-covariant quantum mechanics was accomplished essentially by constructing a representation of the inhomogeneous Lorentz group. Dirac then constructed, in his 1963 paper [4], the Lie algebra of the $O(3, 2)$ de Sitter group by using the algebra of two harmonic oscillators. This is a direct consequence of Heisenberg's uncertainty commutation relations. It is important then to be able to derive from $O(3, 2)$ the Lie algebra of the inhomogeneous Lorentz group. For this purpose, we use the group-contraction procedure of Inönü and Wigner [7] which was discussed in detail in chapter 3 [1].

Therefore, the commutation relations of equations (10.3), (10.5), (10.6), (10.12), (10.14), and (10.15) constitute a closed set of commutation relations for ten generators. This closed set is commonly called the Lie algebra of the Poincaré group or the inhomogeneous Lorentz group [4]. Indeed, this set states Heisenberg's uncertainty principle in the Lorentzian world.

The three rotation and the three translation generators are contained in or derivable from the Heisenberg commutation relations, and the Schrödinger equation provides the time translation operator. They are all Hermitian operators corresponding to dynamical variables. On the other hand, the three boost generators of equation (10.13) are not derivable from the Heisenberg commutation relations. Furthermore, they do not appear to correspond to any observable dynamical variables [3].

The purpose here is to show that the Lie algebra of the Poincaré symmetry is derivable from the Heisenberg commutation relations. To accomplish this we first examine the symmetry of the Heisenberg commutation relation using the Wigner function for the single harmonic oscillator in the two-dimensional phase space. As was noted in chapter 8, the single-variable relation contains the symmetry of the $O(2, 1)$ Lorentz group applicable to two space-like and one time-like dimensions.

In his 1963 paper [4], Dirac noted that two coupled harmonic oscillators lead to the symmetry of the $O(3, 2)$ de Sitter group which is the Lorentz group applicable to three space-like and two time-like dimensions. In this group, there are three additional boost generators with respect to the second time variable, and one rotation generator between the first and second time variables. We are thus led to the question of whether it is possible to convert these additional four generators into the four translation generators of the Poincaré group. For this purpose, we shall use the group contraction procedure discussed in chapter 3. This procedure is illustrated in figure 10.1.

10.2 Symmetries of the single oscillator

For a single Cartesian variable, Heisenberg's uncertainty relation takes the form

$$[x, P] = i, \tag{10.16}$$

with

$$P = -i\frac{\partial}{\partial x}. \tag{10.17}$$

It is frequently more convenient to use the annihilation and creation operators as defined in equation (1.65):

$$a = \frac{1}{\sqrt{2}}(x + iP) \quad \text{and} \quad a^{\dagger} = \frac{1}{\sqrt{2}}(x - iP) \tag{10.18}$$

with

$$[a, a^{\dagger}] = 1. \tag{10.19}$$

This aspect is well known. Using a and $a^{\dagger}$, this representation forms the harmonic oscillator representation of the uncertainty relation. Additionally, this representation forms the basic language for the Fock space for particle numbers and thus is the basic language for quantum optics.

If we next consider the quadratic forms of aa, $a^{\dagger}a^{\dagger}$, $aa^{\dagger}$, and $a^{\dagger}a$, the linear combination of these operators then gives

$$aa^{\dagger} - a^{\dagger}a = 1, \tag{10.20}$$

in accordance with the uncertainty relation. Thus, using the remaining three independent quadratic forms, it is possible to write the following two-by-two matrix:

$$\begin{pmatrix} (aa^{\dagger} + a^{\dagger}a)/2 & aa \\ a^{\dagger}a^{\dagger} & (aa^{\dagger} + a^{\dagger}a)/2 \end{pmatrix}. \tag{10.21}$$

From this matrix we can derive the following three independent operators:

$$J_2 = \frac{1}{2}(aa^{\dagger} + a^{\dagger}a), \quad K_1 = \frac{1}{2}(a^{\dagger}a^{\dagger} + aa), \quad \text{and} \quad K_3 = \frac{i}{2}(a^{\dagger}a^{\dagger} - aa), \tag{10.22}$$

which produce the following closed set of commutation relations:

$$[J_2, K_1] = -iK_3, \quad [J_2, K_3] = iK_1, \quad \text{and} \quad [K_1, K_3] = iJ_2. \tag{10.23}$$

This set forms the Lie algebra of the group commonly called $Sp(2)$, that is like (locally isomorphic to) the $O(2, 1)$ Lorentz group applicable to two space and one time coordinates.

Using the Wigner function for the ground-state oscillator is perhaps the best way to study the symmetry property of these operators [5, 6, 8, 10]. This Wigner function takes the form

$$W(x, p) = \frac{1}{\pi}\exp[-(x^2 + p^2)]. \tag{10.24}$$

Since this distribution is concentrated in the circular region around the origin, a circle can be defined as

$$x^2 + p^2 = 1. \tag{10.25}$$

The area of this circle can be used in the phase space of x and p for the minimum uncertainty which is preserved in the phase space under rotations and squeezing. We can write these transformations from equations (8.12) and (8.13) as:

$$\begin{pmatrix} \cos\alpha/2 & -\sin\alpha/2 \\ \sin\alpha/2 & \cos\alpha/2 \end{pmatrix}\begin{pmatrix} x \\ p \end{pmatrix} \quad \text{and} \quad \begin{pmatrix} e^{\eta/2} & 0 \\ 0 & e^{-\eta/2} \end{pmatrix}\begin{pmatrix} x \\ p \end{pmatrix}, \tag{10.26}$$

respectively. The rotation and the squeeze are generated by

$$J_2 = -i\left(x\frac{\partial}{\partial p} - p\frac{\partial}{\partial x}\right) \quad \text{and} \quad K_1 = -i\left(x\frac{\partial}{\partial p} + p\frac{\partial}{\partial x}\right). \tag{10.27}$$

The commutation relation of these two operators is:

$$[J_2, K_1] = -iK_3, \tag{10.28}$$

with

$$K_3 = -i\left(x\frac{\partial}{\partial x} - p\frac{\partial}{\partial p}\right). \tag{10.29}$$

Indeed, as we have seen before, these three generators form the set of closed commutation relations forming the Lie algebra of the $Sp(2)$ group, isomorphic to the $O(2, 1)$ Lorentz group. This isomorphic correspondence is illustrated in figure 10.2 which is the same as figure 8.1 used to illustrate the two-dimensional phase space.

In the Minkowski space of (t, z, x, y), three four-by-four matrices that satisfy the Lie algebra of equation (10.23) can be written as:

$$J_2 = \begin{pmatrix} 0 & 0 & 0 & 0 \\ 0 & 0 & -i & 0 \\ 0 & i & 0 & 0 \\ 0 & 0 & 0 & 0 \end{pmatrix}, \quad K_3 = \begin{pmatrix} 0 & i & 0 & 0 \\ i & 0 & 0 & 0 \\ 0 & 0 & 0 & 0 \\ 0 & 0 & 0 & 0 \end{pmatrix}, \quad \text{and} \quad K_1 = \begin{pmatrix} 0 & 0 & i & 0 \\ 0 & 0 & 0 & 0 \\ i & 0 & 0 & 0 \\ 0 & 0 & 0 & 0 \end{pmatrix}. \tag{10.30}$$

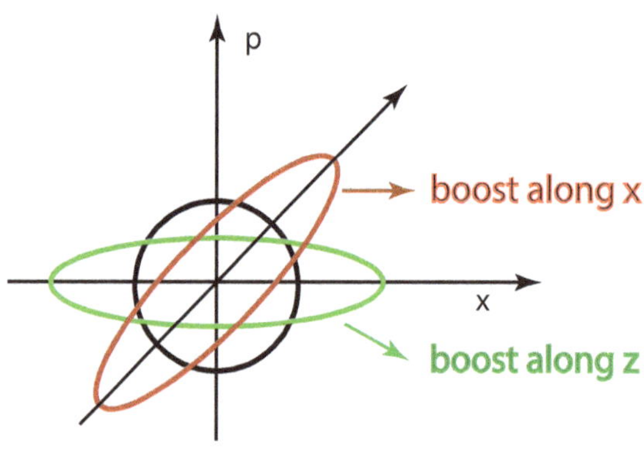

Figure 10.2. The transformations of $Sp(2)$ produce these rotations and squeezes. The Lorentz boost along the z-direction corresponds to the squeeze along the x-direction. The rotation by 90° around the y-axis corresponds to the rotation by 45° while the boost along the x-direction corresponds to the squeeze along the 45° angle [1].

The fourth row and column of each of these matrices all have zero elements. Thus, they are essentially three-by-three matrices applicable to the space–time of (t, z, x). The single oscillator cannot give the symmetry applicable to the full Minkowski space of (t, z, x, y). We shall thus consider two oscillators in section 10.3.

For the case of $O(2, 1)$, it was noted in chapter 3 that this group can be contracted to the two-dimensional Euclidean group. Let us examine this procedure again starting from $Sp(2)$ for the single oscillator in phase space, as illustrated in figure 10.2. Since the symmetry of $Sp(2)$ is similar to that of the $O(2, 1)$ group, we are interested in contracting $Sp(2)$ to the two-dimensional Euclidean group using the process that was discussed in chapter 3 for $O(2, 1)$. We want to study this procedure again in preparation for section 10.4. For this purpose, we use the squeeze matrix

$$C = \begin{pmatrix} \varepsilon & 0 & 0 & 0 \\ 0 & 1/\varepsilon & 0 & 0 \\ 0 & 0 & 1/\varepsilon & 0 \\ 0 & 0 & 0 & 1/\varepsilon \end{pmatrix}. \tag{10.31}$$

This matrix commutes with J_2. Let us squeeze K_3 first and take the small-ε limit:

$$C\, K_3\, C^{-1} = \begin{pmatrix} 0 & i\varepsilon^2 & 0 & 0 \\ i/\varepsilon^2 & 0 & 0 & 0 \\ 0 & 0 & 0 & 0 \\ 0 & 0 & 0 & 0 \end{pmatrix} \to K_3' = \begin{pmatrix} 0 & 0 & 0 & 0 \\ i/\varepsilon^2 & 0 & 0 & 0 \\ 0 & 0 & 0 & 0 \\ 0 & 0 & 0 & 0 \end{pmatrix}. \tag{10.32}$$

In the limit of small ε, the element on the top row becomes zero. After this procedure, we can squeeze back with $C^{-1}\, K_3'\, C$. Finally, we have

$$K_3 \to P_3 = \begin{pmatrix} 0 & 0 & 0 & 0 \\ i & 0 & 0 & 0 \\ 0 & 0 & 0 & 0 \\ 0 & 0 & 0 & 0 \end{pmatrix} \quad \text{and} \quad K_1 \to P_1 = \begin{pmatrix} 0 & 0 & 0 & 0 \\ 0 & 0 & 0 & 0 \\ i & 0 & 0 & 0 \\ 0 & 0 & 0 & 0 \end{pmatrix}, \tag{10.33}$$

and J_2 remains invariant.

These new operators satisfy the commutation relations

$$[P_3, P_1] = 0, \quad [J_2, P_1] = -iP_3, \quad \text{and} \quad [J_2, P_3] = iP_1. \tag{10.34}$$

This closed set of commentators is called the Lie algebra for the two-dimensional Euclidean group, or the group $E(2)$.

As for the four-vector, the squeeze transformation leads to

$$(t\varepsilon,\ z/\varepsilon,\ x/\varepsilon,\ y/\varepsilon). \tag{10.35}$$

In the limit of small ε, t can be changed to one. If this vector is transformed back, it becomes

$$(1,\ z,\ x,\ y). \tag{10.36}$$

This is the contracted four-vector.

The matrices P_3 and P_1 lead to the transformation matrix

$$\exp(-i[aP_3 + bP_1]) = \begin{pmatrix} 1 & 0 & 0 & 0 \\ a & 1 & 0 & 0 \\ b & 0 & 1 & 0 \\ 0 & 0 & 0 & 1 \end{pmatrix}. \tag{10.37}$$

If this matrix is applied to the vector $(1, z, x, y)$ we have,

$$\begin{pmatrix} 1 & 0 & 0 & 0 \\ a & 1 & 0 & 0 \\ b & 0 & 1 & 0 \\ 0 & 0 & 0 & 1 \end{pmatrix} \begin{pmatrix} 1 \\ z \\ x \\ y \end{pmatrix} = \begin{pmatrix} 1 \\ z + a \\ x + b \\ y \end{pmatrix}. \tag{10.38}$$

This is a translation matrix in the two-dimensional Euclidean plane, as illustrated in figure 10.3.

The procedure of eliminating the top row is called the contraction of $Sp(2)$ to $E(2)$ and is very similar to the contraction of $O(2, 1)$ to $E(2)$. Even though this procedure was detailed for $O(2, 1)$ in chapter 3, it is interesting to note that the phase space of a single oscillator produces three different representations sharing the same Lie algebra with the (2 + 1)-dimensional Lorentz group. This is shown in table 10.1.

10.3 Symmetries from two oscillators

It is necessary to have two Heisenberg commutation relations to generate Lorentz transformations applicable to the full Minkowski space. In 1963 [4], Dirac wrote the two uncertainty relations using two harmonic oscillators, where i and j could be 1 or 2, as:

$$\left[a_i, a_j^\dagger\right] = \delta_{ij}, \tag{10.39}$$

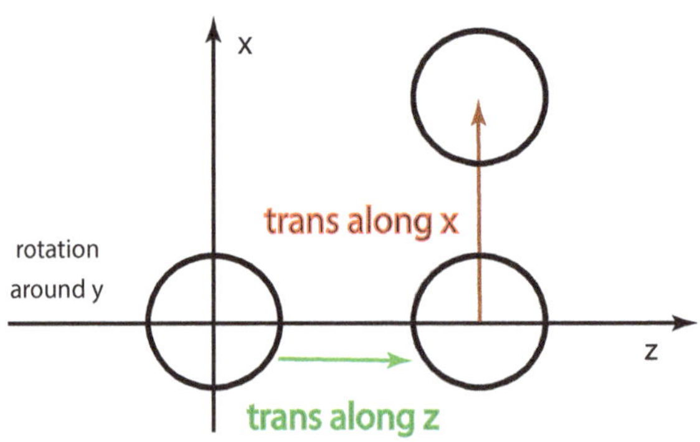

Figure 10.3. Contraction of $Sp(2)$ to $E(2)$. The rotation around the y-direction remains unchanged during this contraction procedure. The boost along the z- and x-directions becomes translations along the z- and x-directions, respectively.

Table 10.1. The generators, and transformation for the Gaussian function in terms of harmonic oscillators, two-dimensional phase space, and the four-dimensional Minkowski space with (t, z, x, y) [1].

Generators	Oscillator	Phase space	Lorentz
J_2	$\frac{1}{2}(aa^\dagger + a^\dagger a)$	$\frac{1}{2}\sigma_2$	$\begin{pmatrix} 0 & 0 & 0 & 0 \\ 0 & 0 & -i & 0 \\ 0 & i & 0 & 0 \\ 0 & 0 & 0 & 0 \end{pmatrix}$
K_3	$\frac{1}{2}(a^\dagger a^\dagger - aa)$	$\frac{i}{2}\sigma_1$	$\begin{pmatrix} 0 & i & 0 & 0 \\ i & 0 & 0 & 0 \\ 0 & 0 & 0 & 0 \\ 0 & 0 & 0 & 0 \end{pmatrix}$
K_1	$\frac{1}{2i}(a^\dagger a^\dagger + aa)$	$\frac{i}{2}\sigma_3$	$\begin{pmatrix} 0 & 0 & i & 0 \\ 0 & 0 & 0 & 0 \\ i & 0 & 0 & 0 \\ 0 & 0 & 0 & 0 \end{pmatrix}$

with

$$a_i = \frac{1}{\sqrt{2}}(x_i + ip_i), \qquad a_i^\dagger = \frac{1}{\sqrt{2}}(x_i - ip_i), \tag{10.40}$$

and

$$x_i = \frac{1}{\sqrt{2}}(a_i + a_i^\dagger), \qquad p_i = \frac{i}{\sqrt{2}}(a_i^\dagger - a_i). \tag{10.41}$$

Similarly to the two-by-two matrix given in equation (10.21), if the oscillators are not coupled the following four-by-four block-diagonal matrix can be written:

$$\begin{pmatrix} \frac{(a_1 a_1^\dagger + a_1^\dagger a_1)}{2} & a_1 a_1 & 0 & 0 \\ a_1^\dagger a_1^\dagger & \frac{(a_1 a_1^\dagger + a_1^\dagger a_1)}{2} & 0 & 0 \\ 0 & 0 & \frac{(a_2 a_2^\dagger + a_2^\dagger a_2)}{2} & a_2 a_2 \\ 0 & 0 & a_2^\dagger a_2^\dagger & \frac{(a_2 a_2^\dagger + a_2^\dagger a_2)}{2} \end{pmatrix}. \tag{10.42}$$

This matrix has six generators.

We wish now to couple them by filling in the off-diagonal blocks. The following two-by-two matrix, together with the Hermitian conjugate, is the most general form for this block:

$$\begin{pmatrix} a_1^{\dagger}a_2 & a_1a_2 \\ a_1^{\dagger}a_2^{\dagger} & a_1a_2^{\dagger} \end{pmatrix}. \tag{10.43}$$

This has four independent generators. Thus we obtain the following four-by-four matrix with ten (6 + 4) generators (table 10.2):

$$\begin{pmatrix} \frac{(a_1a_1^{\dagger} + a_1^{\dagger}a_1)}{2} & a_1a_1 & a_1^{\dagger}a_2 & a_1a_2 \\ a_1^{\dagger}a_1^{\dagger} & \frac{(a_1a_1^{\dagger} + a_1^{\dagger}a_1)}{2} & a_1^{\dagger}a_2^{\dagger} & a_1a_2^{\dagger} \\ a_1a_2^{\dagger} & a_1a_2 & \frac{(a_2a_2^{\dagger} + a_2^{\dagger}a_2)}{2} & a_2a_2 \\ a_1^{\dagger}a_2^{\dagger} & a_1^{\dagger}a_2 & a_2^{\dagger}a_2^{\dagger} & \frac{(a_2a_2^{\dagger} + a_2^{\dagger}a_2)}{2} \end{pmatrix}. \tag{10.44}$$

Table 10.2. Generators for the two-oscillator and phase space systems.

Generators	Two Oscillators	Phase Space
J_1	$\frac{1}{2}\left(a_1^{\dagger}a_2 + a_2^{\dagger}a_1\right)$	$-\frac{1}{2}\begin{pmatrix} 0 & I \\ I & 0 \end{pmatrix}\sigma_2$
J_2	$\frac{1}{2i}\left(a_1^{\dagger}a_2 - a_2^{\dagger}a_1\right)$	$\frac{i}{2}\begin{pmatrix} 0 & -I \\ I & 0 \end{pmatrix}I$
J_3	$\frac{1}{2}\left(a_1^{\dagger}a_1 - a_2^{\dagger}a_2\right)$,	$\frac{1}{2}\begin{pmatrix} -I & 0 \\ 0 & I \end{pmatrix}\sigma_2$
S_3	$\frac{1}{2}\left(a_1^{\dagger}a_1 + a_2a_2^{\dagger}\right)$,	$\frac{1}{2}\begin{pmatrix} I & 0 \\ 0 & I \end{pmatrix}\sigma_2$
K_1	$-\frac{1}{4}\left(a_1^{\dagger}a_1^{\dagger} + a_1a_1 - a_2^{\dagger}a_2^{\dagger} - a_2a_2\right)$	$\frac{i}{2}\begin{pmatrix} I & 0 \\ 0 & -I \end{pmatrix}\sigma_1$
K_2	$+\frac{i}{4}\left(a_1^{\dagger}a_1^{\dagger} - a_1a_1 + a_2^{\dagger}a_2^{\dagger} - a_2a_2\right)$	$\frac{i}{2}\begin{pmatrix} I & 0 \\ 0 & I \end{pmatrix}\sigma_3$
K_3	$\frac{1}{2}\left(a_1^{\dagger}a_2^{\dagger} + a_1a_2\right)$	$-\frac{i}{2}\begin{pmatrix} 0 & I \\ I & 0 \end{pmatrix}\sigma_1$
Q_1	$-\frac{i}{4}\left(a_1^{\dagger}a_1^{\dagger} - a_1a_1 - a_2^{\dagger}a_2^{\dagger} + a_2a_2\right)$	$-\frac{i}{2}\begin{pmatrix} I & 0 \\ 0 & -I \end{pmatrix}\sigma_3$
Q_2	$-\frac{1}{4}\left(a_1^{\dagger}a_1^{\dagger} + a_1a_1 + a_2^{\dagger}a_2^{\dagger} + a_2a_2\right)$	$\frac{i}{2}\begin{pmatrix} I & 0 \\ 0 & I \end{pmatrix}\sigma_1$
Q_3	$\frac{i}{2}\left(a_1^{\dagger}a_2^{\dagger} - a_1a_2\right)$	$\frac{1}{2}\begin{pmatrix} I & 0 \\ 0 & I \end{pmatrix}\sigma_2$

We can now construct, with these ten elements, the following four rotation-like generators:

$$J_1 = \frac{1}{2}\left(a_1^\dagger a_2 + a_2^\dagger a_1\right), \qquad J_2 = \frac{1}{2i}\left(a_1^\dagger a_2 - a_2^\dagger a_1\right),$$
$$J_3 = \frac{1}{2}\left(a_1^\dagger a_1 - a_2^\dagger a_2\right), \qquad S_3 = \frac{1}{2}\left(a_1^\dagger a_1 + a_2 a_2^\dagger\right), \tag{10.45}$$

and six squeeze-like generators:

$$K_1 = -\frac{1}{4}\left(a_1^\dagger a_1^\dagger + a_1 a_1 - a_2^\dagger a_2^\dagger - a_2 a_2\right),$$
$$K_2 = +\frac{i}{4}\left(a_1^\dagger a_1^\dagger - a_1 a_1 + a_2^\dagger a_2^\dagger - a_2 a_2\right), \tag{10.46}$$
$$K_3 = +\frac{1}{2}\left(a_1^\dagger a_2^\dagger + a_1 a_2\right),$$

and

$$Q_1 = -\frac{i}{4}\left(a_1^\dagger a_1^\dagger - a_1 a_1 - a_2^\dagger a_2^\dagger + a_2 a_2\right),$$
$$Q_2 = -\frac{1}{4}\left(a_1^\dagger a_1^\dagger + a_1 a_1 + a_2^\dagger a_2^\dagger + a_2 a_2\right), \tag{10.47}$$
$$Q_3 = +\frac{i}{2}(a_1^\dagger a_2^\dagger - a_1 a_2).$$

These ten operators from equations (10.45), (10.46), and (10.47), as noted by Dirac in 1963 [4], satisfy the following Lie algebra:

$$[J_i, J_j] = i\epsilon_{ijk}J_k, \quad [J_i, K_j] = i\epsilon_{ijk}K_k,$$
$$[J_i, Q_j] = i\epsilon_{ijk}Q_k, \quad [K_i, K_j] = [Q_i, Q_j] = -i\epsilon_{ijk}J_k,$$
$$[K_i, Q_j] = -i\delta_{ij}S_3, \quad [J_i, S_3] = 0, \tag{10.48}$$
$$[K_i, S_3] = -iQ_i, \quad [Q_i, S_3] = iK_i.$$

This set, which Dirac noted, is the same as the Lie algebra for the $O(3, 2)$ de Sitter group. These ten generators constitute the Lorentz group which is applicable to the three-dimensional space with two time variables and plays a very important role in space–time symmetries.

The five-by-five matrices which represent the ten generators for the $O(3, 2)$ group are applicable to (s, t, z, x, y), where s and t are time-like variables. They are explicitly given in table 8.3. The six matrices, $(J_1, J_2, J_3, K_1, K_2, K_3)$, which generate rotations and boosts in the (t, z, x, y) space, have zero elements on their first rows and first columns. This indicates that they leave the s coordinates invariant. Essentially, they are four-by-four matrices of the familiar Lorentz group $O(3, 1)$ in Minkowski space.

The three generators, Q_i, are for boosts along the s variable, and matrix S_3, generates rotations between the two time variables t and s. These matrices have elements only in the first row and first column. We are now interested in converting the Q_i and S_3 into the translation generators P_i and P_0 which are applicable to (t, z, x, y).

These four matrices can be rewritten in the following manner:

$$S_3 = \begin{pmatrix} 0 & -i & 0 & 0 & 0 \\ i & 0 & 0 & 0 & 0 \\ 0 & 0 & 0 & 0 & 0 \\ 0 & 0 & 0 & 0 & 0 \\ 0 & 0 & 0 & 0 & 0 \end{pmatrix} \rightarrow P_0 = \begin{pmatrix} 0 & 0 & 0 & 0 & 0 \\ i & 0 & 0 & 0 & 0 \\ 0 & 0 & 0 & 0 & 0 \\ 0 & 0 & 0 & 0 & 0 \\ 0 & 0 & 0 & 0 & 0 \end{pmatrix},$$
$$Q_3 = \begin{pmatrix} 0 & 0 & i & 0 & 0 \\ 0 & 0 & 0 & 0 & 0 \\ i & 0 & 0 & 0 & 0 \\ 0 & 0 & 0 & 0 & 0 \\ 0 & 0 & 0 & 0 & 0 \end{pmatrix} \rightarrow P_3 = \begin{pmatrix} 0 & 0 & 0 & 0 & 0 \\ 0 & 0 & 0 & 0 & 0 \\ i & 0 & 0 & 0 & 0 \\ 0 & 0 & 0 & 0 & 0 \\ 0 & 0 & 0 & 0 & 0 \end{pmatrix},$$
$$Q_1 = \begin{pmatrix} 0 & 0 & 0 & i & 0 \\ 0 & 0 & 0 & 0 & 0 \\ 0 & 0 & 0 & 0 & 0 \\ i & 0 & 0 & 0 & 0 \\ 0 & 0 & 0 & 0 & 0 \end{pmatrix} \rightarrow P_1 = \begin{pmatrix} 0 & 0 & 0 & 0 & 0 \\ 0 & 0 & 0 & 0 & i \\ 0 & 0 & 0 & 0 & 0 \\ i & 0 & 0 & 0 & 0 \\ 0 & 0 & 0 & 0 & 0 \end{pmatrix},$$
$$Q_2 = \begin{pmatrix} 0 & 0 & 0 & 0 & i \\ 0 & 0 & 0 & 0 & 0 \\ 0 & 0 & 0 & 0 & 0 \\ 0 & 0 & 0 & 0 & 0 \\ i & 0 & 0 & 0 & 0 \end{pmatrix} \rightarrow P_2 = \begin{pmatrix} 0 & 0 & 0 & 0 & 0 \\ 0 & 0 & 0 & 0 & 0 \\ 0 & 0 & 0 & 0 & 0 \\ 0 & 0 & 0 & 0 & 0 \\ i & 0 & 0 & 0 & 0 \end{pmatrix}. \tag{10.49}$$

These translation generators have zero elements on their first rows. We shall discuss this procedure in section 10.4.

10.4 Contraction of $O(3, 2)$ to the inhomogeneous Lorentz group

We can contract $O(3, 2)$ to the inhomogeneous Lorentz group. The procedure for group contraction was introduced in 1953 by Inönü and Wigner [7]. In chapter 3, their procedure was formulated in terms of squeeze transformations. In this section, we use those squeeze techniques to eliminate the elements in the first row of the matrices Q_i and S_3 of equation (10.49), in order to convert them to the translation generators P_i and P_0.

We introduce the squeeze matrices:

$$C(\varepsilon) = \begin{pmatrix} \varepsilon & 0 & 0 & 0 & 0 \\ 0 & 1/\varepsilon & 0 & 0 & 0 \\ 0 & 0 & 1/\varepsilon & 0 & 0 \\ 0 & 0 & 0 & 1/\varepsilon & 0 \\ 0 & 0 & 0 & 0 & 1/\varepsilon \end{pmatrix} \quad \text{and} \quad C^{-1}(\varepsilon) = \begin{pmatrix} 1/\varepsilon & 0 & 0 & 0 & 0 \\ 0 & \varepsilon & 0 & 0 & 0 \\ 0 & 0 & \varepsilon & 0 & 0 \\ 0 & 0 & 0 & \varepsilon & 0 \\ 0 & 0 & 0 & 0 & \varepsilon \end{pmatrix}, \tag{10.50}$$

where ε can become very small. Their transformations are illustrated in figure 10.4.

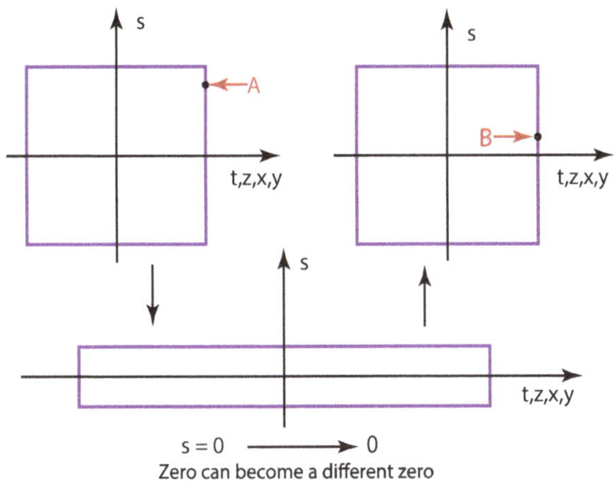

Figure 10.4. The Inönü–Wigner contraction procedure interpreted as squeeze transformations [7]. The square becomes a narrow rectangle during the squeeze process. When the rectangle becomes narrow enough, the point A can be moved to point B along the vertical axis. Then, the inverse squeeze brings back the rectangle to the original shape, but point A is moved to point B.

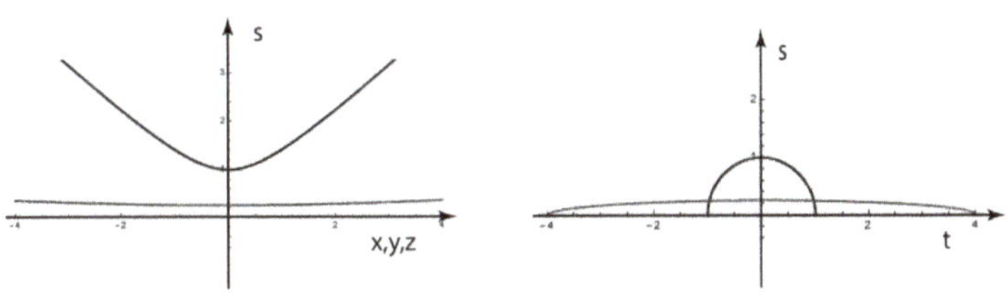

Figure 10.5. The Inönü–Wigner contraction procedure interpreted as squeeze transformations [7]. On the left, the hyperbola becomes straightened and becomes a line on the horizontal axis. It can become another horizontal line during the inverse squeeze with a desired value of s. On the right, the same process is carried out for the circular curve.

These matrices commute with J_i and K_i. The story is different for Q_i and S_3. As illustrated in figure 10.5, they can flatten the hyperbola for Q_i and flatten the circle for S_3.

For Q_3,

$$C\,Q_3\,C^{-1} = \begin{pmatrix} 0 & 0 & i\,\varepsilon^2 & 0 & 0 \\ 0 & 0 & 0 & 0 & 0 \\ i/\varepsilon^2 & 0 & 0 & 0 & 0 \\ 0 & 0 & 0 & 0 & 0 \\ 0 & 0 & 0 & 0 & 0 \end{pmatrix}, \tag{10.51}$$

which, in the limit of small ε, becomes

$$Q_3' = \begin{pmatrix} 0 & 0 & 0 & 0 & 0 \\ 0 & 0 & 0 & 0 & 0 \\ i/\varepsilon^2 & 0 & 0 & 0 & 0 \\ 0 & 0 & 0 & 0 & 0 \\ 0 & 0 & 0 & 0 & 0 \end{pmatrix}. \tag{10.52}$$

We then make the inverse squeeze transformation:

$$C^{-1}\,Q_3'\,C = \begin{pmatrix} 0 & 0 & 0 & 0 & 0 \\ 0 & 0 & 0 & 0 & 0 \\ i & 0 & 0 & 0 & 0 \\ 0 & 0 & 0 & 0 & 0 \\ 0 & 0 & 0 & 0 & 0 \end{pmatrix}. \tag{10.53}$$

Since there is only one element in this matrix, we can write this contraction procedure as

$$P_3 = \lim_{\varepsilon\to 0}\left(\varepsilon^2\,C\,Q_3\,C^{-1}\right), \tag{10.54}$$

where the explicit five-by-five matrix for P_3 is given in equation (10.49), and for the contraction procedure in equation (10.51). Likewise

$$P_1 = \lim_{\varepsilon\to 0}(\varepsilon^2\,C\,Q_1\,C^{-1}),\; P_2 = \lim_{\varepsilon\to 0}(\varepsilon^2\,C\,Q_2\,C^{-1}),\; P_0 = \lim_{\varepsilon\to 0}(\varepsilon^2\,C\,S_3\,C^{-1}). \tag{10.55}$$

These four contracted generators lead to the five-by-five transformation matrix, as can be seen from

$$\exp\{-i(aP_0 + bP_3 + cP_1 + dP_2)\} \tag{10.56}$$

performing translations in the four-dimensional Minkowski space:

$$\begin{pmatrix} 1 & 0 & 0 & 0 & 0 \\ a & 1 & 0 & 0 & 0 \\ b & 0 & 1 & 0 & 0 \\ c & 0 & 0 & 1 & 0 \\ d & 0 & 0 & 0 & 1 \end{pmatrix}\begin{pmatrix} 1 \\ t \\ z \\ x \\ y \end{pmatrix} = \begin{pmatrix} 1 \\ t+a \\ z+b \\ x+c \\ y+d \end{pmatrix}. \tag{10.57}$$

This contraction procedure is illustrated in figure 10.4 and 10.5. There, both the hyperbolic and circular curves can be flattened to horizontal lines with $s = 1$. This means, as illustrated in figure 10.1, that the group $O(3, 2)$, derivable from Heisenberg's uncertainty relations, becomes the inhomogeneous Lorentz group governing the Poincaré symmetry for quantum mechanics and quantum field theory.

Let us translate this into the language of differential operators with the translation operators:

$$P_0 = i\frac{\partial}{\partial t}, \quad P_z = -i\frac{\partial}{\partial z}, \quad P_x = -i\frac{\partial}{\partial x}, \quad P_y = -i\frac{\partial}{\partial y}. \tag{10.58}$$

Among the ten generators of Dirac's $O(3, 2)$, J_i and K_i are tabulated in table 10.3. In addition to the six quadratic forms given in table 10.3, Dirac constructed four additional quadratic forms. They correspond to the differential operators given in table 10.4.

Table 10.3. Three rotation generators and three boost generators among Dirac's ten quadratic forms. These six generators satisfy the Lie algebra of the Lorentz group applicable to (t, z, x, y).

Dirac Oscillators	Differential
$J_3 = \frac{1}{2}\left(a_1^\dagger a_1 - a_2^\dagger a_2\right)$	$-i\left(x\frac{\partial}{\partial y} - y\frac{\partial}{\partial x}\right)$
$J_1 = \frac{1}{2}\left(a_1^\dagger a_2 + a_2^\dagger a_1\right)$	$-i\left(y\frac{\partial}{\partial z} - z\frac{\partial}{\partial y}\right)$
$J_2 = \frac{1}{2i}\left(a_1^\dagger a_2 - a_2^\dagger a_1\right)$	$-i\left(z\frac{\partial}{\partial x} - x\frac{\partial}{\partial z}\right)$
$K_3 = \frac{1}{2}\left(a_1^\dagger a_2^\dagger + a_1 a_2\right)$	$-i\left(z\frac{\partial}{\partial t} + t\frac{\partial}{\partial z}\right)$
$K_1 = -\frac{1}{4}\left(a_1^\dagger a_1^\dagger + a_1 a_1 - a_2^\dagger a_2^\dagger - a_2 a_2\right)$	$-i\left(x\frac{\partial}{\partial t} + t\frac{\partial}{\partial x}\right)$
$K_2 = +\frac{i}{4}\left(a_1^\dagger a_1^\dagger - a_1 a_1 + a_2^\dagger a_2^\dagger - a_2 a_2\right)$	$-i\left(y\frac{\partial}{\partial t} + t\frac{\partial}{\partial y}\right)$

Table 10.4. Four additional quadratic forms in Dirac's paper of 1963 [4]. They correspond to the generators operating on the second time variable.

Dirac Oscillators	Differential	Contracted to
$S_3 = \frac{1}{2}\left(a_1^\dagger a_1 + a_2 a_2^\dagger\right)$	$-i\left(t\frac{\partial}{\partial s} - s\frac{\partial}{\partial t}\right)$	$i\frac{\partial}{\partial t}$
$Q_3 = \frac{i}{2}\left(a_1^\dagger a_2^\dagger - a_1 a_2\right)$	$-i\left(z\frac{\partial}{\partial s} + s\frac{\partial}{\partial z}\right)$	$-i\frac{\partial}{\partial z}$
$Q_1 = -\frac{i}{4}\left(a_1^\dagger a_1^\dagger - a_1 a_1 - a_2^\dagger a_2^\dagger + a_2 a_2\right)$	$-i\left(x\frac{\partial}{\partial s} + s\frac{\partial}{\partial x}\right)$	$-i\frac{\partial}{\partial x}$
$Q_2 = -\frac{1}{4}\left(a_1^\dagger a_1^\dagger + a_1 a_1 + a_2^\dagger a_2^\dagger + a_2 a_2\right)$	$-i\left(y\frac{\partial}{\partial s} + s\frac{\partial}{\partial y}\right)$	$-i\frac{\partial}{\partial y}$

The differential operators in table 10.4 do not depend on the time variable t, but depend only on the second time variable s. We are now interested in converting these differential forms in this table into four translation generators, using the group contraction procedure outlined in this section.

This contraction procedure tells us to fix the s time variable and set $s = 1$ for the generators given in table 10.4 [1, 9]. They then become contracted to the translation generators given also in table 10.4. Indeed, the six generators of the Lorentz group given in table 10.3 together with the four translation generators constitute the ten generators of the inhomogeneous Lorentz group or Einstein's Lorentzian system of space and time.

This contraction procedure produces four translation generators corresponding to the energy–momentum four-vector in the Lorentzian system. They lead to

$$-p_0^2 + p_3^2 + p_1^2 + p_2^2 = \text{constant}. \qquad (10.59)$$

Table 10.5. Three different systems. In the traditional Galilean system, there are three rotational and three translational degrees of freedom. In the Lorentzian system, four boost operations are possible and one additional translation, namely along the time direction. In 1963, Dirac constructed a space–time symmetry from two harmonic oscillators [4]. This was discussed chapter 9.

Systems	Galilean	Lorentzian	Two Oscillators
Rotations	$J_z, J_x, J_y.$	$J_z, J_x, J_y.$	$J_z, J_x, J_y.$
Boosts	None	$K_z, K_x, K_y.$	$K_z, K_x, K_y.$
Translations	$P_z, P_z, P_y.$	$P_t, P_z, P_x, P_y.$	S_3, Q_z, Q_x, Q_y contracted to $P_t, P_z, P_x, P_y.$

This energy–momentum relation is widely known as Einstein's $E = mc^2$ [1, 2].

We can now compare the three different systems, namely, Galilean, Lorentz, and Dirac's two-oscillator systems which leads to the system of the Poincaré or the inhomogeneous Lorentz group. They are compared in table 10.5.

References

[1] Başkal S, Kim Y S, and Noz M E 2019 Einstein's $E = mc^2$ derivable from Heisenberg's uncertainty relations *Quant. Rep.* **1** 236–51

[2] Başkal S, Kim Y S, and Noz M E 2019 Poincaré symmetry from Heisenberg's uncertainty relations *Symmetry* **11** 409

[3] Dirac P A M 1949 Forms of relativistic dynamics *Rev. Mod. Phys.* **21** 392–9

[4] Dirac P A M 1963 A remarkable representation of the 3 + 2 de Sitter group *J. Math. Phys.* **4** 901–9

[5] Dodonov V V and Man'ko V I 2003 *Theory of Nonclassical States of Light* (London; New York: Taylor & Francis) (ISBN 978-0-415-28413-4, OCLC: ocm49550380)

[6] Han D, Kim Y S, and Noz M E 1988 Linear canonical transformations of coherent and squeezed states in the Wigner phase space *Phys. Rev.* A **37** 807–14

[7] Inönü E and Wigner E P 1953 On the contraction of groups and their representations *Proc. Natl. Acad. Sci.* **39** 510–24

[8] Kim Y S and Noz M E 1991 *Phase Space Picture of Quantum Mechanics: Group Theoretical Approach* (*Lecture Notes in Physics* vol 40) (Singapore; Hackensack, NJ: World Scientific Publishing Co.) (ISBN 978-981-02-0360-3 978-981-02-0361-0)

[9] Kim Y S and Noz M E 2020 Integration of Dirac's efforts to construct a quantum mechanics which is Lorentz-covariant *Symmetry* **12** 1270

[10] Kim Y S and Wigner E P 1990 Canonical transformation in quantum mechanics *Am. J. Phys.* **58** 439–48

IOP Publishing

Physics of the Lorentz Group (Second Edition)
Beyond high-energy physics and optics
Sibel Başkal, Young S Kim, and Marilyn E Noz

Chapter 11

Lorentz group in ray optics

While the Lorentz group serves as the basic language for Einstein's special theory of relativity, it is possible to consider it as the essential mathematical instrument in optical sciences, both classical and quantum. This applies particularly in ray and polarization optics. The Lorentz group is based on four-by-four matrices. However, as discussed in chapter 1, it shares the same algebraic properties as $SL(2, c)$ with four complex elements or six real independent parameters. This group is the underlying language for the two-by-two beam transfer matrix, commonly called the *ABCD* matrix in optics. Here the group of *ABCD* matrices is shown to be a two-by-two representation of the Lorentz group. Thus, it is possible to study this group of *ABCD* matrices in terms of the mathematical device developed for studying Wigner's little groups [21], that dictate the internal space–time symmetries of particles in the Lorentz-covariant world. Wigner's little groups have been discussed extensively in chapters 2 and 4.

If from $SL(2, c)$, we choose the transformation matrices with real parameters, they form a subgroup $Sp(2)$ that was discussed in section 8.1. This subgroup shares the same algebraic property as the three-dimensional Lorentz group, $O(2, 1)$, applicable to two space-like and one time-like coordinates. The primary advantage of $Sp(2)$ is the mathematical simplicity and richness of mathematical content. It is easy to follow the logic based on two-by-two matrices, so $Sp(2)$ has become the standard language in classical and quantum optics.

Since the geometrical optics of laser cavities is a form of lens optics, laser optics is basically multilens lens optics. The problem here, though, is how to get a simple mathematical expression for the system of a large number of the same lens separated by equal distance [8]. In multilayer optics, two optical rays are moving in opposite directions, which also is a two-by-two matrix formalism. Similarly to laser cavities, the problem is that many matrix chains need to be multiplied together.

Here, we apply the concepts presented in the earlier chapters to the study of ray optics. The group of *ABCD* matrices is used to illustrate the similarities between ray optics and Wigner's little groups. In section 11.1 we show how the *ABCD* matrix,

doi:10.1088/978-0-7503-3607-9ch11

applied to ray optics, can be derived from Wigner's little groups. In section 11.2 we demonstrate the methods that can be used to render the *ABCD* matrices equi-diagonal. We must do this as the matrices for Wigner's little groups are equi-diagonal for massive, massless, and imaginary-mass particles. We use our knowledge of the Bargmann and Wigner decompositions in section 11.3, to decompose the *ABCD* matrices. We then discuss in section 11.4 lenses and how the *ABCD* matrices allow us to understand optical cavities. We look at the recomposition of the *ABCD* matrices, in section 11.5, using lens and translation matrices. In section 11.6, we illustrate how Wigner's little groups can be used to derive many characteristics of multilayer optical systems. Finally, in section 11.7, we develop camera optics in terms of the things we have learned in the preceding sections. We see that the camera forms an analogy to the tangential continuity that unites the representations of Wigner's little groups, as was done in section 4.6 and 4.7.

11.1 The group of *ABCD* matrices applied to ray optics

The Lorentz group for particle physics is generated by the six matrices given in table 1.1. Some of the generators are composed of real elements, which means the two-by-two transformation matrices derived from them have imaginary elements. However, the elements of the *ABCD* matrices are always real and unimodular (determinant = 1). Referring to table 1.1, we see that the elements of generators J_2, K_3, and K_1 are pure imaginary and lead to two-by-two transformation matrices with all real elements. These generators satisfy the closed set of commutation relations

$$[J_2, K_1] = -iK_3, \quad [J_2, K_3] = iK_1, \quad \text{and} \quad [K_1, K_3] = -iJ_2, \tag{11.1}$$

where

$$J_2 = \frac{i}{2}\begin{pmatrix} 0 & -1 \\ 1 & 0 \end{pmatrix}, \quad K_3 = \frac{i}{2}\begin{pmatrix} 1 & 0 \\ 0 & -1 \end{pmatrix}, \quad \text{and} \quad K_1 = \frac{i}{2}\begin{pmatrix} 0 & 1 \\ 1 & 0 \end{pmatrix}. \tag{11.2}$$

They form the Lie algebra for the $Sp(2)$ group [6, 10, 20].

The representation of the $Sp(2)$ group generated by these commutation relations consists of a rotation about the origin, a squeeze along the x-direction and another squeeze along an axis rotated by 45°, respectively. These generators are equivalent to another set of generators which form a representation of the $Sp(2)$ group consisting of two shear transformations and a squeeze transformation [2, 3]. Therefore we shall use a rotation, a squeeze, and a shear matrix of the form [4]:

$$R(\theta) = \begin{pmatrix} \cos(\theta/2) & -\sin(\theta/2) \\ \sin(\theta/2) & \cos(\theta/2) \end{pmatrix}, \quad B(\eta) = \begin{pmatrix} e^{\eta/2} & 0 \\ 0 & e^{-\eta/2} \end{pmatrix}, \tag{11.3}$$

and

$$T(\gamma) = \begin{pmatrix} 1 & -\gamma \\ 0 & 1 \end{pmatrix}. \tag{11.4}$$

The last matrix can also be viewed as an optical filter [13], or a translation matrix [3].

The traces of these matrices are smaller than 2, greater than 2, and equal to 2, respectively. In section 4.6, we discussed how to make transitions from one to another.

Since the *ABCD* matrix has real elements and a determinant of one, it has three independent parameters. The elements of the *ABCD* matrix are determined by optical materials and how they are arranged. The purpose of this chapter is to explore the mathematical properties of the *ABCD* matrix, which can address more fundamental issues in physics.

11.2 Equi-diagonalization of the *ABCD* matrix

In dealing with matrices, it is a routine process to diagonalize them. However, it is not always possible. The triangular matrix in equation (11.4) cannot be diagonalized. On the other hand, as the two diagonal elements are equal, we say that this matrix is equi-diagonal.

The *ABCD* matrix with inputs from optical instruments is not always equi-diagonal. Thus, we have to make the diagonal elements of the matrix equal; in other words we must equi-diagonalize the matrix. Let us start from the matrix

$$[ABCD] = \begin{pmatrix} A & B \\ C & D \end{pmatrix}, \tag{11.5}$$

where A and D are not necessarily equal to each other. Then the transformation

$$B(\eta)\ [ABCD]\ B(\eta) = \begin{pmatrix} e^{\eta/2} & 0 \\ 0 & e^{-\eta/2} \end{pmatrix} \begin{pmatrix} A & B \\ C & D \end{pmatrix} \begin{pmatrix} e^{\eta/2} & 0 \\ 0 & e^{-\eta/2} \end{pmatrix} \tag{11.6}$$

will lead to the equi-diagonal form

$$\begin{pmatrix} \sqrt{AD} & B \\ C & \sqrt{AD} \end{pmatrix}, \tag{11.7}$$

with

$$e^{\eta} = \sqrt{D/A}\,. \tag{11.8}$$

The diagonal matrix $B(\eta)$ is given in equation (11.3). This form of equi-diagonalization will be useful in camera optics discussed in section 11.7.

Although the transformation of equation (11.6) is a unimodular (determinant-preserving) transformation, it is not a similarity transformation. Applying the rotation matrix also given in equation (11.3) to the *ABCD* matrix we obtain:

$$\begin{aligned} R(\theta)\ [ABCD]\ R(-\theta) \ = & \begin{pmatrix} \cos(\theta/2) & -\sin(\theta/2) \\ \sin(\theta/2) & \cos(\theta/2) \end{pmatrix} \begin{pmatrix} A & B \\ C & D \end{pmatrix} \\ & \begin{pmatrix} \cos(\theta/2) & \sin(\theta/2) \\ -\sin(\theta/2) & \cos(\theta/2) \end{pmatrix}. \end{aligned} \tag{11.9}$$

This leads to the matrix

$$\begin{pmatrix} A' & B' \\ C' & D' \end{pmatrix}, \tag{11.10}$$

with

$$\begin{aligned} A' &= \frac{1}{2}[A(1 + \cos\theta) + D(1 - \cos\theta) - (C + B)\sin\theta], \\ D' &= \frac{1}{2}[A(1 - \cos\theta) + D(1 + \cos\theta) + (C + B)\sin\theta]. \end{aligned} \tag{11.11}$$

If these two diagonal elements are to be equal,

$$\tan\theta = \frac{A - D}{A + B}. \tag{11.12}$$

We now have two different ways of transforming the $ABCD$ matrix into an equi-diagonal form. Is it possible to make these into one mathematical transformation? Let M be an arbitrary element of the $Sp(2)$ group. We can define the *Hermitian transformation* of the $ABCD$ matrix as

$$M\ [ABCD]\ M^{\dagger}, \tag{11.13}$$

where $M^{\dagger}$ is the Hermitian conjugate of M.

The Hermitian transformation of equation (11.13) is like the Lorentz transformation on the four-vector discussed in chapter 8. If M is Hermitian and is anti-symmetric, its Hermitian conjugate is its inverse. Thus the transformation is a similarity transformation. If M is symmetric, its Hermitian conjugate is not its inverse. Thus the transformation is not a similarity transformation.

The rotation matrix of equation (11.3) is anti-symmetric and its Hermitian conjugate is its inverse. Thus, the transformation of equation (11.13) with the rotation matrix is a similarity transformation. The squeeze matrix, such as $B(\eta)$ is symmetric, and it is invariant under Hermitian conjugation. The Hermitian transformation of equation (11.13) with the squeeze matrix is not a similarity transformation.

Once the $ABCD$ matrix is equi-diagonalized, it can be brought to one of the forms [4, 5]:

$$\begin{aligned} &\begin{pmatrix} \cos(\theta/2) & -e^{\eta}\sin(\theta/2) \\ e^{-\eta}\sin(\theta/2) & \cos(\theta/2) \end{pmatrix}, \\ &\begin{pmatrix} 1 & -e^{\eta}\gamma \\ 0 & 1 \end{pmatrix}, \\ &\begin{pmatrix} \cosh(\lambda/2) & -e^{\eta}\sinh(\lambda/2) \\ -e^{-\eta}\sinh(\lambda/2) & \cosh(\lambda/2) \end{pmatrix}. \end{aligned} \tag{11.14}$$

We shall use the notation W' for these three equi-diagonal matrices. It is then possible to use all the mathematical instruments developed for the two-by-two representation of Wigner's little groups in chapter 4. The expressions for these two-by-two matrices are given in table 4.2. These three matrices, like the ones defined in equations (11.3) and (11.4) form different classes with different traces. These traces are smaller than, equal to, and greater than 2, respectively. However, it is possible to make continuations from one to another through tangential continuity, as noted in section 4.6.

11.3 Decomposition of the *ABCD* matrix

The equi-diagonal matrices of equation (11.14) can now be written as

$$B(\eta)\ W\ [B(\eta)]^{-1}, \tag{11.15}$$

where W is one of the three single-parameter matrices:

$$\begin{aligned} R(\theta) &= \begin{pmatrix} \cos(\theta/2) & -\sin(\theta/2) \\ \sin(\theta/2) & \cos(\theta/2) \end{pmatrix}, \\ T(\gamma) &= \begin{pmatrix} 1 & -\gamma \\ 0 & 1 \end{pmatrix}, \\ S(\lambda) &= \begin{pmatrix} \cosh(\lambda/2) & \sinh(\lambda/2) \\ \sinh(\lambda/2) & \cosh(\lambda/2) \end{pmatrix}. \end{aligned} \tag{11.16}$$

These matrices are given in table 4.1, but they perform different physical operations. The transformation of equation (11.15) is not a Hermitian transformation. It is a similarity transformation.

We have seen above that the two-by-two *ABCD* matrix can be written as a similarity transformation of one of the three possible W matrices given in equation (11.16). Thus, if the *ABCD* matrix is equi-diagonalized by a rotation, it can be written as a similarity transformation

$$[ABCD] = R(\sigma)B(\eta)\ W\ B(-\eta)R(-\sigma) = [R(\sigma)B(\eta)]\ W\ [R(\sigma)B(\eta)]^{-1}. \tag{11.17}$$

Thus, the repeated application of the *ABCD* matrix becomes

$$[ABCD]^N = [R(\sigma)B(\eta)]\ W^N\ [R(\sigma)B(\eta)]^{-1}. \tag{11.18}$$

There is another form of decomposition known as the Bargmann decomposition discussed in section 4.4. This procedure combines all three different classes of equation (11.14) into one analytic expression:

$$W_B = R(\alpha)S(-2\chi)R(\alpha), \tag{11.19}$$

where the forms of the rotation matrix R and the squeeze matrix S are given in equation (11.16). If we carry out the matrix multiplication, the W_B matrix becomes

$$\begin{pmatrix} (\cosh\chi)\cos\alpha & -\sinh\chi - (\cosh\chi)\sin\alpha \\ -\sinh\chi + (\cosh\chi)\sin\alpha & (\cosh\chi)\cos\alpha \end{pmatrix}. \tag{11.20}$$

This matrix also has two independent variables α and χ. We can write these parameters in terms of the η, θ, and γ parameters of the matrices given in equation (11.14) by comparing the matrix elements.

If the off-diagonal elements have different signs, with $(\cosh\chi)\sin\alpha > \sinh\chi$, the diagonal elements become

$$\cos(\theta/2) = (\cosh\chi)\cos\alpha. \tag{11.21}$$

The off-diagonal elements lead to

$$e^{2\eta} = \frac{(\cosh\chi)\sin\alpha + \sinh\chi}{(\cosh\chi)\sin\alpha - \sinh\chi}. \tag{11.22}$$

If the off-diagonal elements have the same sign, the diagonal elements become

$$\cosh(\lambda/2) = (\cosh\chi)\cos\alpha, \tag{11.23}$$

with

$$e^{2\eta} = \frac{(\cosh\chi)\sin\alpha + \sinh\chi}{\sinh\chi - (\cosh\chi)\sin\alpha}. \tag{11.24}$$

If the lower left element of equation (11.20) vanishes, the matrix becomes

$$\begin{pmatrix} 1 & -2\sinh\chi \\ 0 & 1 \end{pmatrix}. \tag{11.25}$$

The matrix of equation (11.20) is useful when we make a transition from one form to another among the three matrices given in equation (11.14). We will need this procedure when we discuss multilayer optics.

11.4 Laser cavities

As the first example of the periodic system, let us consider the laser cavity consisting of two identical concave mirrors separated by a distance d, as shown in figure 11.1. Then the $ABCD$ matrix for a round trip of one beam is

$$\begin{pmatrix} 1 & 0 \\ -2/R & 1 \end{pmatrix}\begin{pmatrix} 1 & d \\ 0 & 1 \end{pmatrix}\begin{pmatrix} 1 & 0 \\ -2/R & 1 \end{pmatrix}\begin{pmatrix} 1 & d \\ 0 & 1 \end{pmatrix}, \tag{11.26}$$

where the matrices

$$\begin{pmatrix} 1 & 0 \\ -2/R & 1 \end{pmatrix} \quad \text{and} \quad \begin{pmatrix} 1 & d \\ 0 & 1 \end{pmatrix} \tag{11.27}$$

are the mirror and translation matrices, respectively. The parameters R and d are the radius of the mirror and the mirror separation, respectively. This form is quite familiar to us from the laser literature [14, 15, 22].

The question then is what happens when this process is repeated. We are thus led to the question of whether the chain of matrices in equation (11.26) can be brought to an equi-diagonal form and eventually to a form of the Wigner decomposition also

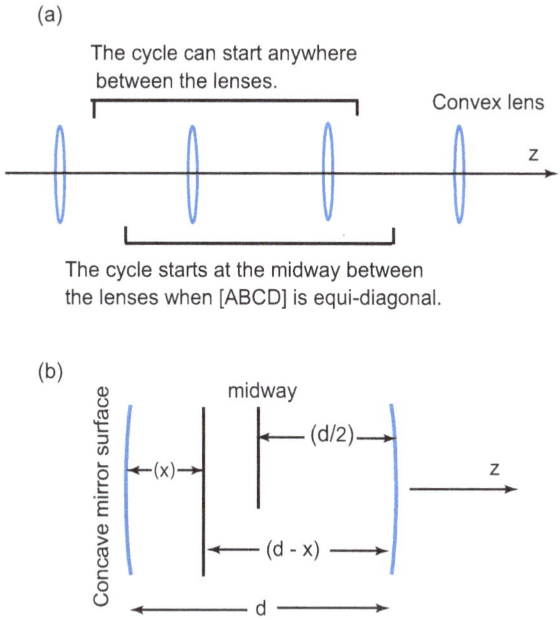

Figure 11.1. Optical rays in the laser cavity. The reflected beams are like those going through a series of convex lenses, as illustrated in (a). How the beams are reflected from concave mirror surfaces is described in (b) [6].

discussed in section 4.4. For this purpose, we rewrite the matrix of equation (11.26) as

$$\begin{pmatrix} 1 & -d/2 \\ 0 & 1 \end{pmatrix}\begin{pmatrix} 1 & d/2 \\ 0 & 1 \end{pmatrix}\begin{pmatrix} 1 & 0 \\ -2/R & 1 \end{pmatrix}\begin{pmatrix} 1 & d/2 \\ 0 & 1 \end{pmatrix}^2 \times \begin{pmatrix} 1 & 0 \\ -2/R & 1 \end{pmatrix}\begin{pmatrix} 1 & d/2 \\ 0 & 1 \end{pmatrix}\begin{pmatrix} 1 & d/2 \\ 0 & 1 \end{pmatrix}. \tag{11.28}$$

In this way, we translate the system by $-d/2$ using a translation matrix given in equation (11.27), and write the *ABCD* matrix of equation (11.26) as

$$\begin{pmatrix} 1 & -d/2 \\ 0 & 1 \end{pmatrix}\left[\begin{pmatrix} 1 - d/R & d - d^2/2R \\ -2/R & 1 - d/R] \end{pmatrix}\right]^2\begin{pmatrix} 1 & d/2 \\ 0 & 1 \end{pmatrix}. \tag{11.29}$$

We are thus led to concentrate on the matrix in the middle

$$\begin{pmatrix} 1 - d/R & d - d^2/2R \\ -2/R & 1 - d/R \end{pmatrix}, \tag{11.30}$$

which can be written as

$$\begin{pmatrix} \sqrt{d} & 0 \\ 0 & 1/\sqrt{d} \end{pmatrix}\begin{pmatrix} 1 - d/R & 1 - d/2R \\ -2d/R & 1 - d/R \end{pmatrix}\begin{pmatrix} 1/\sqrt{d} & 0 \\ 0 & \sqrt{d} \end{pmatrix}. \tag{11.31}$$

It is then possible to decompose the $ABCD$ matrix into

$$E \;\; C^2 \;\; E^{-1}, \tag{11.32}$$

with

$$C = \begin{pmatrix} 1 - d/R & 1 - d/2R \\ -2d/R & 1 - d/R \end{pmatrix},$$
$$E = \begin{pmatrix} 1 & -d/2 \\ 0 & 1 \end{pmatrix} \begin{pmatrix} \sqrt{d} & 0 \\ 0 & 1/\sqrt{d} \end{pmatrix}. \tag{11.33}$$

The C matrix now contains only dimensionless numbers, and it can be written as

$$C = \begin{pmatrix} \cos(\gamma/2) & e^{\eta}\sin(\gamma/2) \\ -e^{-\eta}\sin(\gamma/2) & \cos(\gamma/2) \end{pmatrix}, \tag{11.34}$$

with

$$\cos(\gamma/2) = 1 - \frac{d}{R} \quad \text{and} \quad e^{\eta} = \sqrt{\frac{2R - d}{4d}}. \tag{11.35}$$

Here both d and R are positive, and the restriction on them is that d be smaller than $2R$. This is the stability condition frequently mentioned in the literature [14, 15].

Thus, the equi-diagonal matrix is C^2 and takes the form

$$\begin{pmatrix} \cos(\gamma) & e^{\eta}\sin(\gamma) \\ -e^{-\eta}\sin(\gamma) & \cos(\gamma) \end{pmatrix}, \tag{11.36}$$

which is

$$B(\eta)R(2\gamma)B(-\eta). \tag{11.37}$$

The $ABCD$ matrix takes the form

$$[ABCD] = [E \;\; B(\eta)] \;\; R(2\chi) \;\; [E \;\; B(\eta)]^{-1}. \tag{11.38}$$

Thus, the repeated application becomes

$$[ABCD]^N = [E \;\; B(\eta)] \;\; R(2N\chi) \;\; [E \;\; B(\eta)]^{-1}. \tag{11.39}$$

11.5 Composition of lens and translation matrices

In section 11.3, we decomposed the $ABCD$ matrix. Now we would like to recompose it in terms of lens and translation matrices. To best demonstrate this, let us use the following physical example. In para-axial lens optics, the lens and translation matrices take the form

$$L(f) = \begin{pmatrix} 1 & 0 \\ -1/f & 1 \end{pmatrix} \quad \text{and} \quad T(d) = \begin{pmatrix} 1 & d \\ 0 & 1 \end{pmatrix}. \tag{11.40}$$

The focal length of the lens is f, while the distance between lenses is given by d.

These matrices apply to the two-dimensional space of $(y, m)^T$. Here the height of the ray from the propagation axis is measured by y, and the slope of the ray is given by m. Multiplications are converted into additions by both L and T. We obtain the following computations if we restrict ourselves to L-type or T-type matrices:

$$L(f_1 + f_2) = L(f_1)\,L(f_2) \quad \text{and} \quad T(d_1 + d_2) = T(d_1)\,T(d_2). \tag{11.41}$$

For a one-lens system, the chain consists of TLT, while the form $TLTLT$ is for a two-lens system. As more lenses are included, the chain becomes longer. The net result will be, however, one $ABCD$ matrix with three real independent parameters. Previously we saw that this constitutes a representation of the $Sp(2)$ group. We can express the most general form of this as $W''(\phi, \rho, \eta) = R(\phi)B(\eta)R(\rho)$. It can be more explicitly written as:

$$W''(\phi, \rho, \eta) = \begin{pmatrix} \cos\phi & -\sin\phi \\ \sin\phi & \cos\phi \end{pmatrix} \begin{pmatrix} e^{\eta} & 0 \\ 0 & e^{-\eta} \end{pmatrix} \begin{pmatrix} \cos\rho & -\sin\rho \\ \sin\rho & \cos\rho \end{pmatrix}. \tag{11.42}$$

Here we consider the Bargmann decomposition given in section 4.4. It is possible to note that $R(\rho)$ can be decomposed as $R(\rho) = R(-\phi)R(\theta)$. Here $R(\theta)$ is as in table 1.1. Thus in terms of ϕ and θ, the last matrix in equation (11.42) becomes:

$$R(\rho) = \begin{pmatrix} \cos\phi & \sin\phi \\ -\sin\phi & \cos\phi \end{pmatrix} \begin{pmatrix} \cos\theta & -\sin\theta \\ \sin\theta & \cos\theta \end{pmatrix}, \tag{11.43}$$

with $\rho = \theta - \phi$. Therefore, we have

$$W''(\phi, \theta, \eta) = R(\phi)B(\eta)R(-\phi)R(\theta) \tag{11.44}$$

where now θ is an independent parameter, instead of the angle ρ.

The W'' matrix can then be obtained from the product of one symmetric matrix, S, and the other an orthogonal matrix, R. Explicitly, $W''(\phi, \theta, \eta) = S(\phi, \eta)R(\theta)$, with $S(\phi, \eta) = R(\phi)B(\eta)R(-\phi)$. We then write the symmetric matrix S in the form [13]:

$$S(\phi, \eta) = \begin{pmatrix} \cosh\eta + (\sinh\eta)\cos(2\phi) & (\sinh\eta)\sin(2\phi) \\ (\sinh\eta)\sin(2\phi) & \cosh\eta - (\sinh\eta)\cos(2\phi) \end{pmatrix}. \tag{11.45}$$

Next, we proceed to write S and R separately as L and T chains. To accomplish this, we consider first the rotation matrix. This can be obtained through the multiplication of L-type and T-type matrices as [16]:

$$R(\theta) = \begin{pmatrix} 1 & -\tan(\theta/2) \\ 0 & 1 \end{pmatrix} \begin{pmatrix} 1 & 0 \\ \sin\theta & 1 \end{pmatrix} \begin{pmatrix} 1 & -\tan(\theta/2) \\ 0 & 1 \end{pmatrix}. \tag{11.46}$$

The above expression is written in the form of TLT, but it can also have the form of LTL. By taking the transpose and changing the sign of θ, we obtain R as:

$$R'(\theta) = \begin{pmatrix} 1 & 0 \\ \tan(\theta/2) & 1 \end{pmatrix} \begin{pmatrix} 1 & -\sin\theta \\ 0 & 1 \end{pmatrix} \begin{pmatrix} 1 & 0 \\ \tan(\theta/2) & 1 \end{pmatrix}. \tag{11.47}$$

Both R and R' are the same matrix, but they have been decomposed in different ways.

The two-parameter symmetric matrix of equation (11.45) can be started with the form $LTLT$ [2]:

$$S(a, b) = \begin{pmatrix} 1 & 0 \\ b & 1 \end{pmatrix}\begin{pmatrix} 1 & a \\ 0 & 1 \end{pmatrix}\begin{pmatrix} 1 & 0 \\ a & 1 \end{pmatrix}\begin{pmatrix} 1 & b \\ 0 & 1 \end{pmatrix}. \tag{11.48}$$

We can then combine this into one symmetric matrix:

$$S(a, b) = \begin{pmatrix} 1 + a^2 & b(1 + a^2) + a \\ b(1 + a^2) + a & 1 + 2ab + b^2(1 + a^2) \end{pmatrix}. \tag{11.49}$$

If equations (11.45) and (11.49) are compared, then the parameters a and b can be computed in terms of η and ϕ. The result we get is:

$$\begin{aligned} a &= \pm\sqrt{(\cosh\eta - 1) + (\sinh\eta)\cos(2\phi)}, \\ b &= \frac{(\sinh\eta)\sin(2\phi) \mp \sqrt{(\cosh\eta - 1) + (\sinh\eta)\cos(2\phi)}}{\cosh\eta + (\sinh\eta)\cos(2\phi)}. \end{aligned} \tag{11.50}$$

We can also write this matrix in a $TLTL$ form:

$$S'(a', b') = \begin{pmatrix} 1 & b' \\ 0 & 1 \end{pmatrix}\begin{pmatrix} 1 & 0 \\ a' & 1 \end{pmatrix}\begin{pmatrix} 1 & a' \\ 0 & 1 \end{pmatrix}\begin{pmatrix} 1 & 0 \\ b' & 1 \end{pmatrix}. \tag{11.51}$$

The parameters a' and b' then can be written as:

$$\begin{aligned} a' &= \pm\sqrt{(\cosh\eta - 1) - (\sinh\eta)\cos(2\phi)}, \\ b' &= \frac{(\sinh\eta)\sin(2\phi) \mp \sqrt{(\cosh\eta - 1) - (\sinh\eta)\cos(2\phi)}}{\cosh\eta - (\sinh\eta)\cos(2\phi)}. \end{aligned} \tag{11.52}$$

The sign of the parameter η is the only difference between the two sets of parameters ab and $a'b'$. The meaning of this sign change is that the squeeze operation is in the direction perpendicular to the original direction. When we choose either ab or $a'b'$, it is also necessary to make sure the sign of the quantity inside the square root is positive. Both sets are acceptable when $\cos(2\phi)$ is sufficiently small, but if the absolute value of $(\sinh\eta)\cos(2\phi)$ is greater than $(\cosh\eta - 1)$, only one of ab or $a'b'$ is valid.

If we now combine the S and R matrices, it is possible to construct the $ABCD$ matrix. By doing this, we reduce the number of matrices by one as:

$$\begin{aligned} SR = &\begin{pmatrix} 1 & 0 \\ b & 1 \end{pmatrix}\begin{pmatrix} 1 & a \\ 0 & 1 \end{pmatrix}\begin{pmatrix} 1 & 0 \\ a & 1 \end{pmatrix}\begin{pmatrix} 1 & b - \tan(\theta/2) \\ 0 & 1 \end{pmatrix} \\ &\times \begin{pmatrix} 1 & 0 \\ \sin\theta & 1 \end{pmatrix}\begin{pmatrix} 1 & -\tan(\theta/2) \\ 0 & 1 \end{pmatrix}. \end{aligned} \tag{11.53}$$

We can combine as well the product $S'R'$, which becomes

$$\begin{aligned} S'R' &= \begin{pmatrix} 1 & b' \\ 0 & 1 \end{pmatrix} \begin{pmatrix} 1 & 0 \\ a' & 1 \end{pmatrix} \begin{pmatrix} 1 & a' \\ 0 & 1 \end{pmatrix} \begin{pmatrix} 1 & 0 \\ b' + \tan(\theta/2) & 1 \end{pmatrix} \\ &\times \begin{pmatrix} 1 & -\sin\theta \\ 0 & 1 \end{pmatrix} \begin{pmatrix} 1 & 0 \\ \tan(\theta/2) & 1 \end{pmatrix}. \end{aligned} \tag{11.54}$$

Two adjoining T matrices were combined into one T matrix to reduce SR of equation (11.53). To simplify $S'R'$ of equation (11.54), two L matrices were combined into one. We see in both combinations that there are six matrices, which consist of three T and three L matrices. The conclusion we have come to is that for the most general form of the $ABCD$ matrix, we require a minimum number of L and T-type matrices, which is six.

In para-axial optics, special forms of the $ABCD$ matrix are often encountered. For instance, for pure magnification [12], the squeeze matrix $B(2\eta)$ in equation (11.42) is required. This has $\phi = 0$, and is thus a special case of the decomposition given for S and S' in equations (11.49) and (11.51), respectively. The set $a'b'$ is not acceptable if η is positive because the quantity in the square root in equation (11.52) becomes negative. For the ab set,

$$a = \pm(e^{\eta} - 1)^{1/2} \quad \text{and} \quad b = \mp e^{-\eta}(e^{\eta} - 1)^{1/2}. \tag{11.55}$$

Mentioned in reference to equation (4.4), the L and T matrices of equation (11.40) are products of Iwasawa decompositions. By implementing the inverse process of the Iwasawa decomposition, using solely the products of L and T matrices, we obtained the most general form of the $ABCD$ matrix. Therefore, we have the recomposition of the $ABCD$ matrix.

11.6 Optical beam propagation through multilayers

Here, we take into account an optical beam propagating through a cascaded medium, comprising two individual successive layers, with each layer having its own refractive index n_1 and n_2. Partial reflection and partial transmission are observed while the beam goes from the first medium to the second one. We define the electric fields as

$$\begin{aligned} E_1^{(\pm)} &= \frac{1}{\sqrt{n_1}} \exp(\pm ik_1 z - \omega t), \\ E_2^{(\pm)} &= \frac{1}{\sqrt{n_2}} \exp(\pm ik_2 z - \omega t) \end{aligned} \tag{11.56}$$

for the optical beams in the first and second media respectively. It is clear that, characterized in this form, the Poynting vector of the field is kept continuous. The incoming and the reflected rays are denoted by the superscripts (+) and (-), respectively.

The two-by-two $ABCD$ matrix,

$$\begin{pmatrix} E_2^{(+)} \\ E_2^{(-)} \end{pmatrix} = \begin{pmatrix} A & B \\ C & D \end{pmatrix} \begin{pmatrix} E_1^{(+)} \\ E_1^{(-)} \end{pmatrix} \tag{11.57}$$

associates the incoming and the reflected beams of the second medium to those of the first. The components of this matrix are expressed in terms of the transmission coefficients and the phase shifts of the beams as they propagate through the media [1, 10]. We consider the boundary matrix

$$S(\sigma) = \begin{pmatrix} \cosh(\sigma/2) & \sinh(\sigma/2) \\ \sinh(\sigma/2) & \cosh(\sigma/2) \end{pmatrix} \tag{11.58}$$

to deal with the behavior of the beams at the interface, introduced by Azzam and Bashara [1] and Monzón and Sánchez-Soto [9, 17, 18]. The σ parameter can be expressed in terms of the refractive indices as

$$\cosh\left(\frac{\sigma}{2}\right) = \frac{n_1 + n_2}{2\sqrt{n_1 n_2}} \quad \text{and} \quad \sinh\left(\frac{\sigma}{2}\right) = \frac{n_1 - n_2}{2\sqrt{n_1 n_2}}. \tag{11.59}$$

Accordingly, the boundary matrix for the beam originating from the second medium becomes $S(-\sigma)$.

Moreover, beams experience phase shifts as they propagate; so we bring in the phase-shift matrix

$$P(\delta_1) = \begin{pmatrix} e^{-i\delta_1/2} & 0 \\ 0 & e^{i\delta_1/2} \end{pmatrix}, \tag{11.60}$$

for the beam going through the first medium. The expression of the phase-shift matrix for the second medium takes a similar form as $P(\delta_2)$, where the value of δ depends on the wave-number and on the thickness of the medium.

Now, instead of the beam starting from a randomly located input plane to propagate through some periodic media, we shall adopt the following approach, which will render the behavior of the beam, passing through a cascaded system composed of a large number of periodic media, tractable. Consider the beam originating from an input plane located at the midpoint of medium two and reaching again to medium two after traversing medium one in between, while experiencing the effects of the boundaries and phase shifts. This makes one complete cycle, represented by the sequence of matrices as follows:

$$M_I = P(\delta_2/2)S(\sigma)P(\delta_1)S(-\sigma)P(\delta_2/2), \tag{11.61}$$

as can be seen from figure 11.2.

In order to deal with N cycles, we have to check whether the result of the above matrix multiplication will be in an equi-diagonal form to accept Wigner and Bargmann decompositions. To this end, the matrices in equation (11.61) should be converted into matrices with real elements, which can be done by way

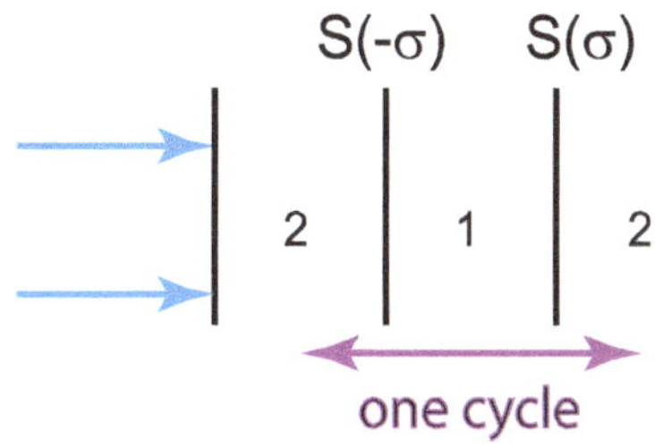

Figure 11.2. Multilayers. The cycle starts from the midpoint of medium two so that the ABCD matrix becomes equi-diagonal [10, 11].

of two successive similarity transformations. The first transformation to be done is

$$C_1 \ P(\delta)S(\sigma) \ C_1^{-1}, \tag{11.62}$$

with

$$C_1 = \frac{1}{\sqrt{2}}\begin{pmatrix} 1 & i \\ i & 1 \end{pmatrix}. \tag{11.63}$$

Thus, the chain in equation (11.62) yields

$$R(\delta)S(\sigma), \tag{11.64}$$

where

$$R(\delta) = \begin{pmatrix} \cos(\delta/2) & -\sin(\delta/2) \\ \sin(\delta/2) & \cos(\delta/2) \end{pmatrix}. \tag{11.65}$$

This rotation matrix has already been employed in section 11.3. We have to bring in another similarity transformation, using

$$C_2 = \frac{1}{\sqrt{2}}\begin{pmatrix} 1 & 1 \\ -1 & 1 \end{pmatrix}, \tag{11.66}$$

which transforms $S(\sigma)$ into $B(\sigma)$, while keeping $R(\delta)$ unaltered. Here

$$B(\sigma) = \begin{pmatrix} e^{\sigma/2} & 0 \\ 0 & e^{-\sigma/2} \end{pmatrix}, \tag{11.67}$$

and again is in the same form as the $B(\eta)$ matrix treated in section 11.3.

Thus the final form of the similarity transformation matrix is [10]

$$C = C_2 C_1 = \frac{1}{\sqrt{2}}\begin{pmatrix} e^{i\pi/4} & e^{i\pi/4} \\ -e^{-i\pi/4} & e^{-i\pi/4} \end{pmatrix}, \tag{11.68}$$

with

$$C^{-1} = \frac{1}{\sqrt{2}}\begin{pmatrix} e^{-i\pi/4} & -e^{i\pi/4} \\ e^{-i\pi/4} & e^{i\pi/4} \end{pmatrix}. \tag{11.69}$$

When this similarity transformation is applied to M_I of equation (11.61), it yields another chain

$$M_R = R(\delta_2/2)B(\sigma)R(\delta_1)B(-\sigma)R(\delta_2/2), \tag{11.70}$$

where all the matrices are now real.

Our main purpose is to express M_I, initially governing one whole cycle, as an equi-diagonal matrix. Since there is this rotation matrix $R(\delta_2/2)$ at the beginning and at end of the chain in equation (11.70), it may not be possible to write M_R as a similarity transformation. In order to circumvent this problem, we first note that the three middle matrices can be expressed in a familiar pattern

$$\begin{aligned} M_D &= B(\sigma)R(\delta_1)B(-\sigma) \\ &= \begin{pmatrix} \cos(\delta_1/2) & -e^{\sigma}\sin(\delta_1/2) \\ e^{-\sigma}\sin(\delta_1/2) & \cos(\delta_1/2) \end{pmatrix}, \end{aligned} \tag{11.71}$$

which can in return be written as a Bargmann decomposition

$$R(\alpha)S(-2\chi)R(\alpha), \tag{11.72}$$

whose explicit expression is given in equation (11.20). The parameters α and χ are related to σ and δ_1 by

$$\begin{aligned} \cos(\delta_1/2) &= (\cosh\chi)\cos\alpha, \\ e^{2\eta} &= \frac{(\cosh\chi)\sin\alpha + \sinh\chi}{(\cosh\chi)\sin\alpha - \sinh\chi}. \end{aligned} \tag{11.73}$$

Now the entire chain of equation (11.61) can be put into the form

$$M_B = R(\alpha + \delta_2/2)S(-2\chi)R(\alpha + \delta_2/2), \tag{11.74}$$

which is easily seen to be another Bargmann decomposition. Finally, M_B can be expressed as a Wigner decomposition [11]

$$M_W = B(\eta)R(\theta)B(-\eta), \tag{11.75}$$

with

$$\begin{aligned} \cos(\theta/2) &= (\cosh\chi)\cos(\alpha + \delta_2/2), \\ e^{2\eta} &= \frac{(\cosh\chi)\sin(\alpha + \delta_2/2) + \sinh\chi}{(\cosh\chi)\sin(\alpha + \delta_2/2) - \sinh\chi}. \end{aligned} \tag{11.76}$$

Thanks to the above Wigner decomposition, now the repeated applications of M_W can easily be dealt with

$$(M_W)^N = B(\eta)R(N\theta)B(-\eta) \tag{11.77}$$

to be applied to a light beam traveling through a periodic system of multilayers.

11.7 Camera optics

The basic optical arrangement for the camera consists of a lens with focal length f and the propagation of the ray by an amount d [19]. The lens matrix is given by

$$\begin{pmatrix} 1 & 0 \\ -1/f & 1 \end{pmatrix}, \tag{11.78}$$

and a translation of the ray is expressed by the matrix

$$\begin{pmatrix} 1 & d \\ 0 & 1 \end{pmatrix}. \tag{11.79}$$

If the object and the image are d_1 and d_2 distances away from the lens, respectively, the system is described by

$$\begin{pmatrix} 1 & d_2 \\ 0 & 1 \end{pmatrix}\begin{pmatrix} 1 & 0 \\ -1/f & 1 \end{pmatrix}\begin{pmatrix} 1 & d_1 \\ 0 & 1 \end{pmatrix}. \tag{11.80}$$

The multiplication of these matrices leads to the camera matrix of the form

$$\begin{pmatrix} 1 - d_2/f & d_1 + d_2 - d_1 d_2/f \\ -1/f & 1 - d_1/f \end{pmatrix}. \tag{11.81}$$

The image becomes focused when the upper right element of this matrix vanishes [3], i.e.,

$$\frac{1}{d_1} + \frac{1}{d_2} = \frac{1}{f}. \tag{11.82}$$

For the camera optics, both d_1 and d_2 are longer than f. It is then more convenient to deal with the negative of the matrix given in equation (11.81) with positive diagonal elements. In addition, let us use the dimensionless variables

$$x_1 = d_1/f \quad \text{and} \quad x_2 = d_2/f. \tag{11.83}$$

Then the camera matrix becomes

$$\begin{pmatrix} x_2 - 1 & (x_1 - 1)(x_2 - 1) - 1 \\ 1 & x_1 - 1 \end{pmatrix}, \tag{11.84}$$

with the focal condition

$$\frac{1}{x_1} + \frac{1}{x_2} = 1. \tag{11.85}$$

If we wish to study the formula of the camera matrix equation (11.84) as a representation of the Lorentz group, the first step is obtain its equi-diagonal form. However, if the focal condition of equation (11.85) is to be preserved, the off-diagonal elements should remain invariant. We thus have to resort to the Hermitian transformation given in equation (11.6). For this purpose, we can use the transformation matrix

$$\begin{pmatrix} e^{\eta/2} & 0 \\ 0 & e^{-\eta/2} \end{pmatrix} \tag{11.86}$$

with

$$e^{\eta} = \sqrt{\frac{1 - x_2}{1 - x_1}}. \tag{11.87}$$

The diagonal elements become

$$\sqrt{(1 - x_1)(1 - x_2)}. \tag{11.88}$$

If the diagonal elements are smaller than one, the camera matrix should take the form

$$\begin{pmatrix} \cos(\theta/2) & -e^{\eta}\sin(\theta/2) \\ e^{-\eta}\sin(\theta/2) & \cos(\theta/2) \end{pmatrix} \tag{11.89}$$

with

$$e^{-\eta}\sin(\theta/2) = 1. \tag{11.90}$$

Thus the matrix becomes

$$\begin{pmatrix} \cos(\theta/2) & -\sin^2(\theta/2) \\ 1 & \cos(\theta/2) \end{pmatrix}. \tag{11.91}$$

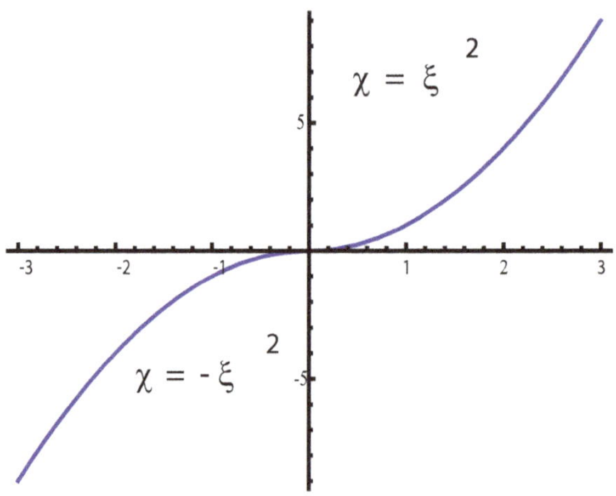

Figure 11.3. Tangential continuity in the camera focusing. The vertical axis χ of this graph is the upper-right element of the camera matrix and is $-\sin^2(\theta/2)$ for $\xi < 0$, and $\sinh^2(\lambda/2)$ for $\xi > 0$ [7].

If the diagonal elements are greater than one, the camera matrix becomes

$$\begin{pmatrix} \cosh(\lambda/2) & e^{\eta}\sinh(\lambda/2) \\ e^{-\eta}\sinh(\lambda/2) & \cosh(\lambda/2) \end{pmatrix} \tag{11.92}$$

with

$$e^{-\eta}\sinh(\lambda/2) = 1, \tag{11.93}$$

leading to

$$\begin{pmatrix} \cosh(\lambda/2) & \sinh^2(\lambda/2) \\ 1 & \cosh(\lambda/2) \end{pmatrix}. \tag{11.94}$$

Thus, the focusing process is the transition from

$$\begin{pmatrix} 1-\xi^2/2 & -\xi^2 \\ 1 & 1-\xi^2/2 \end{pmatrix} \quad \text{to} \quad \begin{pmatrix} 1+\xi^2/2 & \xi^2 \\ 1 & 1+\xi^2/2 \end{pmatrix}, \tag{11.95}$$

via $\xi = 0$. This transition is illustrated in figure 11.3. Indeed, this is the tangential continuity discussed extensively in section 4.6 and 4.7.

References

[1] Azzam R M A-G and Bashara N M 1999 *Ellipsometry and Polarized Light. North-Holland Personal Library* (Amsterdam: Elsevier) 4. Impression, paperback edition (ISBN 978-0-444-87016-2, Originally published 1977; OCLC: 247501433)

[2] Başkal S and Kim Y S 2001 Shear representations of beam transfer matrices *Phys. Rev.* E **63** 056606-1–6

[3] Başkal S and Kim Y S 2003 Lens optics as an optical computer for group contractions *Phys. Rev.* E **67** 056601-1–8

[4] Başkal S and Kim Y S 2009 ABCD matrices as similarity transformations of Wigner matrices and periodic systems in optics *J. Opt. Soc. Am. A: Opt. Image Sci. Vis.* **26** 2049–54

[5] Başkal S and Kim Y S 2010 One analytic form for four branches of the *ABCD* matrix *J. Mod. Opt.* **57** 1251–9

[6] Başkal S and Kim Y S 2013 Lorentz group in ray and polarization optics *Mathematical Optics: Classical, Quantum and Computational Methods* ed V Lakshminarayanan, M L Calvo, and T Alieva (Boca Raton, FL: Taylor and Francis) pp 303–49 (ISBN 978-1-4398-6961-1)

[7] Başkal S and Kim Y S 2014 Lens optics and the continuity problems of the *ABCD* matrix *J. Mod. Opt.* **61** 161–6

[8] Başkal S and Kim Y S 2002 Wigner rotations in laser cavities *Phys. Rev.* E **66**

[9] Dragoman D 2010 Polarization optics analogy of quantum wavefunctions in graphene *J. Opt. Soc. Am.* B **27** 1325–31

[10] Georgieva E and Kim Y S 2001 Iwasawa effects in multilayer optics *Phys. Rev.* E **64** 26602-1–26602-6

[11] Georgieva E and Kim Y S 2003 Slide-rule-like property of Wigner's little groups and cyclic *S* matrices for multilayer optics *Phys. Rev.* E **68** 026606-1–12

[12] Gerrard A and Burch J M 1994 *Introduction to Matrix Methods in Optics* (New York: Dover) (ISBN 978-1-62198-651-5, Originally published 1975 New York: Wiley)
[13] Han D, Kim Y S, and Noz M E 1999 Wigner rotations and Iwasawa decompositions in polarization optics *Phys. Rev.* E **60** 1036–41
[14] Haus H A 1984 *Waves and Fields in Optoelectronics. Prentice-Hall Series in Solid State Physical Electronics* (Englewood Cliffs, NJ: Prentice-Hall) (ISBN 978-0-13-946053-1)
[15] Hawkes J and Latimer I 1995 *Lasers: Theory and Practice. Prentice-Hall International Series in Optoelectronics* (Upper Saddle River, NJ: Prentice Hall) (ISBN 978-0-13-521493-0)
[16] Lohmann A W 1993 Image rotation, Wigner rotation, and the fractional Fourier transform *J. Opt. Soc. Am.* A **10** 2181
[17] Monzón J J and Sánchez-Soto L L 1999 Fully relativistic-like formulation of multilayer optics *J. Opt. Soc. Am.* A **16** 2013–8
[18] Monzón J J and Sánchez-Soto L L 2000 Fresnel formulas as Lorentz transformations *J. Opt. Soc. Am.* A **17** 1475–81
[19] Saleh B E A and Carl Teich M 2007 *Fundamentals of Photonics: Wiley Series in Pure and Applied Optics* 2nd edn (Hoboken, NJ: Wiley Interscience) (ISBN 978-0-471-35832-9)
[20] Sánchez-Soto L L, Monzón J J, Barriuso A G, and Cariñena J F 2012 The transfer matrix: a geometrical perspective *Phys. Rep.* **513** 191–227
[21] Wigner E P 1939 On unitary representations of the inhomogeneous Lorentz group *Ann. Math.* **40** 149–204
[22] Yariv A 1989 *Quantum Electronics* 3rd edn (Hoboken, NJ: Wiley) (ISBN 978-0-471-60997-1, Originally published 1975)

IOP Publishing

Physics of the Lorentz Group (Second Edition)
Beyond high-energy physics and optics
Sibel Başkal, Young S Kim, and Marilyn E Noz

Chapter 12

Polarization optics

In studying polarized light propagating along the z-direction, the traditional approach is to consider the x- and y-components of the electric fields. Their amplitude ratio and phase difference determine the state of polarization. Thus, we can change the polarization either by adjusting the amplitudes, by changing the relative phases, or both. For convenience, we call the optical device which changes amplitudes an *attenuator* and the device which changes the relative phase a *phase shifter*.

It is thus possible to describe the polarization of light as two-component column vectors, and two-by-two matrices applicable to those vectors. The two-by-two representation of the Lorentz group was discussed in section 1.2. We will examine the polarization of light using the methods developed in the early chapters of this book. Although the mathematics of the Lorentz group was originally developed for understanding Einstein's special relativity, it is interesting to note that the same set of mathematical tools can be used for studying polarization optics. In this chapter and the next, we will study the Jones vectors, Mueller matrices, and Stokes parameters, all of which constitute appropriate representations of the Lorentz group [4, 5, 7, 9, 22, 23, 26].

While the abstract group theoretical ideas make two-by-two matrix calculations more systematic and transparent in optics, optical instruments can act as analog computers for Lorentz transformations in special relativity. It is gratifying to note that special relativity as well as ray and polarization optics can be formulated as the physics of two-by-two matrices.

In section 12.1 we introduce the Jones matrix formalism and the Jones vector. We derive the attenuation and phase shift matrices. In section 12.2 we apply the matrices derived in the previous section to the Jones vector. In section 12.3 we show the combined effect of the attenuation and phase shift matrices on the Jones vector. In section 12.4 we delineate the $SL(2, c)$ content contained in polarization optics. Then we turn to a discussion of the optical activities in section 12.5. Here in section 12.6 we show optical instruments can be used to illustrate Wigner's little groups.

doi:10.1088/978-0-7503-3607-9ch12

Finally, we discuss, in section 12.7, how some possible optical filters can be useful in computing and as components in future computers.

12.1 Jones vectors

When studying the two-component light vector, the traditional method is to use the Jones-matrix formalism. This is covered in standard optics textbooks [13, 24]. In this formalism, it is usual to combine the two transverse components of the electric fields into one column matrix. By using the exponential form for the sinusoidal function, the column vector can be written as:

$$\begin{pmatrix} E_x \\ E_y \end{pmatrix} = \begin{pmatrix} a \exp\left\{i(kz - \omega t + \phi_1)\right\} \\ b \exp\left\{i(kz - \omega t + \phi_2)\right\} \end{pmatrix}. \tag{12.1}$$

This column matrix is commonly called the Jones vector [14, 15].

The projection operator

$$\begin{pmatrix} 1 & 0 \\ 0 & 0 \end{pmatrix}, \tag{12.2}$$

is the usual way the Jones-matrix formalism [12] starts. This can be found in existing textbooks [13] and is applicable to the Jones vector of equation (12.1). This operator completely eliminates the y-component, while keeping the x-component of the electric field. Thus, the attenuation factor in the y-direction is always considered to be much greater than that along the x-direction. This is an oversimplification of the real world. We shall replace this projection operator by an attenuation matrix which is closer to the real world.

Here the projection operator of equation (12.2) is replaced by a squeeze matrix. This matrix contains two transverse directions which are perpendicular to each other. In one transverse direction, the absorption coefficient could therefore, differ from that along the other direction. This can be described by the two-by-two matrix [4, 5, 7, 9, 22, 25, 26]

$$\begin{pmatrix} e^{-\eta_1} & 0 \\ 0 & e^{-\eta_2} \end{pmatrix} = e^{-(\eta_1+\eta_2)/2} \begin{pmatrix} e^{\eta/2} & 0 \\ 0 & e^{-\eta/2} \end{pmatrix}, \tag{12.3}$$

with $\eta = \eta_2 - \eta_1$. If we look at the projection operator of equation (12.2), we see that physically, it means that along the y-direction, the absorption coefficient is much greater than it is along the x-direction. If η_1 is very close to zero and η_2 becomes infinitely large, then this absorption matrix given in equation (12.3), becomes the projection matrix. Hence we see that the projection operator of equation (12.2) is just a special case of the attenuation matrix given in equation (12.3).

What the above attenuation matrix tells us is that there are two different attenuation rates associated with the electric fields. Since both components in the exponential factor $e^{-(\eta_1+\eta_2)/2}$ are reduced at the same rate, the state of polarization is not affected. The polarization effect is solely determined by the squeeze matrix

$$B(\eta) = \begin{pmatrix} e^{\eta/2} & 0 \\ 0 & e^{-\eta/2} \end{pmatrix}. \tag{12.4}$$

The diagonal matrix of this type served many different purposes in earlier chapters of this book. It is a key element of the two-by-two representation of the Lorentz group, as seen in table 1.1. Here, this squeeze matrix will play the role of the attenuator.

The optical filter, is another basic element. It has different values of the index of refraction along the two orthogonal directions. We show the effect of this filter by writing it as

$$\begin{pmatrix} e^{-i\delta_1} & 0 \\ 0 & e^{-i\delta_2} \end{pmatrix} = e^{-i(\delta_1+\delta_2)/2}\begin{pmatrix} e^{-i\delta/2} & 0 \\ 0 & e^{i\delta/2} \end{pmatrix}. \tag{12.5}$$

Here we have $\delta = \delta_1 - \delta_2$. Since the overall phase factor $e^{-i(\delta_2+\delta_2)/2}$ cannot be detected in measurement processes it can be deleted. As solely determined by the matrix

$$P(\delta) = \begin{pmatrix} e^{-i\delta/2} & 0 \\ 0 & e^{i\delta/2} \end{pmatrix}, \tag{12.6}$$

the polarization effect of the filter leads to a phase difference of δ between the x- and y-components. We shall call this phase shift matrix a phase shifter. The matrix in equation (12.6) has the same form as a rotation around the negative z-direction as seen from table 1.1, but has a completely different role here.

Since the x- and y-axes are not always the polarization axes, we need the rotation matrix

$$R(\theta) = \begin{pmatrix} \cos(\theta/2) & -\sin(\theta/2) \\ \sin(\theta/2) & \cos(\theta/2) \end{pmatrix}. \tag{12.7}$$

This matrix is also familiar to us from the Lorentz group.

The three components given in equations (12.2), (12.6), and (12.7), when systematically combined, traditionally constitute Jones-matrix formalism. However, in this chapter, we shall use the squeeze matrix of equation (12.4) instead of the projection operator of equation (12.2). Then they become the starters of the two-by-two representation of the Lorentz group.

12.2 Squeeze transformation and phase shift

The effect of the squeeze matrix given in equation (12.4) on the Jones vector is not addressed adequately in the literature. However, the effect of the phase shift matrix $P(\delta)$ of equation (12.6) is well known. The combined effect of these two matrices will now be discussed. From the equations we can see that both are diagonal and that they commute with each other.

The Jones vector is effected in a straightforward manner by the squeeze matrix. Applying the squeeze matrix of equation (12.4) to the Jones vector, the net result is

$$\begin{pmatrix} e^{\eta/2} & 0 \\ 0 & e^{-\eta/2} \end{pmatrix}\begin{pmatrix} E_x \\ E_y \end{pmatrix} = \begin{pmatrix} e^{\eta/2}E_x \\ e^{-\eta/2}E_y \end{pmatrix}. \tag{12.8}$$

The product of the amplitude, therefore, remains invariant because this squeeze transformation expands one amplitude, while contracting the other.

Starting with the Jones vector of the form

$$\begin{pmatrix} \exp(ikz) \\ \exp[i(kz - \pi/2)] \end{pmatrix}, \tag{12.9}$$

we can illustrate phase shifts. The real part of equation (12.9) is

$$\mathrm{Re}\begin{pmatrix} x \\ y \end{pmatrix} = \begin{pmatrix} \cos(kz) \\ \sin(kz) \end{pmatrix}. \tag{12.10}$$

This corresponds to a circular polarization with

$$x^2 + y^2 = 1. \tag{12.11}$$

Suppose now, we apply the phase shift matrix. Then the resulting vector is given by

$$\begin{pmatrix} x \\ y \end{pmatrix} = \begin{pmatrix} \cos(kz + \delta/2) \\ \sin(kz - \delta/2) \end{pmatrix}. \tag{12.12}$$

This can then be rewritten as

$$\begin{pmatrix} x \\ y \end{pmatrix} = \begin{pmatrix} \cos(kz - \pi/4 + \alpha) \\ \cos(kz - \pi/4 - \alpha) \end{pmatrix}, \tag{12.13}$$

where we use

$$\alpha = \frac{\delta}{2} + \frac{\pi}{4}. \tag{12.14}$$

Then we obtain

$$\begin{aligned} x + y &= 2(\cos\alpha)\cos(kz - \pi/4), \\ x - y &= -2(\sin\alpha)\sin(kz - \pi/4), \end{aligned} \tag{12.15}$$

which can also be written as

$$\frac{(x+y)^2}{4(\cos\alpha)^2} + \frac{(x-y)^2}{4(\sin\alpha)^2} = 1. \tag{12.16}$$

This is easily seen to be an elliptic polarization.

Since the squeeze operation of equation (12.4) is seen to change the amplitudes, and also to commute with the phase shift matrix it is relatively simple. Figure 12.1 illustrates the combined effect of squeeze transformation and phase shift.

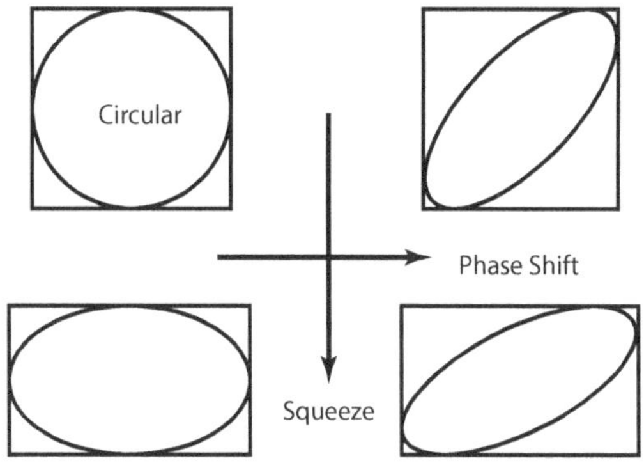

Figure 12.1. Squeeze transformation and phase shift. We see that the squeeze transformation and phase shift are both elliptic deformations, but they are done differently [4].

12.3 Rotation of the polarization axes

The circumstances under which the electric field components take the form of equation (12.1) are dependent upon the polarization coordinate system being the same as the xy coordinate system. Then the phase shifter of equation (12.6) and the attenuator of equation (12.4) can be directly applied to the column matrix of equation (12.1). If the matrix

$$R(\theta) = \begin{pmatrix} \cos(\theta/2) & -\sin(\theta/2) \\ \sin(\theta/2) & \cos(\theta/2) \end{pmatrix}, \tag{12.17}$$

is used to rotate the polarization coordinate system through an angle of $\theta/2$, the result is the phase shifter taking the form

$$P(\theta, \delta) = R(\theta)P(\delta)R(-\theta) \tag{12.18}$$

$$= \begin{pmatrix} \cos(\delta/2) + i\sin(\delta/2)\cos\theta & i\sin(\delta/2)\sin\theta \\ i\sin(\delta/2)\sin\theta & \cos(\delta/2) - i\sin(\delta/2)\cos\theta \end{pmatrix}. \tag{12.19}$$

If, however, we use an angle of 45° to rotate the polarization coordinate system, we obtain the phase shifter matrix

$$Q(\delta) = \begin{pmatrix} \cos(\delta/2) & i\sin(\delta/2) \\ i\sin(\delta/2) & \cos(\delta/2) \end{pmatrix}, \tag{12.20}$$

which is a rotation around the x-axis, as given in table 1.1.

Let us start with the circularly polarized wave

$$\begin{pmatrix} x \\ y \end{pmatrix} = \begin{pmatrix} 1 \\ -i \end{pmatrix} e^{(ikz - i\omega t)}, \tag{12.21}$$

to illustrate what this matrix does to polarized beams. The real part of equation (12.21) is

$$\mathrm{Re}\begin{pmatrix} x \\ y \end{pmatrix} = \begin{pmatrix} \cos(kz - \omega t) \\ \sin(kz - \omega t) \end{pmatrix}. \tag{12.22}$$

This leads to the familiar equation for the circle

$$x^2 + y^2 = 1. \tag{12.23}$$

If we apply the phase shifter of equation (12.20) to the Jones vector of equation (12.1), we obtain the result

$$\begin{pmatrix} [\cos(\delta/2) + \sin(\delta/2)]\cos(kz - \omega t) \\ i[\sin(\delta/2) - \cos(\delta/2)]\sin(kz - \omega t) \end{pmatrix}. \tag{12.24}$$

Using

$$\begin{aligned} \cos(\delta/2) &= \cos([\delta/2 + \pi/4] - \pi/4), \\ -\sin(\delta/2) &= \cos([\delta/2 + \pi/4] + \pi/4), \end{aligned} \tag{12.25}$$

we obtain

$$\begin{aligned} \cos(\delta/2) - \sin(\delta/2) &= \sqrt{2}\cos(\delta/2 + \pi/4), \\ \cos(\delta/2) + \sin(\delta/2) &= \sqrt{2}\sin(\delta/2 + \pi/4). \end{aligned} \tag{12.26}$$

After this phase shift, the Jones vector can now be written as

$$\begin{pmatrix} [\sqrt{2}\cos\alpha]\cos(kz - \omega t) \\ [\sqrt{2}\sin\alpha]\sin(kz - \omega t) \end{pmatrix}, \tag{12.27}$$

with

$$\alpha = \frac{\delta}{2} + \frac{\pi}{4}. \tag{12.28}$$

This is an elliptic polarization because the x- and y-components now satisfy the equation

$$\frac{x^2}{(\sqrt{2}\cos\alpha)^2} + \frac{y^2}{(\sqrt{2}\sin\alpha)^2} = 1. \tag{12.29}$$

Figure 12.2 illustrates these steps.

Next, by considering rotations of the squeeze matrix

$$B(\theta, \eta) = R(\theta)B(\eta)R(-\theta), \tag{12.30}$$

we are led to

$$B(\theta, \eta) = \begin{pmatrix} \cosh(\eta/2) + \sinh(\eta/2)\cos\theta & \sinh(\eta/2)\sin\theta \\ \sinh(\eta/2)\sin\theta & \cosh(\eta/2) - \sinh(\eta/2)\cos\theta \end{pmatrix}. \tag{12.31}$$

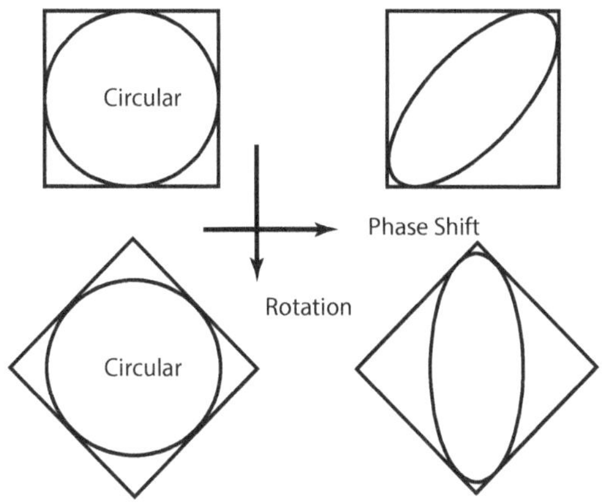

Figure 12.2. Phase shift and rotation. They are now both rotated by 45° [4].

We are familiar with this squeeze operation from section 11.3. This changes the amplitudes. If the squeeze matrix is rotated by 45°, we use the notation $S(\eta)$ for this special rotation angle, and

$$S(\eta) = \begin{pmatrix} \cosh(\eta/2) & \sinh(\eta/2) \\ \sinh(\eta/2) & \cosh(\eta/2) \end{pmatrix}. \tag{12.32}$$

We can then ask the question about what happens if there are two squeeze transformations made in two different directions. Would this result in another squeeze? The answer is no. The result is another squeeze matrix followed by a rotation, which can be written as [3]

$$B(\theta, \lambda)B(0, \eta) = B(\phi, \xi)R(\omega), \tag{12.33}$$

where

$$\begin{aligned} \cosh\xi &= \cosh\eta \ \cosh\lambda + \sinh\eta \ \sinh\lambda \ \cos\theta, \\ \tan\phi &= \frac{\sin\theta[\sinh\lambda + \tanh\eta(\cosh\lambda - 1)\cos\theta]}{\sinh\lambda\cos\theta + \tanh\eta[1 + (\cosh\lambda - 1)\cos^2\theta]}, \\ \tan\omega &= \frac{2(\sin\theta)[\sinh\lambda\sinh\eta + C_-\cos\theta]}{C_+ + C_-\cos(2\theta) + 2\sinh\lambda\sinh\eta\cos\theta}, \end{aligned} \tag{12.34}$$

with

$$C_\pm = (\cosh\lambda \pm 1)(\cosh\eta \pm 1). \tag{12.35}$$

Indeed, we write equation (12.33) as

$$R(\omega) = B(\phi, -\xi) \ B(\theta, \lambda) \ B(0, \eta), \tag{12.36}$$

where it can be observed that three squeeze transformations lead to one rotation.

If the angle θ is 90°, the calculation becomes simpler, and

$$B(\lambda)B(\eta) = B(\phi, \xi)R(\omega), \tag{12.37}$$

where

$$\begin{aligned}\cosh \xi &= \cosh \eta \ \cosh \lambda, \\ \tan \phi &= \frac{\sinh \lambda}{\tanh \eta}, \\ \tan \omega &= \frac{\sinh \lambda \sinh \eta}{\cosh \eta + \cosh \lambda}.\end{aligned} \tag{12.38}$$

This calculation has been done using the kinematics of Lorentz transformations. However, it does not seem to be possible to perform experiments using relativistic particles. However, we note that by using polarization optics this experiment can be realized in laboratories.

12.4 The $SL(2, c)$ group content of polarization optics

The attenuator or squeeze operator of equation (12.4) can be written in the exponential form:

$$B(\eta) = \exp(-i\eta K_3) \quad \text{with} \quad K_3 = \frac{i}{2}\begin{pmatrix} 1 & 0 \\ 0 & -1 \end{pmatrix}. \tag{12.39}$$

The phase shifter of equation (12.6) can be written as

$$P(\delta) = \exp(-i\delta J_1) \quad \text{with} \quad J_1 = \frac{1}{2}\begin{pmatrix} 0 & 1 \\ 1 & 0 \end{pmatrix}. \tag{12.40}$$

The rotation operator of equation (12.7) takes the form:

$$R(\theta) = \exp(-i\theta J_2) \quad \text{with} \quad J_2 = \frac{1}{2}\begin{pmatrix} 0 & -i \\ i & 0 \end{pmatrix}. \tag{12.41}$$

However, the generators J_1, J_2, and K_3 do not form a closed set of commutation relations. If we take their commutators among themselves, they produce three more generators. These six generators then result in the full set of commutation relations necessary to form the Lie algebra of the Lorentz group, as given in equations (1.5) and (1.10).

The squeeze matrix of equation (12.31) needs the generators K_3 and J_2. If these generators are repeatedly applied to the attenuation matrix, this leads to the commutation relation

$$[J_2, K_3] = iK_1. \tag{12.42}$$

Indeed, as discussed previously in section 8.1, J_2, K_1, and K_3 form a closed set of commutation relations for the $Sp(2)$ group which is a subgroup of $SL(2, c)$ [21]. The group $Sp(2)$ is a three-parameter group and it has been extensively discussed in connection with squeezed states of light [18, 28].

Let us now consider only the phase shifters. The mathematics consists of basically repeated applications of J_1 and J_2. This also gives an equivalent result in the repeated applications of J_3 with either of J_1 or J_2. The explicit two-by-two matrix forms of these generators are given in table 1.1. Thus, the phase-shift filters form an $SU(2)$ or $O(3)$-like little group. This is again a subgroup of $SL(2, c)$.

The use of both the attenuators and phase shifters results in the full $SL(2, c)$ group with six parameters. The transformation matrix was previously given in equation (1.18). If two matrices of this kind are applied successfully, the result is

$$\begin{pmatrix} \alpha_2 & \beta_2 \\ \gamma_2 & \delta_2 \end{pmatrix} \begin{pmatrix} \alpha_1 & \beta_1 \\ \gamma_1 & \delta_1 \end{pmatrix} = \begin{pmatrix} \alpha_2\alpha_1 + \beta_2\gamma_1 & \alpha_2\beta_1 + \beta_2\delta_1 \\ \gamma_2\alpha_1 + \delta_2\gamma_1 & \gamma_2\beta_1 + \delta_2\delta_1 \end{pmatrix}. \tag{12.43}$$

The application of this algebra to the column matrix of equation (12.1) is the most general form of the polarization transformation.

The generators of the rotation and the phase-shifters are Hermitian, hence forming a unitary subset of the G matrices of equation (1.18). If the phase shifters are repeatedly applied, the result is a representation of the rotation-like subgroup. From table 1.1 we see that the generators J_2, K_1 and K_3 have all imaginary elements. However, the transformation matrices they generate have all real elements and form the $Sp(2)$ group. Thus, the real subset of the G matrices represents the attenuation filters and their repeated applications.

12.5 Optical activities

It is convenient for this discussion to change the parameters θ and η as

$$\theta = 2\alpha z \quad \text{and} \quad \eta = 2\beta z. \tag{12.44}$$

Then the $R(\theta)$ matrix can be written as

$$R(\alpha z) = \begin{pmatrix} \cos(\alpha z) & -\sin(\alpha z) \\ \sin(\alpha z) & \cos(\alpha z) \end{pmatrix}. \tag{12.45}$$

It is now plain that, as the beam propagates along the z-direction, the rotation angle increases. This is the well-known version of optical activity.

Let us now consider the squeeze operation

$$S(-\beta z) = \begin{pmatrix} \cosh(\beta z) & -\sinh(\beta z) \\ -\sinh(\beta z) & \cosh(\beta z) \end{pmatrix}. \tag{12.46}$$

As the beam moves, it can be seen that the squeeze parameter increases. For convenience, we use a negative sign for β.

If we follow this squeeze by the rotation of equation (12.45), we get a net effect of

$$\begin{pmatrix} \cosh(\beta z) & -\sinh(\beta z) \\ -\sinh(\beta z) & \cosh(\beta z) \end{pmatrix} \begin{pmatrix} \cos(\alpha z) & -\sin(\alpha z) \\ \sin(\alpha z) & \cos(\alpha z) \end{pmatrix}. \tag{12.47}$$

Here, z is in a macroscopic scale, and could perhaps be measured in centimeters, but this does not accurately describe the optical process.

When this happens in a microscopic scale of z/N, after the N repetitions, it can become accumulated into the macroscopic scale of z. Here, N is a very large number. We are thus led to the transformation matrix of the form [16]

$$M(\alpha, \beta, z) = [S(-\beta z/N)R(\alpha z/N)]^N. \tag{12.48}$$

When we take the limit of large N, we see that this quantity can be written as

$$\left[\begin{pmatrix} 1 & -\beta z/N \\ -\beta z/N & 1 \end{pmatrix}\begin{pmatrix} 1 & -\alpha z/N \\ \alpha z/N & 1 \end{pmatrix}\right]^N. \tag{12.49}$$

As $\alpha z/N$ and $\beta z/N$ are very small, we can write M in the following manner:

$$M(\alpha, \beta, z) = \left[\begin{pmatrix} 1 & 0 \\ 0 & 1 \end{pmatrix} + \begin{pmatrix} 0 & -(\alpha + \beta) \\ (\alpha - \beta) & 0 \end{pmatrix}\frac{z}{N}\right]^N. \tag{12.50}$$

If N becomes very large, this matrix can be written as

$$M(\alpha, \beta, z) = \exp(Gz), \tag{12.51}$$

with

$$G = \begin{pmatrix} 0 & -(\alpha + \beta) \\ (\alpha - \beta) & 0 \end{pmatrix}. \tag{12.52}$$

Using the procedure developed in section 11.3 this matrix can be computed. If α is greater than β, G becomes

$$G = \alpha'\begin{pmatrix} 0 & -\exp(\eta) \\ \exp(-\eta) & 0 \end{pmatrix}, \tag{12.53}$$

where

$$\alpha' = \sqrt{\alpha^2 - \beta^2} \quad \text{and} \quad \exp(\eta) = \sqrt{\frac{\alpha + \beta}{\alpha - \beta}}. \tag{12.54}$$

Then the M matrix of equation (12.51) has the form

$$\begin{pmatrix} \cos(\alpha' z) & -e^{\eta}\sin(\alpha' z) \\ e^{-\eta}\sin(\alpha' z) & \cos(\alpha' z) \end{pmatrix}. \tag{12.55}$$

On the other hand if α is less than β, then the off-diagonal elements will have the same sign. In this case G can be written as

$$G = -\beta'\begin{pmatrix} 0 & \exp(\eta) \\ \exp(-\eta) & 0 \end{pmatrix}, \tag{12.56}$$

with

$$\beta' = \sqrt{\beta^2 - \alpha^2} \quad \text{and} \quad \exp(\eta) = \sqrt{\frac{\beta + \alpha}{\beta - \alpha}}, \tag{12.57}$$

and the M matrix of equation (12.51) becomes

$$\begin{pmatrix} \cosh(\beta' z) & -e^{\eta}\sinh(\beta' z) \\ -e^{-\eta}\sinh(\beta' z) & \cosh(\beta' z) \end{pmatrix}. \tag{12.58}$$

However, the lower-left element of the G matrix must vanish if $\alpha = \beta$. Thus the G matrix becomes

$$G = \begin{pmatrix} 0 & -2\alpha \\ 0 & 0 \end{pmatrix}. \tag{12.59}$$

The M matrix then takes the triangular form

$$\begin{pmatrix} 1 & -2\alpha z \\ 0 & 1 \end{pmatrix}. \tag{12.60}$$

We know that rotations of the polarization axis can be provided by optical material. This additional effect of squeeze which we have seen is much more interesting.

12.6 Correspondence to space–time symmetries

We have seen that the Lorentz group forms a basic mathematical framework for ray and polarization optics. Now we wish to see how the two-by-two matrices discussed in the previous section on optical activities can be translated into four-by-four Lorentz transformation matrices. These matrices are applicable to the Minkowski space of (t, z, x, y). As before, the energy–momentum four-vector is (E, p_z, p_x, p_y) in this space. We will translate these two-by-two matrices into the language of Wigner's little groups given in chapter 2.

If the particle is massive, there is a Lorentz frame in which it is at rest. As we saw in equation (2.3), the energy–momentum four-vector is proportional to

$$P = (1, 0, 0, 0), \tag{12.61}$$

which is invariant under rotations around the z-axis. To Lorentz boost the energy–momentum four-vector of equation (12.61) along the z-direction we can use the Lorentz boost matrix

$$B(\beta) = \begin{pmatrix} \cosh\beta & -\sinh\beta & 0 & 0 \\ -\sinh\beta & \cosh\beta & 0 & 0 \\ 0 & 0 & 1 & 0 \\ 0 & 0 & 0 & 1 \end{pmatrix}. \tag{12.62}$$

This Lorentz boost matrix corresponds to the squeeze matrix of equation (12.46) [2, 10, 16]. Here $\beta = \eta/2$. The energy–momentum four-vector of equation (12.61) is also invariant under the rotation matrix

$$R(\alpha z) = \begin{pmatrix} 1 & 0 & 0 & 0 \\ 0 & \cos(\alpha z) & -\sin(\alpha z) & 0 \\ 0 & \sin(\alpha z) & \cos(\alpha z) & 0 \\ 0 & 0 & 0 & 1 \end{pmatrix}. \tag{12.63}$$

Here $\alpha = \theta/2$. Hence we see that matrix

$$B(\beta)R(\alpha z)B(-\beta) \tag{12.64}$$

leaves the energy–momentum four-vector of equation (12.61) invariant.

The matrix given in equation (12.63) performs a rotation around the y-axis in the particle's rest frame. However, we can also rotate the energy–momentum four-vector around the z-axis without changing it. For the particle with mass m, this is what Wigner's little group describes. It is clear that the matrix of equation (12.64) does not change the momentum. Instead, the two-by-two counterpart of this matrix rotates the direction of the spin of the particle in the rest frame. This allows us to conclude that the little group is not a trivial mathematical device. Since the rotation matrix of equation (12.63) corresponds to the rotation matrix of equation (12.45) [10, 16], this two-by-two rotation matrix together with the squeeze matrix of equation (12.46), generates the little group for particles with non-zero mass.

For the particle where the mass is imaginary, we start with the energy–momentum four-vector proportional to

$$P = (E, p, 0, 0). \tag{12.65}$$

If E is smaller than p, the momentum can be brought to the Lorentz frame where the energy–momentum four-vector becomes

$$P = (0, 1, 0, 0). \tag{12.66}$$

This was given in equation (2.32). The Lorentz boost matrix which effects this transformation takes the same form as equation (12.62), with

$$\tanh(2\eta) = \frac{E}{p}. \tag{12.67}$$

Here, $P^2 = -1$, which is a negative number. This indeed means the particle mass is imaginary. The energy–momentum four-vector of equation (12.66) is invariant under the Lorentz boost

$$S(\beta z) = \begin{pmatrix} \cosh(\beta z) & 0 & -\sinh(\beta z) & 0 \\ 0 & 1 & 0 & 0 \\ -\sinh(\beta z) & 0 & \cosh(\beta z) & 0 \\ 0 & 0 & 0 & 1 \end{pmatrix} \tag{12.68}$$

along the x-direction. The energy–momentum four-vector of equation (12.66) is, of course, invariant under rotations around the z-axis.

The four-by-four matrix of equation (12.68) corresponds to the two-by-two squeeze matrix of equation (12.46). This matrix is applicable to optical activities [2, 10, 16]. This squeeze matrix, together with the squeeze matrix along the y-direction and the rotation matrix in the z-direction, generates the group $O(2, 1)$-like little group, which we saw in section 2.4, is the little group for imaginary-mass particles.

Lastly, let us consider a massless particle. The energy–momentum four-vector for this particle, given in equation (2.8), is proportional to

$$(1, 1, 0, 0). \tag{12.69}$$

As always, this energy–momentum four-vector is invariant under the rotation around the z-axis. Additionally, Wigner [27] observed it to be invariant under the transformation given in table 2.2:

$$\begin{pmatrix} 1+\alpha^2 & -\alpha^2 & 2\alpha & 0 \\ \alpha^2 & 1-\alpha^2 & 2\alpha & 0 \\ 2\alpha & -2\alpha & 1 & 0 \\ 0 & 0 & 0 & 1 \end{pmatrix}. \tag{12.70}$$

Here we have used 2α for γ. There is a controversial history [17, 19] associated with this four-by-four matrix. The bottom line is that it corresponds to the triangular matrix of equation (12.59), and the variable 2α performs gauge transformations.

Thus it is interesting to note, as shown here, that optical activities can indeed act as computational devices for the internal space–time symmetries of elementary particles. This means that we can perform high energy experiments using optical instruments.

12.7 More optical filters from $E(2)$-like groups

Now we turn our attention to deriving a set of optical filters from another three-parameter subgroup of the Lorentz group. We know of course that the Lorentz group has many sets of three-parameter subgroups. The subgroup we shall use is the two-dimensional Euclidean group, $E(2)$. Here we will consider the group, defined in section 2.3, which is generated by the four-by-four matrices given in equation (2.10). We derived the two-by-two form in equation (5.10). This $E(2)$-like little group consists of the generators J_3, N_1, and N_2, with

$$N_1 = K_1 - J_2 \quad \text{and} \quad N_2 = K_2 + J_1. \tag{12.71}$$

The matrices N_1 and N_2 take the form given in equation (5.10):

$$N_1 = \begin{pmatrix} 0 & i \\ 0 & 0 \end{pmatrix} \quad \text{and} \quad N_2 = \begin{pmatrix} 0 & 1 \\ 0 & 0 \end{pmatrix}. \tag{12.72}$$

These matrices constitute a closed set of commutation relations:

$$[J_3, N_1] = iN_2, \quad [J_3, N_2] = -iN_1, \quad \text{and} \quad [N_1, N_2] = 0. \tag{12.73}$$

As shown in sections 2.3 and 5.1, these commutation relations are like those of the two-dimensional Euclidean group which consists of two translations and one rotation around the origin. Earlier in this book, this group was studied extensively in connection with the space–time symmetries of massless particles. It was shown that J_3 and the two N_i generators correspond to the helicity and gauge degrees of freedom, respectively [11]. The physics of J_3 is well known through the phase shifter given in equation (12.6). The phase shifter, if the angle δ is $\pi/2$, becomes a quarter-wave shifter. We can write this quarter-wave shifter as:

$$Q = P(\pi/2) = \begin{pmatrix} e^{-i\pi/4} & 0 \\ 0 & e^{i\pi/4} \end{pmatrix}. \tag{12.74}$$

Then the quarter-wave conjugates of J_1 and K_2 are J_2 and K_1, respectively:

$$J_2 = QJ_1Q^{-1} \quad \text{and} \quad K_1 = -QK_2Q^{-1}. \tag{12.75}$$

Consequently,

$$N_2 = QN_1Q^{-1}. \tag{12.76}$$

The N_i generators, therefore, lead to the following transformation matrices:

$$\begin{aligned} T_1(\tau) &= \exp(-i\tau N_1) = \begin{pmatrix} 1 & \tau \\ 0 & 1 \end{pmatrix}, \\ T_2(\tau) &= \exp(-i\tau N_2) = \begin{pmatrix} 1 & -i\tau \\ 0 & 1 \end{pmatrix}. \end{aligned} \tag{12.77}$$

It is clear that T_1 forms the quarter-wave conjugate of T_2, hence it is possible to concentrate on the transformation matrix T_1.

If we apply T_1 to the column matrix of equation (12.1) we obtain:

$$\begin{pmatrix} 1 & \tau \\ 0 & 1 \end{pmatrix}\begin{pmatrix} E_x \\ E_y \end{pmatrix} = \begin{pmatrix} E_x + \tau E_y \\ E_y \end{pmatrix}. \tag{12.78}$$

This new filter, while it leaves the y-component invariant, superimposes the y-component of the electric field to the x-component with an appropriate constant.

We have achieved this because the generator N_1 consists of J_2 which generates rotations around the y-axis. Additionally, N_1 consists of K_1 which generates a squeeze along the 45°-axis. Physically, J_2 generates optical activities. Hence, this new filter consists of a suitable combination of these two operations. We have to take into account the overall attenuation factor, in both cases. The attenuation of the y-component, which is not affected by the symmetry operation of equation (12.78), can be used to measure this factor.

The question then is whether optical filters of this kind can be produced? If we start from an optically active material, either mechanical or electrical methods can be used to introduce an asymmetric attenuation into the material. It is also possible to produce optical filters by piling up alternative layers of the J_2-type and K_1-type. It is interesting to note that, by using either of these two methods, a special effect can be produced, predictable from the Lorentz group.

Transformations generated by J_1 are included in the $E(2)$-like symmetry. Additionally, the phase shifter matrix given in equation (12.6) and the physics associated with it is well understood. We apply now this phase shift matrix to T_1 from left and from right. Then

$$P(\delta)T_1(\tau) = \begin{pmatrix} e^{-i\delta/2} & e^{-i\delta/2}\tau \\ 0 & e^{i\delta/2} \end{pmatrix},$$
$$T_1(\tau)P(\delta) = \begin{pmatrix} e^{-i\delta/2} & e^{i\delta/2}\tau \\ 0 & e^{i\delta/2} \end{pmatrix}. \tag{12.79}$$

This leads to

$$P(-\delta)T_1(\tau)P(\delta) = T_1(\delta, \tau) = \begin{pmatrix} 1 & e^{i\delta}\tau \\ 0 & 1 \end{pmatrix}. \tag{12.80}$$

By rotating the $P(\delta)$ filter around the z-axis by 90°we can obtain the $P(-\delta)$ filter. We see then that it is possible, using phase shifters of the type $P(\delta)$, to add a phase factor to the τ variable.

Like the three-dimensional rotation group which occupies an important place in many different branches of physics, the group $Sp(2)$ is useful in many fields, which include optics [1, 8, 17, 18]. Thus, in constructing analog computers which perform the symmetry operations of the rotation and $Sp(2)$ groups, a useful tool might be the traditional attenuation and phase-shift filters.

We saw in chapter 2 that the group $E(2)$, though somewhat new in optics [20], deals with translations and rotations on a flat surface. It is possible then that this filter might be useful as a computational device for recording and reading two-dimensional maps. First we rewrite the T_2 matrix of equation (12.77) as:

$$\begin{pmatrix} 1 & \beta \\ 0 & 1 \end{pmatrix}. \tag{12.81}$$

An interesting algebraic property of this matrix is:

$$\begin{pmatrix} 1 & \beta_1 \\ 0 & 1 \end{pmatrix}\begin{pmatrix} 1 & \beta_2 \\ 0 & 1 \end{pmatrix} = \begin{pmatrix} 1 & \beta_1 + \beta_2 \\ 0 & 1 \end{pmatrix}. \tag{12.82}$$

It is apparent that this matrix, like the logarithmic function, can therefore be used for converting multiplication into addition. This is one of the most basic operations in computational machines.

For a more immediate application, we could consider lens optics. A trivial laboratory operation consists of rotating a given filter around the z-axis by 90°. If the matrix of equation (12.81) is rotated, the resulting matrix has the form

$$\begin{pmatrix} 1 & 0 \\ \beta & 1 \end{pmatrix}. \tag{12.83}$$

These two forms of equations (12.81) and (12.83) together, can serve in para-axial optics as lens and translation matrices, respectively. We saw this in section 11.5. We have seen, therefore, that, for a multilens system, an analog computer can be composed of a system of polarization filters. Additionally, a matrix of the form given in equation (12.81) can represent a *shear* transformation. This is one of the basic deformations [6] in engineering applications.

Thus, the Lorentz group, while it was introduced to physics as the basic language for space–time symmetries of elementary particles [27], is becoming increasingly prominent in many branches of physics and engineering which include classical and quantum optics. Optical filters may provide excellent calculational tools for the Lorentz group. Hence, these filters may be useful components in future computers.

References

[1] Abraham R and Marsden J E 2008 *Foundations of Mechanics* 2nd edn (Providence, RI: AMS Chelsea Pub./American Mathematical Society) (ISBN 978-0-8218-4438-0, Originally published 1978; OCLC: ocn191847156)

[2] Başkal S and Kim Y S 2003 Lens optics as an optical computer for group contractions *Phys. Rev.* E **67** 056601 1–8

[3] Başkal S and Kim Y S 2005 Rotations associated with Lorentz boosts *J. Phys. A: Math. General* **38** 6545–56

[4] Başkal S and Kim Y S 2013 Lorentz group in ray and polarization optics *Mathematical Optics: Classical, Quantum and Computational Methods* ed V Lakshminarayanan, M L Calvo, and T Alieva (Boca Raton, FL: Taylor and Francis) pp 303–49 (ISBN 978-1-4398-6961-1)

[5] Ben-Aryeh Y 2005 Nonunitary squeezing and biorthogonal scalar products in polarization optics *J. Opt. B: Quant. Semiclass. Opt.* **7** S452–7

[6] Chen B and Kaufman A 2000 3D volume rotation using shear transformations *Graph. Models* **62** 308–22

[7] Franssens G R 2015 Relativistic kinematics formulation of the polarization effects of Jones–Mueller matrices *J. Opt. Soc. Am.* A **32** 164–72

[8] Guillemin V and Sternberg S 2001 *Symplectic Techniques in Physics* reprinted edition (Cambridge: Cambridge University Press) (ISBN 978-0-521-38990-7, Originally published 1984; OCLC: 248721606)

[9] Han D, Kim Y S, and Noz M E 1997 Jones-matrix formalism as a representation of the Lorentz group *J. Opt. Soc. Am.* A **14** 2290–8

[10] Han D, Kim Y S, and Noz M E 1997 Stokes parameters as a Minkowskian four-vector *Phys. Rev.* E **56** 6065–76

[11] Han D, Kim Y S, and Son D 1982 E(2)-like little group for massless particles and neutrino polarization as a consequence of gauge invariance *Phys. Rev.* D **26** 3717–25

[12] Hecht E 1970 Note on an operational definition of the Stokes parameters *Am. J. Phys.* **38** 1156–8

[13] Hecht E 2002 *Optics* 4th edn (Reading, MA: Addison-Wesley) (ISBN 978-0-8053-8566-3)

[14] Jones R C 1941 A new calculus for the treatment of optical systems. I. Description and discussion of the calculus *J. Opt. Soc. Am.* **31** 488–93

[15] Jones R C 1947 A new calculus for the treatment of optical systems. V. A more general formulation, and description of another calculus *J. Opt. Soc. Am.* **37** 107–10

[16] Kim Y S 2010 Optical activities as computing resources for space–time symmetries *J. Mod. Opt.* **57** 17–22
[17] Kim Y S and Noz M E 1986 *Theory and Applications of the Poincaré Group* (Dordrecht: Springer) (ISBN 978-94-010-8526-7 978-94-009-4558-6)
[18] Kim Y S and Noz M E 1991 *Phase Space Picture of Quantum Mechanics: Group Theoretical Approach.* (*Lecture Notes in Physics* vol 40) (Singapore; Hackensack, NJ: World Scientific Publishing Co.) (ISBN 978-981-02-0360-3 978-981-02-0361-0)
[19] Kim Y S and Wigner E P 1990 Space–time geometry of relativistic particles *J. Math. Phys.* **31** 55–60
[20] Kim Y S and Yeh L 1992 E(2)-symmetric two-mode sheared states *J. Math. Phys.* **33** 1237–46
[21] Kitano M and Yabuzaki T 1989 Observation of Lorentz-group Berry phases in polarization optics *Phys. Lett.* A **142** 321–5
[22] Opatrny` T and Peřina J 1993 Non-image-forming polarization optical devices and Lorentz transformations—an analogy *Phys. Lett.* A **181** 199–202
[23] Red'kov V M 2011 Lorentz group theory and polarization of the light *Adv. Appl. Clifford Algebras* **21** 203–20
[24] Saleh B E A and Carl Teich M 2007 *Fundamentals of Photonics* (*Wiley Series in Pure and Applied Optics*) 2nd edn (Hoboken, NJ: Wiley Interscience) (ISBN 978-0-471-35832-9)
[25] Tudor T 2010 Vectorial Pauli algebraic approach in polarization optics. I. Device and state operators *Optik—Int. J. Light Electron.* **121** 1226–35
[26] Tudor T 2015 On a quasi-relativistic formula in polarization theory *Opt. Lett.* **40** 693
[27] Wigner E P 1939 On unitary representations of the inhomogeneous Lorentz group *Ann. Math.* **40** 149–204
[28] Yuen H P 1976 Two-photon coherent states of the radiation field *Phys. Rev.* A **13** 2226–43

IOP Publishing

Physics of the Lorentz Group (Second Edition)

Beyond high-energy physics and optics

Sibel Başkal, Young S Kim, and Marilyn E Noz

Chapter 13

Poincaré sphere

With the intention of analyzing the polarization of light, Henri Poincaré constructed, in 1892, what is known today as the Poincaré sphere [1, 5–7, 18]. Poincaré meant this sphere as a graphic illustration of the polarization ellipse that was discussed in chapter 12. In this chapter we formulate the Poincaré sphere in terms of two-by-two matrices. We shall show that the matrix elements of the Poincaré sphere consist of the four Stokes parameters which transform like $SL(2, c)$ spinors. Thus, the Poincaré sphere shares the same symmetry property as that of the Lorentz group.

While the Jones vector formalism is a concrete physical example of the two-by-two representation of the Lorentz group, it cannot tell whether two orthogonal components are coherent with each other. In order to address this coherency issue, we need the two-by two coherency matrix consisting of the four Stokes parameters. These four Stokes parameters define the parameters for the three-dimensional Poincaré sphere. The traditional Poincaré sphere needs three parameters like the Euler angles. Then what role does the fourth Stokes parameter play? We shall show in this chapter, the radius of the Poincaré sphere can change.

Since the Stokes parameters are constructed from the two-component Jones vectors, which are transformed like the $SL(2, c)$ spinors, the two-by-two coherency matrix should transform like the two-by-two form of the space–time four-vector discussed extensively in section 1.2 and again in chapter 4. Thus, the Lorentz group should leave the determinant of the coherency matrix invariant. On the other hand the degree of coherency changes the determinant. Hence, the coherency is an extra-Lorentzian variable. We shall study its implications in regard to Einstein's energy–momentum relation where the particle mass is a Lorentz-invariant quantity. To accommodate this extra-Lorentzian variable, the symmetry group has to be extended to the $O(3, 2)$ group discussed in sections 8.3 and 9.5.

In section 13.1 we introduce the problem of decoherence in two beam systems and show how to solve it with the density matrix since this problem cannot be handled by the Jones matrix. In section 13.2 we define the coherency matrix and identify the

doi:10.1088/978-0-7503-3607-9ch13

elements in terms of the Stokes parameters. The Poincaré sphere is formally defined in section 13.3. We look at two concentric Poincaré spheres in section 13.4 and discuss the form of the inner and outer radii and the relationship between them. In section 13.5 the symmetry properties derivable from the Poincaré sphere are applied to the energy–momentum four-vector. In section 13.6 we enlarge the symmetry group of the Poincaré sphere to that of the $O(3, 2)$ group and show the implications for a possible non-Lorentzian variable mass. The entropy problem associated with the coherency matrix is discussed in section 13.7.

13.1 Decoherence in polarization optics

In chapter 12, we gave a concrete physical example of the two-by-two representation of the Lorentz group by using the Jones vector formalism. There, with appropriate transformation matrices of the Lorentz group, we were able to identify the matrix in equation (12.4) as the attenuator, the matrix in equation (12.6) as the phase shifter, and the rotation matrix as that in equation (12.7). This formulation applies to all coherent or partially coherent phases which occur in all two-beam systems, for example interferometers [12]. It is not restricted solely to polarization optics. The lack of coherence between two beams cannot be formulated using the Jones vectors. However, the lack of coherence between beams can be described [3] with the density matrix, the Stokes parameters, and the Poincaré sphere.

If, for example, we start with a pair of complex numbers a and b we can construct the density matrix as:

$$\rho = \begin{pmatrix} aa^* & ab^*e^{-\lambda t} \\ a^*be^{-\lambda t} & bb^* \end{pmatrix}. \tag{13.1}$$

A fundamental role in the decoherence process is played by the decay in the off-diagonal elements of this matrix. This matrix has a determinant of

$$aa^*bb^*(1 - e^{-2\lambda t}). \tag{13.2}$$

If $t = 0$, the determinant is zero and the system is in a pure state. The system looses coherence as the determinant in equation (13.2) increases from zero to aa^*bb^*, as t increases.

Thus, we need the two-by-two coherency matrix to address the issue of coherency between two beams. This takes the form of a density matrix. The elements of this matrix consist of four parameters which can be identified in terms of the Stokes parameters. These four Stokes parameters define the parameters for the three-dimensional Poincaré sphere. Like the Euler angles, however, the traditional Poincaré sphere needs only three parameters. We therefore need to investigate the role of this fourth Stokes parameter. Here we'll show that the radius of the Poincaré sphere can change.

We saw in chapter 12, that the two-component Jones vectors transform like the $SL(2, c)$ spinors. Thus, if we construct the Stokes parameters from these vectors, the two-by-two coherency matrix should transform like the two-by-two form of the space–time four-vector discussed extensively in section 1.2 and chapter 4. From this we know that the Lorentz group should leave the determinant of the coherency matrix invariant. However, the determinant of the coherency matrix changes according to the degree of coherency. Thus, we can draw the conclusion that the coherency is an extra-Lorentzian variable and study its implications in Einstein's energy–momentum relation. In this formulation the particle mass is a Lorentz-invariant quantity. It is then necessary to extend the symmetry group to the $O(3, 2)$ group which was discussed in section 8.3 and 9.5.

13.2 Coherency matrix

In order to address the question of coherency between the two orthogonal electric fields, we study the coherency matrix defined as [6, 7]

$$C = \begin{pmatrix} S_{11} & S_{12} \\ S_{21} & S_{22} \end{pmatrix}, \tag{13.3}$$

with

$$\langle \psi_i^* \psi_j \rangle = \frac{1}{T} \int_0^T \psi_i^*(t + \tau) \psi_j(t) dt, \tag{13.4}$$

where T for a sufficiently long time interval, is much larger than τ. Then, those four elements become [11]

$$\begin{aligned} S_{11} &= \langle \psi_1^* \psi_1 \rangle = a^2, & S_{12} &= \langle \psi_1^* \psi_2 \rangle = ab\, e^{-(\lambda t + i\phi)}, \\ S_{21} &= \langle \psi_2^* \psi_1 \rangle = ab\, e^{-(\lambda t - i\phi)}, & S_{22} &= \langle \psi_2^* \psi_2 \rangle = b^2. \end{aligned} \tag{13.5}$$

The diagonal elements are the absolute values of ψ_1 and ψ_2, respectively. The off-diagonal elements could be smaller than the product of ψ_1 and ψ_2, if the two transverse components are not completely coherent. The t parameter specifies the degree of decoherence. The decoherence of the system is minimal if $t = 0$ and reaches a maximum as the value of $t \to \infty$.

If we start with the Jones vector of the form of equation (12.1), the coherency matrix becomes

$$C = \begin{pmatrix} a^2 & ab\, e^{-(\lambda t + i\phi)} \\ ab\, e^{-(\lambda t - i\phi)} & b^2 \end{pmatrix}. \tag{13.6}$$

We are interested in the symmetry properties of this matrix. Since the transformation matrix applicable to the Jones vector is the two-by-two representation of the Lorentz group, we are particularly interested in the transformation matrices applicable to this coherency matrix.

The determinant and trace of the above coherency matrix are

$$\begin{aligned} \det(C) &= (ab)^2(1 - e^{-2\lambda t}), \\ \mathrm{trace}(C) &= a^2 + b^2. \end{aligned} \tag{13.7}$$

Since $e^{-\lambda t}$ is always equal to or smaller than one, we can introduce an angle χ defined as

$$\cos \chi = e^{-\lambda t}, \tag{13.8}$$

and call it the *decoherence angle*. If $\chi = 0$, the decoherence is minimum, and it is maximum when $\chi = 90°$. We can then write the coherency matrix of equation (13.6) as

$$C = \begin{pmatrix} a^2 & ab(\cos\chi)e^{-i\phi} \\ ab(\cos\chi)e^{i\phi} & b^2 \end{pmatrix}. \tag{13.9}$$

The degree of polarization is defined as [17]

$$f(t) = \sqrt{1 - \frac{4\,\det(C)}{(\mathrm{trace}(C))^2}} = \sqrt{1 - \frac{4(ab)^2\sin^2\chi}{(a^2+b^2)^2}}. \tag{13.10}$$

This degree is one if $\chi = 0$. When $\chi = 90°$, it becomes

$$f(t) = \frac{a^2 - b^2}{a^2 + b^2}. \tag{13.11}$$

Without loss of generality, we can assume that a is greater than b. If they are equal, this minimum degree of polarization is zero.

Under the influence of the Lorentz transformation defined for the Jones vectors of chapter 12, this coherency matrix is transformed as

$$\begin{aligned} C' = G\,C\,G^{\dagger} &= \begin{pmatrix} S'_{11} & S'_{12} \\ S'_{21} & S'_{22} \end{pmatrix} \\ &= \begin{pmatrix} \alpha & \beta \\ \gamma & \delta \end{pmatrix}\begin{pmatrix} S_{11} & S_{12} \\ S_{21} & S_{22} \end{pmatrix}\begin{pmatrix} \alpha^* & \gamma^* \\ \beta^* & \delta^* \end{pmatrix}. \end{aligned} \tag{13.12}$$

Indeed, this is the Lorentz transformation defined in section 1.2.

Using equations (1.23) and (1.24), it is then possible, from the components of the coherency matrix, to write

$$\begin{aligned} S_0 &= \frac{1}{2}(S_{11} + S_{22}), & S_3 &= \frac{1}{2}(S_{11} - S_{22}), \\ S_1 &= \frac{1}{2}(S_{12} + S_{21}), & S_2 &= \frac{i}{2}(S_{12} - S_{21}), \end{aligned} \tag{13.13}$$

as a four-vector. These four parameters are called Stokes parameters, and the four-by-four transformations applicable to these parameters are widely known as Mueller matrices [1, 7, 16]. These Mueller matrices give the relationship between the incident and the emerging state of polarization of light after it has passed through any number and any type of optical elements, which include scatterers, polarizers, and waveplates. It is by employing the Mueller matrices that Lorentz transformations can be performed on the four Stokes parameters. Thus, we see that the Stokes parameters do indeed behave like the components of a four-vector.

We can construct the Jones vector of equation (12.1) by making Lorentz transformations on the simpler form

$$\begin{pmatrix} \psi_1 \\ \psi_2 \end{pmatrix} = \begin{pmatrix} a \exp\{i(kz - \omega t)\} \\ a \exp\{i(kz - \omega t)\} \end{pmatrix}. \tag{13.14}$$

Since the amplitudes are now equal, we can now drop the amplitude a and work with the coherency matrix of the form

$$C = \begin{pmatrix} 1 & e^{-i\phi}\cos\chi \\ e^{i\phi}\cos\chi & 1 \end{pmatrix}. \tag{13.15}$$

The Stokes parameters then become

$$\begin{aligned} S_0 &= 1, & S_3 &= 0, \\ S_1 &= (\cos\chi)\cos\phi, & S_2 &= (\cos\chi)(\sin\phi). \end{aligned} \tag{13.16}$$

Alternatively, by using equations (13.9) and (13.13), this four vector can be expressed as:

$$\begin{pmatrix} S_0 \\ S_1 \\ S_2 \\ S_3 \end{pmatrix} = \frac{1}{2}\begin{pmatrix} a^2 + b^2 \\ 2ab(\cos\phi)e^{-\lambda t} \\ 2ab(\sin\phi)e^{-\lambda t} \\ a^2 - b^2 \end{pmatrix}. \tag{13.17}$$

We can also now succinctly construct the coherency matrix from the Stokes parameters and the Pauli spin matrices as

$$C = \sum_{i=0}^{3} \sigma_i\, S_i, \tag{13.18}$$

where σ_0 is the two-by-two identity matrix.

13.3 Poincaré sphere

The Poincaré sphere is defined by S_1, S_2, and S_3. The radius of the sphere is defined by:

$$r = \sqrt{S_1^2 + S_2^2 + S_3^2}. \tag{13.19}$$

Since $S_3 = 0$ in equation (13.16), the sphere collapses into a circle of radius

$$r = \sqrt{S_1^2 + S_2^2} = \cos\chi. \tag{13.20}$$

Let us go back to the two-by-two matrix for the energy–momentum four-vector. Its determinant is m^2 and remains invariant under Lorentz transformations defined by equation (1.27). Likewise, the determinant of the coherency matrix of equation (13.15) should also remain invariant. The determinant in this case is

$$S_0^2 - r^2 = \sin^2\chi. \tag{13.21}$$

Table 13.1. The same mathematics is shared by polarization optics and special relativity. Each two-by-two transformation matrix has a role in both optics and relativity. The determinants, whether these matrices are obtained from the Stokes vector or from the energy–momentum four-vector, remain invariant under Lorentz transformations. Thus it is apparent that the mass (most fundamental) in particle physics corresponds to the decoherence parameter (least fundamental) in optics.

Polarization Optics	Transformation Matrix	Particle Symmetry
Phase shift ϕ	$\begin{pmatrix} e^{-i\phi/2} & 0 \\ 0 & e^{i\phi/2} \end{pmatrix}$	Rotation around z
Rotation around z	$\begin{pmatrix} \cos(\theta/2) & -\sin(\theta/2) \\ \sin(\theta/2) & \cos(\theta/2) \end{pmatrix}$	Rotation around y
Squeeze along x and y	$\begin{pmatrix} e^{\eta/2} & 0 \\ 0 & e^{-\eta/2} \end{pmatrix}$	Boost along z
$\sin^2\chi$	Determinant	$(\text{mass})^2$

However, this quantity depends on the angle χ variable which measures decoherence of the two transverse components. This aspect is illustrated in table 13.1.

While the decoherence parameter is not fundamental and is influenced by environment, it plays the same mathematical role as the particle mass which remains as the most fundamental quantity since Isaac Newton, and even after Einstein.

13.4 Two concentric Poincaré spheres

We can combine equations (13.17) and (13.19) and write the radius of the sphere as

$$r = \frac{1}{2}\sqrt{(a^2 - b^2)^2 + 4(ab)^2 e^{-2\lambda t}}. \tag{13.22}$$

This radius is that of the conventional Poincaré sphere. The four-vector in equation (13.17), however, consists not only of the space-like components, but has a time-like component, providing another radius as

$$s = \frac{1}{2}(a^2 + b^2). \tag{13.23}$$

This defines a concentric outer sphere. The quantity $s^2 - r^2$ is Lorentz-invariant. This quantity is equal to the value of the determinant in equation (13.7). The t parameter cannot be changed in the Lorentzian regime. However, this restriction can be released, if a larger Lorentz group is introduced, as we shall see in section 13.6. In that case, when $t = 0$, the inner radius is equal to the outer radius. When t attains very large values the inner radius becomes

$$S_3 = \frac{1}{2}(a^2 - b^2). \tag{13.24}$$

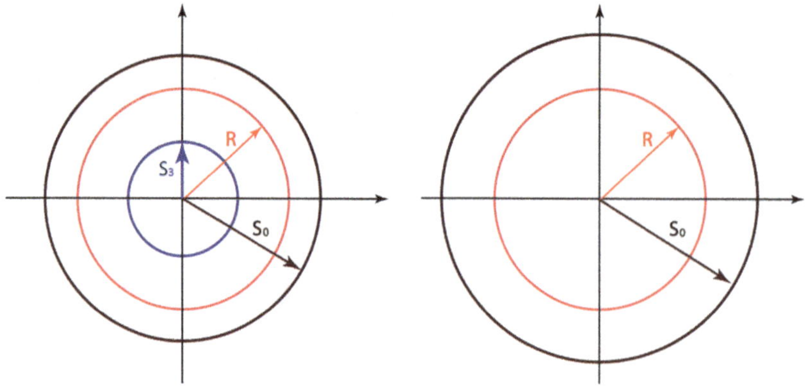

Figure 13.1. The Poincaré sphere has a variable radius. Using equations (13.22) and (13.8), it is apparent that when $\chi = 0°$ the variable radius R takes the maximum value S_0. The radius becomes minimum when this coherency angle reaches 90°. The left figure illustrates that the minimum value of the radius is S_3. When $R = S_0$ the degree of polarization is maximum, and the minimum value occurs when $R = S_3$. Using equation (13.24), it is apparent the S_3 becomes 0 when $a = b$. Thus R attains the minimum value of zero. This is indicated in the right figure. The maximum value of R is still S_0 [13].

We illustrate this in figure 13.1.

A spherical coordinate system can now introduced with

$$\begin{aligned} r_z &= (a^2 - b^2)/2 = r(\cos\theta), \\ r_x &= ab(\cos\phi)e^{-\lambda t} = r(\sin\theta)\cos\phi, \\ r_y &= ab(\sin\phi)e^{-\lambda t} = r(\sin\theta)\sin\phi. \end{aligned} \tag{13.25}$$

The Lorentz symmetry allows rotations. Hence in this three-dimensional scheme, we can use an appropriate rotation to bring the four-vector to:

$$\begin{pmatrix} s \\ r \\ 0 \\ 0 \end{pmatrix}. \tag{13.26}$$

Since the rotations do not change the radii of the outer and inner spheres, s and r remain invariant. However, if we boost the above four-vector along the negative z-axis it becomes

$$\begin{pmatrix} s(\cosh\eta) - r(\sinh\eta) \\ r(\cosh\eta) - s(\sinh\eta) \\ 0 \\ 0 \end{pmatrix}. \tag{13.27}$$

This transformation keeps $(s^2 - r^2)$ invariant, even though it changes the outer and inner radii. This can be seen from

$$[s(\cosh\eta) - r(\sinh\eta)]^2 - [r(\cosh\eta) - s(\sinh\eta)]^2 = s^2 - r^2. \tag{13.28}$$

It is possible to choose the value of η such that

$$r(\cosh\eta) - s(\sinh\eta) = 0. \tag{13.29}$$

This leads to $\tanh\eta = r/s$. When this condition is met, the four-vector of equation (13.27) becomes

$$\begin{pmatrix} \sqrt{s^2 - r^2} \\ 0 \\ 0 \\ 0 \end{pmatrix} = \begin{pmatrix} ab\sqrt{1 - e^{-2\lambda t}} \\ 0 \\ 0 \\ 0 \end{pmatrix}. \tag{13.30}$$

Indeed, within the Lorentz symmetry, the Poincaré sphere is brought to a one-number system. The question then becomes how is it possible to change the value of $(s^2 - r^2)$ in the above expression by changing the time variable t? This expression is clearly an invariant quantity in the $O(3, 1)$ regime. In section 13.6 we shall see that this can be achieved when the symmetry group is enlarged to the Lorentz group $O(3, 2)$, which we studied in section 8.3.

13.5 Symmetries derivable from the Poincaré sphere

It has been demonstrated in section 13.2 that the Poincaré sphere contains the symmetry of the Lorentz group applicable to the energy–momentum four-vector. While the Lorentz group cannot tolerate the variable mass, the sphere has an extra-Lorentz variable which can change the mass. In order to understand this extra variable, we go back to equation (13.15) and exploit the symmetry between $\cos\chi$ and $\sin\chi$, to write another coherency matrix

$$C' = \begin{pmatrix} 1 & e^{-i\phi}\sin\chi \\ e^{i\phi}\sin\chi & 1 \end{pmatrix}, \tag{13.31}$$

with determinant $\cos^2\chi$.

It is possible to diagonalize both C and C' to

$$C = \begin{pmatrix} 1 + \cos\chi & 0 \\ 0 & 1 - \cos\chi \end{pmatrix} \quad \text{and} \quad C' = \begin{pmatrix} 1 + \sin\chi & 0 \\ 0 & 1 - \sin\chi \end{pmatrix}, \tag{13.32}$$

whose determinants are $\sin^2\chi$ and $\cos^2\chi$, respectively. The matrices C and C' share the same physics and the same mathematics. The choice is thus only for convenience.

It is interesting also to note that the

$$\det(C) + \det(C') = 1. \tag{13.33}$$

This is independent of χ.

As we did in equation (1.39), we write the energy–momentum four-vector in the two-by-two matrix format as

$$P = \begin{pmatrix} E + p & 0 \\ 0 & E - p \end{pmatrix}, \tag{13.34}$$

where E and p are the energy and the magnitude of the momentum, respectively. The particle moves along the z-direction.

Let us write $p = E\cos\chi$, and then the matrix P becomes

$$P = E\begin{pmatrix} 1 + \cos\chi & 0 \\ 0 & 1 - \cos\chi \end{pmatrix}. \tag{13.35}$$

This matrix is like C of equation (13.32). If we let $p = E\sin\chi$, the matrix becomes

$$P = E\begin{pmatrix} 1 + \sin\chi & 0 \\ 0 & 1 - \sin\chi \end{pmatrix} \tag{13.36}$$

which is like C'.

Let us pick P of equation (13.36). If $\chi = 0$, the matrix becomes the two-by-two representation of the energy–momentum four-vector of the particle at rest with a mass E. If $\chi = 90°$, the particle becomes massless with an energy–momentum of E. The continuous transition from $\chi = 0$ to $90°$ is illustrated in figure 13.2.

Starting from a massive particle at rest, we are interested in reaching a massless particle with the same energy. This problem is not new in that it was discussed in detail in section 4.6 within the framework of the Lorentz group. However, the transition from the massive case to the massless case is a singular transformation, since the mass remains invariant under Lorentz transformations.

On the other hand, in the case of the Poincaré sphere, this transition is continuous, as indicated in figure 13.3. The question is whether this extra-Lorentzian transformation can be accommodated by a larger symmetry group.

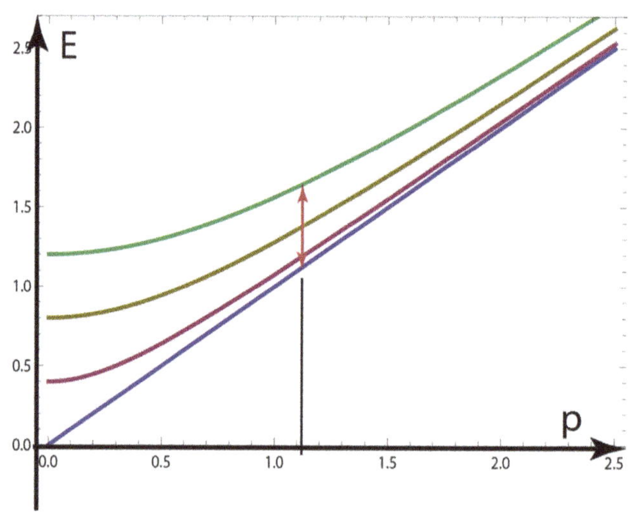

Figure 13.2. Different values of the mass produce different energy–momentum hyperbolas. It is not allowed within the Lorentz group to jump from one hyperbola to another, but within the framework of the $O(3, 2)$ de Sitter symmetry it is possible. The transition, while keeping the magnitude of the momentum constant, is illustrated in this figure.

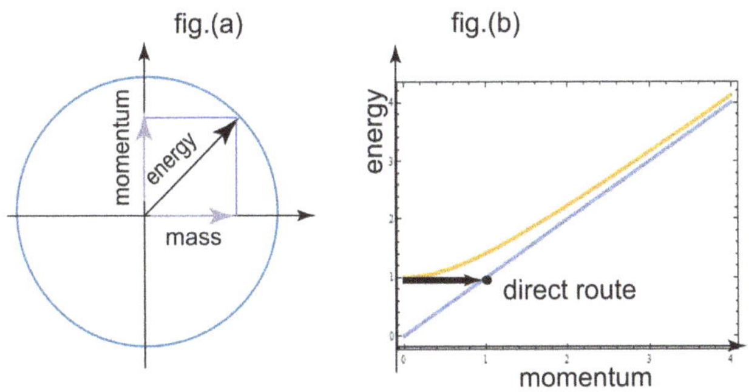

Figure 13.3. In (a) the momentum–mass relation for a fixed energy value is shown. The momentum increases as the angle χ increases, while the mass becomes smaller. At the value $\chi = 90°$ the particle becomes massless. In (b) the traditional energy–momentum plot describes this transition. Within the framework of the Lorentz group this transition is not allowed. The Poincaré sphere does allow this extra-Lorentz transformation [5].

13.6 $O(3, 2)$ symmetry for energy couplings

The group $O(3, 2)$ is the Lorentz group applicable to a five-dimensional space which has three space and two time dimensions. Likewise, in energy–momentum space, there are two energy variables, which lead to a five-component vector

$$\left(E_1, E_2, p_z, p_x, p_y\right). \tag{13.37}$$

In order to study this group, we have to use five-by-five matrices, which are explicitly given in table 8.3. We are, however, interested in its subgroups. First of all, there is a three-dimensional Euclidean space consisting of p_z, p_x, and p_y, to which the $O(3)$ rotation group applies, as in the case of the $O(3, 1)$ Lorentz group.

If the momentum is in the z-direction, this five-vector becomes

$$(E_1, E_2, p, 0, 0). \tag{13.38}$$

As for these two energy variables, they take the form

$$E_1 = \sqrt{p^2 + m_1^2} \quad \text{and} \quad E_2 = \sqrt{p^2 + m_2^2}, \tag{13.39}$$

with [3, 21]

$$m_1 = m\cos\chi \quad \text{and} \quad m_2 = m\sin\chi. \tag{13.40}$$

This results in

$$E_1^2 + E_2^2 = m^2 + 2p^2, \tag{13.41}$$

which remains constant for a fixed value of p^2. There is thus a rotational symmetry in the two-dimensional space of E_1 and E_2. This does, indeed, define the $O(3, 2)$ de Sitter symmetry [3, 4].

For the present purpose, the most important subgroups are two Lorentz subgroups applicable to the Minkowski spaces of

$$(E_1, p, 0, 0) \quad \text{and} \quad (E_2, p, 0, 0). \tag{13.42}$$

Then, in the two-by-two matrix representation, these energy–momentum four-vectors take the form

$$\begin{pmatrix} E_1 + p & 0 \\ 0 & E_1 - p \end{pmatrix} \quad \text{and} \quad \begin{pmatrix} E_2 + p & 0 \\ 0 & E_2 - p \end{pmatrix}, \tag{13.43}$$

with determinants equal to $m^2 \cos^2\chi$ and $m^2 \sin^2\chi$, respectively.

With this understanding, we can now concentrate only on the matrix with E_1. For $\chi = 0$, E_1 takes a maximum value of $\sqrt{p^2 + m^2}$. As χ approaches 90° E_1 becomes equal to p. Thus we have the transition from the massive particle at rest to the massless particle in terms of the Poincaré sphere and Einstein's covariant relativistic equation $E = mc^2$. This fixed-momentum variation is illustrated in figure 13.2. Indeed, the de Sitter symmetry allows us to jump from one mass hyperbola to another.

$O(3, 2)$ or $SO(3, 2)$ appears as the symmetry group in various branches of physics, ranging from quantum mechanics to extended theories of gravity [20, 21]. In section 9.5, we have extensively discussed the quantum mechanical case in the context of the Dirac's coupled harmonic oscillators [8]. In addition to Dirac's coupled oscillators, spin-orbit coupled harmonic oscillators also admit the same symmetry group [10].

In fact, emerging from different viewpoints, attempts to accommodate another time-like dimension to the usual space–time is not new [2, 19]. $SO(3, 2)$ is also the isometry group for the AdS^4 (anti-de Sitter) space–time. It admits closed time-like curves and solves Einstein's equations with a negative cosmological constant accounting for a contracting Universe.

On the other hand, the dS^4 (de Sitter) space–time whose isometry group is $O(4, 1)$, is in contrast to AdS^4 in the sense that it solves Einstein's equations with a positive cosmological constant and accounts for the expansion of the Universe which is what we are observing now in the real cosmos [21]. It is gratifying to note that the Poincaré sphere contains this important symmetry.

13.7 Entropy problem

It is remarkable that the coherency matrix can also serve as the density matrix if it is divided by 2. It is written as

$$\rho(\chi) = \frac{C}{2} = \frac{1}{2}\begin{pmatrix} 1 & e^{-i\phi}\cos\chi \\ e^{i\phi}\cos\chi & 1 \end{pmatrix}. \tag{13.44}$$

The trace of this density matrix is one and its determinant is $\sin^2\chi$.

This density matrix can be diagonalized to

$$\rho(\chi) = \frac{1}{2}\begin{pmatrix} 1 + \cos\chi & 0 \\ 0 & 1 - \cos\chi \end{pmatrix}. \tag{13.45}$$

If we exploit the symmetry between $\cos\chi$ and $\sin\chi$, we can write another density matrix, where we have

$$\rho'(\chi) = \frac{C'}{2} = \frac{1}{2}\begin{pmatrix} 1 & e^{-i\phi}\sin\chi \\ e^{i\phi}\sin\chi & 1 \end{pmatrix}. \tag{13.46}$$

The trace of this density matrix is one and its determinant is $\cos^2\chi$. This matrix has the diagonalized form:

$$\rho'(\chi) = \frac{1}{2}\begin{pmatrix} 1+\sin\chi & 0 \\ 0 & 1-\sin\chi \end{pmatrix}. \tag{13.47}$$

These two matrices are mathematically equivalent. It is interesting to note that the

$$\det(\rho) + \det(\rho') = 1. \tag{13.48}$$

This is independent of χ.

Using the entropy as:

$$S = -Tr(\rho \ln \rho) \tag{13.49}$$

we obtain for the density matrix of equation (13.44) [9, 14, 15]

$$S(\chi) = -\frac{1+\cos\chi}{2}\ln\left(\frac{1+\cos\chi}{2}\right) - \frac{1-\cos\chi}{2}\ln\left(\frac{1-\cos\chi}{2}\right), \tag{13.50}$$

which can be simplified to

$$S(\chi) = -[\cos^2(\chi/2)]\ln[\cos^2(\chi/2)] - [\sin^2(\chi/2)]\ln[\sin^2(\chi/2)]. \tag{13.51}$$

If the entropy is zero the system is minimally decoherent with $\chi = 0$. The entropy takes the maximum value of $\ln(2)$ when the system is totally decoherent with $\chi = 90°$.

In a similar manner, the entropy of equation (13.46) can be written as

$$S'(\chi) = -\frac{1+\sin\chi}{2}\ln\left(\frac{1+\sin\chi}{2}\right) - \frac{1-\sin\chi}{2}\ln\left(\frac{1-\sin\chi}{2}\right). \tag{13.52}$$

The entropy S of equation (13.51) is a monotonically increasing function of χ, while that of S' of equation (13.52) is a decreasing function. In section 6.7, we have discussed Feynman's rest of the Universe in detail within the context of the covariant harmonic oscillator. The first form of the entropy belongs to the part of the Universe in which we make measurements, while the second is in the rest of the Universe. The decrease of entropy in the second can be interpreted as the gain of energy in the first. There is a process of decoherence and recoherence going on here, even though the sum of the entropies is not a conserved quantity of the total system. Nevertheless, if we insist on conserving some physical quantity, we can, using equation (13.48), resort to the sum of the determinants of the two coherency or density matrices, which as we saw in equations (13.33) and (13.48) are independent of the coherency angle χ.

References

[1] Azzam R M A-G and Bashara N M 1999 *Ellipsometry and Polarized Light. North-Holland Personal Library* (Amsterdam: Elsevier) 4 Impression, paperback edition (ISBN 978-0-444-87016-2, Originally published 1977; OCLC: 247501433)

[2] Bars I, Deliduman C, and Andreev O 1998 Gauged duality, conformal symmetry, and spacetime with two times *Phys. Rev.* D **58** 066004-1–066004-10

[3] Başkal S and Kim Y S 2006 de Sitter group as a symmetry for optical decoherence *J. Phys. A: Math. General* **39** 7775–88

[4] Başkal S and Kim Y S 2013 *Mathematical Optics: Classical, Quantum and Computational Methods* (*Lorentz group in ray and polarization optics*) ed V Lakshminarayanan, M L Calvo, and T Alieva (Boca Raton, FL: Taylor and Francis) pp 303–49 (ISBN 978-1-4398-6961-1)

[5] Başkal S, Kim Y S, and Noz M E 2014 Wigner's space–time symmetries based on the two-by-two matrices of the damped harmonic oscillators and the Poincaré sphere *Symmetry* **6** 473–515

[6] Born M and Wolf E 1999 *Principles of Optics: Electromagnetic Theory of Propagation, Interference and Diffraction of Light* 7th expanded edn (Cambridge; New York: Cambridge University Press) (ISBN 978-0-521-64222-4 978-0-521-63921-7, Originally published 1980)

[7] Brosseau C 1998 *Fundamentals of Polarized Light: A Statistical Optics Approach* (New York: Wiley) (ISBN 978-0-471-14302-4)

[8] Dirac P A M 1963 A remarkable representation of the 3 + 2 de Sitter group *J. Math. Phys.* **4** 901–9

[9] Eisert J, Cramer M, and Plenio M B 2010 *Colloquium*: area laws for the entanglement entropy *Rev. Mod. Phys.* **82** 277–306

[10] Haaker S M, Bais F A, and Schoutens K 2014 Noncompact dynamical symmetry of a spin-orbit-coupled oscillator *Phys. Rev.* A **89**

[11] Han D, Kim Y S, and Noz M E 1997 Stokes parameters as a Minkowskian four-vector *Phys. Rev.* E **56** 6065–76

[12] Han D, Kim Y S, and Noz M E 2000 Interferometers and decoherence matrices *Phys. Rev.* E **61** 5907–13

[13] Kim Y S and Noz M E 2013 Symmetries shared by the Poincaré group and the Poincaré sphere *Symmetry* **5** 233–52

[14] Kim Y S and Noz M E 2014 Entropy and temperature from entangled space and time *Phys. Sci. Int. J.* **4** 1015–39

[15] Kim Y S and Wigner E P 1990 Entropy and Lorentz transformations *Phys. Lett.* A **147** 343–7

[16] Mueller H 1943 Memorandum on the polarization optics of the photo elastic shutter. OSRD project OEMsr-576 2, Office of Scientific Research and Development of the United States

[17] Saleh B E A and Carl Teich M 2007 *Fundamentals of Photonics* (*Wiley Series in Pure and Applied Optics*) 2nd edn (Hoboken, NJ: Wiley Interscience) (ISBN 978-0-471-35832-9)

[18] Tudor T 2018 Lorentz transformation, Poincaré vectors and Poincaré sphere in various branches of physics *Symmetry* **10** 52

[19] Wesson P S 2002 Five-dimensional relativity and two times *Phys. Lett.* B **538** 159–63

[20] Wesson P S 2006 *Five-Dimensional Physics: Classical and Quantum Consequences of Kaluza–Klein Cosmology* (Singapore; Hackensack, NJ: World Scientific Publishing Co.) (ISBN 978-981-256-661-4, OCLC: ocm67839346)

[21] Zee A 2013 *Einstein Gravity in a Nutshell (In a Nutshell)* (Princeton, NJ: Princeton University Press) (ISBN 978-0-691-14558-7)

Appendix A

Physics as art of synthesis

Karl Marx was born in Trier (Germany 1818) and died in London (1883). He was buried at London's Highgate Cemetery. On his grave stone, and at the entrance lobby of Humboldt University of Berlin, as seen in figure A.1, Marx is quoted as saying

> *The philosophers have only interpreted the world in various ways. The point however is to change it.*

If we translate this into the language of physics, Marx says

> *Physicists change the world, and the philosophers talk about it.*

Before and during the time of Marx, there were many philosophers who talked about this world. Did they understand physics? The answer is NO. During their lifetimes, physics did not exist as a scholarly subject in higher education. Paul A M Dirac (1902–1984) studied electrical engineering at the University of Cambridge. Eugene Paul Wigner (1902–1995) studied chemical engineering at the Technical University of Berlin.

Physics became a prominent subject after the nuclear explosion on the Japanese city of Hiroshima in 1945. The philosophers can now ask how the physicists changed this world. Before philosophers, physicists should ask how their earlier colleagues changed the world. Thanks to Isaac Newton, we travel with machines instead of horses. Thanks to James Clerk Maxwell, we talk with friends on the opposite side of the Earth without a wire.

Yet, those philosophers wrote down their *opinions* based on how the world was changing during the course of history. Since physics is an experimental science, we are interested in those philosophers who formulated their theories based on what they observed and what they experienced. Among them are David Hume (1711–1776), Immanuel Kant (1724–1804), and Georg Wilhelm Friedrich Hegel (1770–1831). Since they all lived one generation earlier than Marx, it is quite

doi:10.1088/978-0-7503-3607-9ch14

Figure A.1. Karl Marx in Berlin and in London. He said something profound about philosophers in German and in English. (Photos by Y S Kim.)

Figure A.2. David Hume in Edinburgh (Scotland), Immanuel Kant Kaliningrad (Russia; German city of Königsberg until 1945), and Georg Wilhelm Friedrich Hegel in Berlin (Germany). It is safe to assume that physics was not on their scopes. Yet, it is interesting to note that they are talking about how theories in physics are developed. (Photos by Y S Kim.)

possible that Marx included them when he mentioned philosophers. We are thus interested in interpreting the history of physics in terms of their philosophies. Photos of their statues are shown in figure A.2.

During the 20th century, quantum mechanics and relativity were developed. They are quite successful theories. Since they were developed independently, there has been and still is a question about whether these two theories are compatible with each other, that is, if they can be derived from a common set of equations, as is true for electricity and magnetism. This issue was studied in detail in chapter 10.

For this purpose, we shall point out first that physics is an experimental science. We shall review how philosophers organized their 'opinions' or philosophy based on their experiences and the history of the real world.

In section A.1, we mention three philosophers who formulated their work based on their experiences and their study of history. They are David Hume, Immanuel Kant, and Georg Wilhelm Hegel. We use the transparent history of the United States to illustrate their philosophic work.

Einstein was quite interested in the philosophy of Kant during his high-school years. In section A.2, we mention how Kant influenced Einstein to realize that things could appear differently to moving observers.

In section A.3, we discuss the geographical conditions that led Kant to develop his philosophy. It is noted that the same geographical environment led the ancient Chinese to develop their philosophy of Taoism. In Taoism, many become two in harmony, while, according to Kant, many different views have to become one.

In section A.4, Hegel's philosophy is introduced. We then explain how Einstein transformed himself from a Kantianist to Hegelianist while doing physics.

A.1 Illustration of Hume, Kant, and Hegel

Philosophy books are very difficult to read. The best way to understand philosophical ideas is to use examples and illustrations. First of all, humans developed their habits of formulating their abstract ideas from their observations and experiences in the real world. This has been true throughout history.

It was David Hume who said this clearly. Hume was talking about physics which became a major university subject in later years. Physics is an experimental science. We formulate physical laws based on what we observe in the world and apply those laws to other observable phenomena. Newton's apple is a case in point.

Kant's philosophy is also based on observations. He pointed out the same thing could appear differently depending on the observer's environment or state of mind. In the courtroom, witnesses give different stories on the same incident. It is the responsibility of the judge (and jury) to produce one correct verdict. In physics, solutions of the same differential equation take different forms depending on initial and boundary conditions.

Hegel studied history carefully. He observed that two different ideas could be synthesized into one greater idea. He pointed out Christianity as a synthesis of Jewish ethics and Greek logic. In physics, the synthesis of electricity and magnetism led to the present wireless world.

These three philosophers dominate the way humans think these days. Let us look at how Americans constructed their democracy. The history of the United States is short and is transparent. After Columbus, many Europeans came to the New Land without government and institutions. The idea of democracy was developed in Europe before they migrated to the New Land. The Declaration of Independence was written by Thomas Jefferson in 1776. It is known that he was influenced by an English enlightenment philosopher named John Locke (1632–1704).

The basic idea of democracy is to run the country not for the king, but according to the desire of the people. The problem is that there are so many individuals in one country. They all have their own views and desires, and they cannot all be the same.

The solution to this problem was to choose a manageable number of representatives and let them manage those problems.

Yes, it is possible, but the number of representatives is still too big. Thus, Americans divided those many representatives into two groups. These two groups have two different opinions. Yet, it is easier to manage two than many. This is known as the two-party system and bi-partisanship in American democracy. With two different views, the government is able to produce one president and one national policy.

The idea of producing a single national policy is the idea of Kant. There is one, but they are different because people look at it with different perspectives as illustrated in figure A.3. The idea of bi-partisanship was formulated by Hegel as illustrated also in figure A.3.

Kant's idea of making many different views into one is understandable. However, it is seldom practical. Americans thus developed the idea of grouping them into two. From these two groups, Americans produce one president and unified national policies. The Bible does not mention this two-party system. Americans developed this system solely based on their experiences in their New Land.

The idea of bringing many into two opposing groups was developed by the ancient Chinese. After the ice age, many different people came to the bank of their northern river to start constructing their civilization. In order to communicate, they started drawing pictures. This is how Chinese characters were developed. In order to express their strong feelings, they started singing. This is why spoken Chinese has tones even these days.

How about different ideas? They started dividing them into two groups with opposite views. This was the beginning of the concept of Taoism saying that the Universe consists of two opposite elements in harmony. Thus, the happenings in the Universe can be divided into two groups, such as bright and dark, or cold and hot, as well as male and female. The ancient Chinese scholar named Lao Tsu systematically

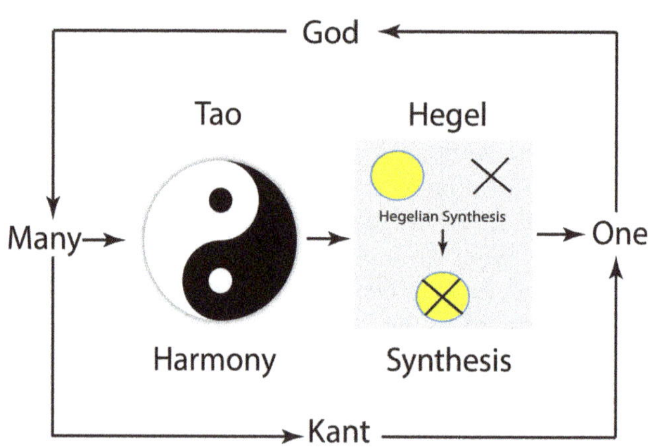

Figure A.3. Integration of Kantianism and Hegelianism. Kant wanted to reduce many to one while Hegel wanted to combine two into one. Thus, we need a many-to-two mechanism in between. This is the place for Taoism. The square for Hegel's synthesis is from Russell's book on western civilization [12].

organized this idea when he wrote his books. This is what Taoist philosophy is all about. The exact dates of his birth and death are unknown. It is believed that he lived during the period of Confucius (550–450 BC).

Again, let us go to figure A.3. During the process of transforming many to one, the Hegelianism idea of combining two into one is more practical and manageable than many into one. However, from where do these two come? Taoism provides the answer to this question.

Let us go back to the American system. Did Americans develop their two-party system according to the Chinese philosophy of Taoism? The answer is clearly no. How about Kant and Hegel? They are strange names even at university campuses in America. Thus, it is quite safe to say that this integrated Kant–Hegel system is the natural way to manage complex problems. This system was developed solely from human experiences, as David Hume clearly pointed out.

The purpose of this chapter is to show that physical theories have been and are being developed according to this integrated Kant–Hegel approach, as illustrated in figure A.3.

A.2 Kant and Einstein

It is known that Einstein was interested in Kant's philosophy during his high school years. His teachers did not like him because he asked so many strange questions. Let us examine how Einstein was influenced by Kant in his development of the concept of relativity.

In order to illustrate Kantianism, let us look at a statue of Toyotomi Hideyoshi (1637–1598) which was displayed at the Osaka Castle which serves as a museum dedicated to his life. Toyotomi is a very important person in Japanese history. He unified Japan by eliminating brutally all competing warlords. However, there are jokes about him in Japan.

He is a monkey (O-Saru-san in Japanese) when humans look at him, but he is a human when monkeys see him. As is shown in figure A.4, his statue shows a monkey-like face. How did he appear to himself? He was the emperor of the world. At his time, China was the center of the world, and he appointed himself as the emperor of China. In order to position himself as the emperor he asked Koreans to

Figure A.4. Toyotomi Hideyoshi of Japan and his Osaka Castle. His name is not well known unless you are Japanese, but his castle in Osaka is a popular destination for tourists to Japan. (Photos by Y S Kim.)

provide a safe pass-way to Beijing. Koreans said no, and his troops had to fight in Korea for seven years until he died in 1598.

Indeed, one person appears so differently depending on the observer's environment or state of mind, as Kant pointed out. Let us go to figure A.4 again. He looks differently if viewed from different angles. As a Kantianist, Einstein asked how he would look to a moving observer.

If Toyotomi's statue is too complicated, we can reduce it to one hydrogen atom, and ask how the hydrogen atom looks to moving observers. Einstein did not provide the answer to this question. Indeed, this book addresses this question in chapter 7 on quarks and partons.

During the early years of the 20th century, Einstein was interested in how things look to moving observers, while Niels Bohr was worrying about the discrete energy levels of the electron orbits in the hydrogen atom. Einstein and Bohr met occasionally to discuss physics, but there are no written records of their discussion of this issue. If they did not talk about this problem, it is understandable, because the hydrogen atom moving with relativistic speed was unthinkable at their time. Indeed, it is still not possible to observe the atom moving with relativistic speed. After all, physics is an experimental science.

These days, high-energy accelerators produce many protons with speeds close to that of light. However, those protons are not hydrogen atoms. Yet, thanks to Gell-Mann's quark model [7], the proton is a bound state of more fundamental particles called quarks sharing the same quantum mechanics as the hydrogen atom. This aspect is illustrated in figure A.5. Indeed, we can examine the original Bohr–Einstein issue by looking at protons moving with relativistic speeds.

The proton at rest appears like a hydrogen-like bound state. However, when it is accelerated close to the speed of light it appears like a collection of free particles with wide-spread momentum distribution. This is known as Feynman's parton picture [5, 8, 9]. In chapter 7, we have shown that Gell-Mann's quark model and Feynman's parton picture are two different ways of looking at the same proton or the hydrogen atom.

This is precisely what Kant said in his philosophy. Kant's philosophy does not stop here. We use many differential equations in physics. Their solutions take different forms depending on initial and boundary conditions. Indeed, the solutions are different because we look at the same equation from different conditions.

Matrices in physics are constructed from a finite number of generators satisfying a closed set of commutators called the Lie algebra. The size of those matrices depends on the physical system we want to explain. The sizes are different, but they all come from the same set of commutators. Indeed, Kantianism is everywhere in physics.

A.3 Kantianism and Taoism

The geography of Venice is well known. It is an island within a lagoon at the northern end of the Adriatic Sea. The lagoon provides a stable water level without violent waves. It is therefore a stopping place for the ships navigating in the ocean beyond. Venice was developed as an important trading port, especially between

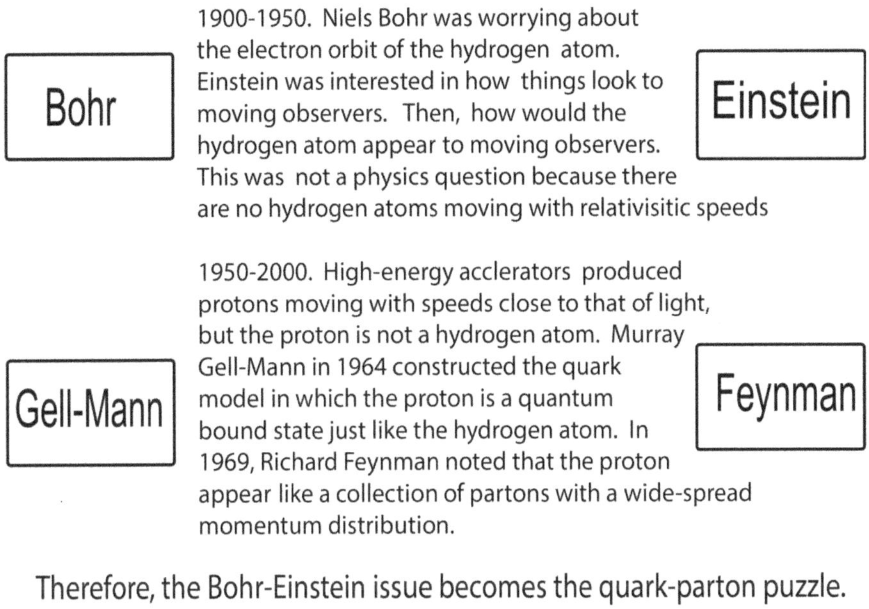

Figure A.5. Bohr and Einstein should have asked how the hydrogen atom appears to a moving observer, or how the atom appears when it moves with a speed comparable with that of light. This is not a physics question because there were and still are no hydrogen atoms moving that fast. This question arises again when the proton becomes a quantum bound state just like the hydrogen atom. There are now many protons moving with a speed close to that of light from particle accelerators.

Europe and Islamic world. It is known that science, especially astronomy, developed in the Islamic world during the Medieval age, came to the Western world through this seaport. The theory of Copernicus is a case in point. The concept of the modern banking system was also developed in Venice because it was a trade port. Shakespeare's *Merchant of Venice* is a case in point. Indeed, new ideas emerge when many people with different views come to one place.

Likewise, Königsberg was a German city in East Prussia until 1945. It is now a Russian city called Kaliningrad. Like Venice, this city is protected by a lagoon. This means that many people from different areas surrounding the Baltic Ocean came to this city with different ideas for the same thing. Immanuel Kant spent the entire 80 years of his life in this city. Thus, his philosophy is about different interpretations of the same thing, which was called *Ding-an-Sich* by Kant.

As was mentioned in section A.1, the oriental philosophy of Taoism is also a product of geography. There were many different opinions and ideas. Instead of transforming them into one, the ancient Chinese scholars divided them into two groups.

Indeed, it is interesting to note that both Kantianism and Taoism were developed from the same root, namely the same geographical environment. They were developed independently, but from similar human experiences. The minor difference is that the end result in Kantianism is one, while it is two in harmony in Taoism.

Then the question is how to make these two into one, as illustrated in figure A.3. This is indeed, Hegel's philosophy of synthesis.

A.4 Einstein and Hegel

While Einstein started as a Kantianist, his best contributions were made along the line of Hegelian synthesis. Even though Einstein's most significant contribution is his $E = mc^2$, he received his Nobel prize in 1921 for his discovery of the photoelectric effect. This effect led to the basic concept of quantum mechanics, namely the synthesis of the particle nature and wave nature of matter. He also synthesized both massive and massless particles.

His $E = mc^2$ means the energy–momentum relations for rest and moving particles. Its expression is of course

$$E = \sqrt{(cp)^2 + m^2c^4}. \tag{A.1}$$

This formula synthesizes the energy–momentum relation $E = p^2/2m$ for slow (massive) particles and $E = pc$ for massless particles.

Indeed, Einstein started as a Kantianist, but he made major contributions according to the Hegelian process of synthesis. In this book, we extended Einstein's special relativity to internal space–time symmetries of particles. We can summarize the result in table A.1. The elements of the middle column synthesize those in the left and right columns.

Throughout history, the physical laws were developed according to this integrated Kant–Hegel process. Physicists like to unite many different events into one formula or one set of formulas. There are many heavenly bodies. They can be divided into two groups, namely comets (with open orbits) and planets with (localized orbits). Isaac Newton synthesized these two groups into one with his second-order differential equation. This is a Hegelian process. James Clerk Maxwell synthesized the

Table A.1. Hegelian view of further contents of Einstein's $E = mc^2$. The first row of this table in well known. Einstein synthesized the energy–momentum relations for massive and massless particles. The second row is for internal space–time symmetries, the photon spin along the direction of momentum remains invariant, but the perpendicular components become one gauge degree of freedom. This aspect was discussed in detail in section 4.6 on one little group with three branches. The third row is for the synthesis of Gell-Mann's quark model and Feynman's parton picture, explained in detail in chapters 6 and 7 on Lorentz-covariant harmonic oscillators and quarks and partons in the Lorentz-covariant world. Dirac 1927, 1945, 1949 means his papers written in 1927 [2], 1945 [3], and 1949 [4].

	Slow Massive	Hegelian Synthesis	Ultra fast Massless
Einstein's $E = mc^2$	$E = p^2/2m$	$E = \sqrt{(cp)^2 + m^2c^4}$	$E = cp$
Wigner's Little Groups	J_3, J_1, J_2	Internal Symmetry	Helicity Gauge Trans.
Integration of Dirac 1927, 1945, 1949	Gell-Mann's Quark Model	Covariant Oscillators	Feyman's Parton Picture

equations governing electricity and those for magnetism into a set of four equations. This created the present-day wireless civilization.

The present form of quantum mechanics is a synthesis of the particle and wave nature of matter. Einstein's special relativity synthesizes massive and massless particles. All these are the integrated processes of Kantianism and Hegelianism. The remaining problem is whether quantum mechanics and special relativity can be synthesized.

In order to answer the question, the present form of quantum field theory is quite compatible with special relativity theory. It is regarded as a representation of the inhomogeneous Lorentz group. As for quantum bound states, we addressed this question in chapter 6 on the Lorentz-covariant oscillators. As shown in figure A.6, it is a Hegelian synthesis of the hydrogen atom and special relativity.

The question then is whether this covariant oscillator formalism can be regarded as a representation of the inhomogeneous Lorentz group. The inhomogeneous Lorentz or Poincaré group is generated by three rotation, three boost, and four translation generators. This aspect was discussed in detail by Kim *et al* in 1979 [11]. Thus, quantum bound-states and quantum scattering problems can now be regarded as two different representations of the same set of commutation relations, namely the Lie algebra of the inhomogeneous Lorentz group, as illustrated in figure A.7.

Then there comes the question of whether quantum mechanics and special relativity with $E = mc^2$ can be unified. Let us go back to figure A.7. It tells us that both scattering and bound states can be synthesized in the Lorentz-covariant world as representations of the same Lie algebra of the inhomogeneous Lorentz group.

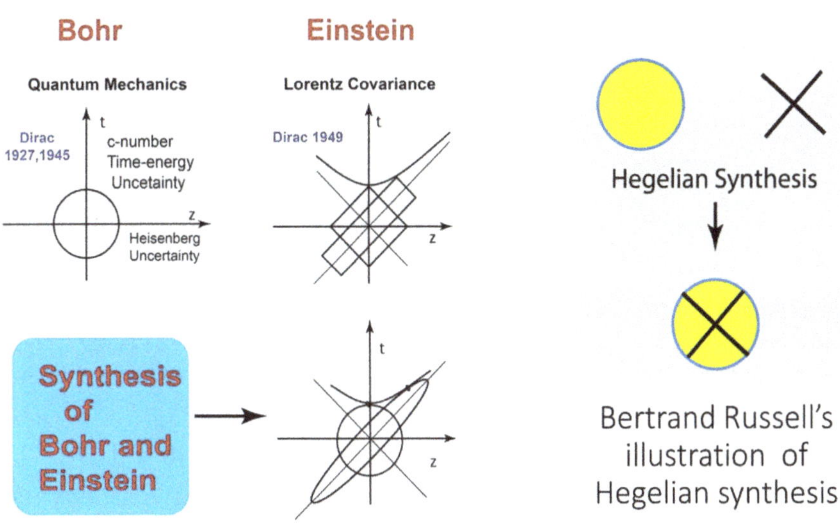

Figure A.6. Synthesis of Bohr and Einstein. Bohr was interested in quantum bound states. Einstein was interested in how things look to moving observers. Paul A M Dirac wrote three papers on this subject. Dirac used a Gaussian wave function for bound states. He also invented the light-cone coordinate system. In this figure, the Lorentz boost and quantum bound state became synthesized according to Hegel. This effect is observable in laboratories, as discussed in chapter 6. The synthesis block in this figure is from Russell's book on western civilization [12].

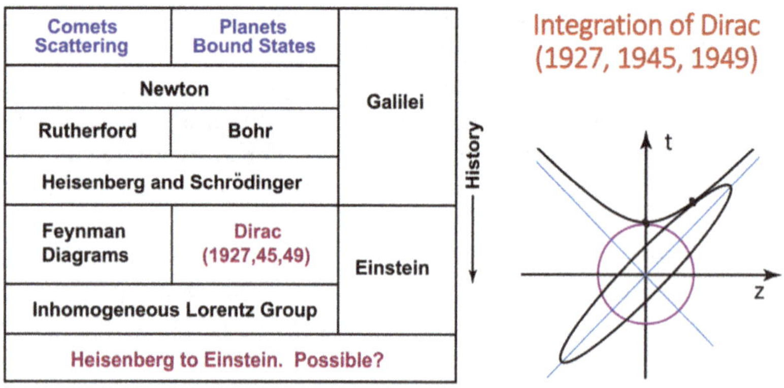

Figure A.7. Synthesis processes in physics. Newton synthesized scattering and bound states with his $f = ma$. Schrödinger and Heisenberg synthesized particle and wave natures to formulate quantum mechanics which can synthesize scattering and bound states in the quantum world. In the Lorentz-covariant world, Feynman diagrams work well for scattering processes. For bound states, the synthesis of Dirac's papers lead to a Lorentz-covariant picture of bound states. It is then possible to say that both Feynman diagrams and covariant bound states are representations that can be constructed from the same closed set of commutation relations, namely the Lie algebra of the inhomogeneous Lorentz group. The synthesis of Dirac's three papers is illustrated in figure A.6.

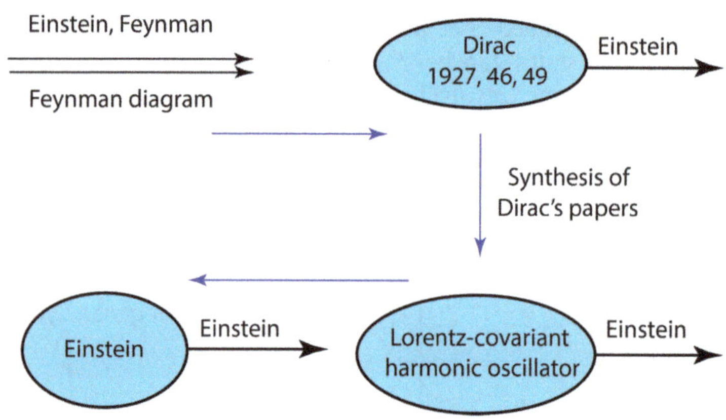

Figure A.8. Bound state in motion. Let us start with two free particles. Their quantum mechanics in the Lorentz-covariant world is well known in terms of Feynman diagrams. The problem arises when they become a bound state. As Feynman *et al* said in their paper [6], harmonic oscillators can be used for this bound-state problem. The question then is how to construct a Lorentz-covariant oscillator system. As in figure A.6, this can be achieved through the integration or synthesis of Dirac's papers.

Next question is whether quantum mechanics and special relativity can be synthesized. In other words, whether they can be derived from the same set of equations or the same closed set of commutators. This issue was addressed in [1, 10], and discussed in detail in chapter 10 of this book and synthesized in figure A.8.

References

[1] Başkal S, Kim Y S, and Noz M E 2019 Einstein's $E = mc^2$ derivable from Heisenberg's uncertainty relations *Quant. Rep.* **1** 236–51

[2] Dirac P A M 1927 The quantum theory of the emission and absorption of radiation *Proc. R. Soc. A: Math. Phys. Eng. Sci.* **114** 243–65

[3] Dirac P A M 1945 Unitary representations of the Lorentz group *Proc. R. Soc. A: Math. Phys. Eng. Sci.* **183** 284–95

[4] Dirac P A M 1949 Forms of relativistic dynamics *Rev. Mod. Phys.* **21** 392–9

[5] Feynman R P 1969 Very high-energy collisions of hadrons *Phys. Rev. Lett.* **23** 1415–7

[6] Feynman R P, Kislinger M, and Ravndal F 1971 Current matrix elements from a relativistic quark model *Phys. Rev.* D **3** 2706–32

[7] Gell-Mann M 1964 A schematic model of baryons and mesons *Phys. Lett.* **8** 214–5

[8] Kim Y S 1989 Observable gauge transformations in the parton picture *Phys. Rev. Lett.* **63** 348–51

[9] Kim Y S and Noz M E 1977 Covariant harmonic oscillators and the parton picture *Phys. Rev.* D **15** 335–8

[10] Kim Y S and Noz M E 2020 Integration of Dirac's efforts to construct a quantum mechanics which is Lorentz-covariant *Symmetry* **12** 1270

[11] Kim Y S, Noz M E, and Oh S H 1979 Representations of the Poincaré group for relativistic extended hadrons *J. Math. Phys.* **20** 1341–4 See also, Physics Auxiliary Publication Service Document No. PAPS JMAPA-20-1336-12.

[12] Russell B and Foulkes P 1989 *Wisdom of the West: A Historical Survey of Western Philosophy in Its Social and Political Setting* (New York: Crescent Books: Distributed by Crown Publishers) (ISBN 978-0-517-69041-3, OCLC: 440104494; Originally published by Doubleday & Co., Garden City, New York, USA 1959)

Index

A
γ matrices, Dirac, 8.5, 9.5, 9.6
ABCD matrix, 11.0–11.4, 11.6
ABCD matrix - decomposition, 11.3
ABCD matrix - equi-diagonalization, 11.2, 11.4
ABCD matrix - recomposition, 11.5
angular momentum, 5.0
annihilation operator, 9.3, 10.2, 1.5.1
area-preserving squeeze transformation, 6.2
area-preserving transformation, 3.1, 8.0, 8.1, 10.0
attenuator, 12.0, 12.1, 12.3, 12.4

B
Bargmann decomposition, 4.4, 4.5, 11.0, 11.3, 11.5, 11.6
Bogoliubov transformation, 1.5.2
Bohr, 4.0, 6.0, 6.7, 7.0, 7.1, A.2
boosted Wigner matrix, 4.3
boundary matrix, 11.6

C
camera matrix, 11.7
camera optics, 11.7
canonical transformation, 8.3, 8.4, 9.0, 9.2, 9.5, 9.6
Cherenkov radiation, 2.4.1
classical mechanics, 2.0, 8.0, 8.1, 9.6, 10.1
c-number time-energy uncertainty relation, 6.1, 6.2, 6.4
coherence, 7.2, 7.4, 7.5
coherence problem, 7.4
coherency angle, 13.4, 13.7
coherency matrix, 13.0–13.2
conformal transformation, 1.3
contravariant transformations, 1.1
coupled harmonic oscillator equation, 1.4
coupled harmonic oscillators, 8.2, 8.6, 9.0, 9.2, 9.6, 10.1, 13.6
covariant harmonic oscillator formalism, 7.2, 7.6
covariant harmonic oscillator model, 7.6
covariant harmonic oscillators, 6.6, 7.3, 9.3
covariant transformations, 1.1
creation operator, 9.3, 10.2, 1.5.1
cylindrical group, 2.3, 3.2
cylindrical transformation, 2.3

D
damped harmonic oscillator, 4.6, 4.7
de Sitter group, 8.0, 10.0, 10.1, 1.5.3
decoherence, 13.2, 13.7
decoherence angle, 13.2
density distribution, 1.6, 8.0
density matrix, 6.7, 7.6, 9.3, 9.5, 9.7, 13.7
dipole behavior, 7.5
dipole cut-off, 7.0, 7.4, 7.5
Dirac equation, 7.3
Dirac spinor, 1.4, 5.1, 5.3
Dirac's γ matrices, 8.5, 9.4–9.6
Dirac's two-oscillator system, 10.4

E
$E(2)$, 2.3, 2.6, 3.2, 3.3, 4.0, 10.2, 12.7, 1.5.2, 2.3.1, 2.4.1
$E(2)$-like little group, 2.3, 2.5, 2.6, 5.4, 12.7
$E(2)$-like symmetry, 12.7
Einstein, 1.0, 2.0, 4.0, 5.0, 6.0, 6.2, 6.7, 7.0, 7.1, 10.0, 10.4, 11.0, 12.0, 13.0, 13.1, 13.3, 13.6, A.0, A.2, A.4
electromagnetic four-potential, 5.0, 5.5.1
electromagnetic four-tensor, 5.0
electromagnetic four-vector, 5.0
electromagnetic tensor, 5.5.2
electromagnetic waves, 5.0

energy-momentum four-vector, 1.2, 1.4, 2.2–2.4, 4.1–4.4, 5.1, 6.3, 7.2–7.4, 10.1, 12.6, 13.3, 13.5
energy-momentum space, 7.3
energy-momentum wave function, 7.2, 7.5
energy-momentum width, 7.2
entropy, 6.7, 9.7, 13.7
equi-diagonal, 11.0, 11.2, 11.4, 11.6, 11.7
Euclidean group, 2.3, 2.6, 3.0, 3.2, 5.1, 10.2
Euler angles, 2.0–2.2, 4.2, 13.0, 13.1
Euler decomposition, 2.1, 4.2

F

Feynman's parton picture, 7.0, 7.2
Feynman's rest of the universe, 6.6, 6.7, 7.6, 9.6, 9.7
Feynman's time dilation, 7.2
focal condition, 11.7
Fock space, 10.2, 1.5.1
four-dimensional phase space, 8.0, 8.2, 9.5, 10.0
four-dimensional special linear group with real elements, 8.4
four-dimensional symplectic group, 8.0, 8.3, 10.0
four-dimensional Wigner function, 9.4

G

Galilean boost matrix, 3.4
Galilean group, 10.1
Galilean system, 10.4
gauge transformation, 2.3, 2.6, 5.1–5.4, 5.5.1–5.5.3
gauge-dependent, 5.3, 5.4
gauge-invariant, 5.3, 5.4, 5.5.2
gauge-invariant four-tensors, 5.0
gauge-invariant spinors, 5.4
Gaussian wave function, 8.0, 10.0
graviton, 5.5.3
group contraction, 3.0, 10.0, 10.4

H

hadron, 1.4, 4.0
harmonic oscillator, 1.5, 6.0–6.2, 7.0–7.4, 7.6, 8.6, 9.3, 9.4, 9.6, 10.1, 10.3, 1.5.1
harmonic oscillator equation, 6.3, 6.4, 9.1
harmonic oscillator force, 6.0
harmonic oscillator potential, 6.3
harmonic oscillator representation, 10.0, 10.2
harmonic oscillator wave functions, 1.5, 6.0–6.2, 6.5, 6.7, 7.0, 7.1, 7.4, 7.6, 9.4, 1.5.1
Hegel, Georg Wilhelm Friedrich, A.0, A.1, A.4
Heisenberg commutation relations, 10.0, 10.1, 10.3
Heisenberg uncertainty relation, 8.0, 8.1, 10.0–10.2, 10.4
helicity, 5.0, 5.1, 12.7
Hermite polynomial, 1.5.1
Hermitian, 1.0, 4.2, 4.5, 10.3, 12.4, 1.2.2, 2.4.1
Hermitian operators, 10.1
Hermitian transformation, 11.2, 11.3, 11.7
Hume, David, A.0, A.1

I

imaginary-mass particle, 2.0, 2.6, 4.3, 5.1, 12.6
inhomogeneous Lorentz group, 1.4, 6.2, 10.1, 10.4
inhomogeneous Lorentz transformations, 1.4
internal space-time coordinate variables, 1.4
internal space-time structure, 2.0
internal space-time symmetry, 4.0, 4.1, 5.0, 6.0, 6.3, 8.5, 11.0, A.4
internal symmetry, 2.6
internal symmetry of photon, 2.6
invariant subgroup, 1.4
Iwasawa decomposition, 4.4, 11.5

J

Jones matrix, 12.0
Jones vector, 12.0–12.3, 13.0–13.2
Jones-matrix formalism, 12.1

K

Kant, Immanuel, A.0–A.3
Klein-Gordon equation, 6.3

L

laser cavity, 11.4
lens matrix, 11.5, 11.7
Lie algebra, 1.0, 1.1, 2.0, 2.2–2.4, 2.6, 3.2, 4.5, 5.4, 8.0–8.4, 8.6, 9.0, 9.4–9.6, 10.1–10.3, 11.1, 12.4, 1.2.1, 1.5.2, 2.4.1, A.4
Lie algebra of the Lorentz group, 1.2
Lie groups, 1.0
light-cone axis, 7.2
light-cone coordinate system, 2.0, 2.6, 6.2
light-cone system, 9.6
light-cone variables, 9.1
linear canonical transformations, 8.0
little groups, 2.0, 2.3, 2.4, 2.6, 4.0–4.4, 4.6, 4.7, 5.0, 5.1, 5.4, 6.0, 6.3, 7.1, 11.0, 11.2, 12.6
Lorentz boost, 1.0, 1.1, 2.0, 2.1, 2.3, 2.4, 2.6, 3.3, 4.5, 4.6, 5.1, 5.2, 5.5, 6.0–6.2, 6.5, 7.0–7.2, 8.6, 9.6, 12.6, 1.2.1, 5.5.2, 5.5.3
Lorentz coherence, 7.0, 7.4, 7.5
Lorentz condition, 2.3, 2.6
Lorentz contraction, 6.6, 7.4
Lorentz covariance, 6.5, 7.2
Lorentz frame, 2.2, 2.5, 4.1, 6.4, 7.2, 7.4, 12.6
Lorentz group, 1.0–1.5, 2.5, 2.6, 3.0, 3.4, 4.0, 5.0, 5.1, 5.4, 6.1, 6.2, 8.0, 8.1, 8.3, 8.4, 8.6, 9.0, 9.3–9.5, 10.0, 10.1, 10.3, 11.0, 11.1, 11.7, 12.0, 12.1, 12.4, 12.6, 12.7, 13.0–13.2, 13.5, 13.6, 1.5.2, 2.4.1, 5.5.1, A.4
Lorentz invariant, 1.0, 6.0, 6.1
Lorentz system, 10.4
Lorentz transformation, 1.0, 1.2, 1.4, 2.5, 4.1, 6.3, 6.4, 8.1, 9.2, 9.6, 10.1, 11.2, 13.2, 13.3, 13.5, 5.5.2
Lorentz transformations, 1.4, 5.1, 6.2, 9.0, 12.0, 12.3, 12.6, 13.2, 5.5.1, 5.5.3
Lorentz-boost, 2.0
Lorentz-covariant, 1.0, 5.0, 5.1, 6.0–6.2, 7.0, 7.2, 5.5.3, A.4
Lorentz-covariant model, 7.0
Lorentz-covariant world, 5.5, 6.6, 6.7, 7.0, 11.0
Lorentz-squeeze, 6.5, 7.2
Lorentz-squeezed Gaussian distribution, 6.2
Lorentz-squeezed hadron, 7.6
Lorentz-squeezed wave function, 7.2, 7.5

M

Majorana representation, 9.0
massive particle, 2.6, 4.3, 5.0–5.3, 12.6, 2.4.1, 5.5.1, 5.5.2, A.4
massless neutrino, 5.4
massless particle, 2.5, 2.6, 4.3, 5.0–5.3, 12.6, 12.7, 5.5.1–5.5.3, A.4
Maxwell, A.0
Minkowski space, 1.0, 1.2, 1.4, 3.2, 8.3, 10.1–10.4, 12.6, 13.6, 1.2.1, 1.2.2
mirror matrix, 11.4
momentum wave function, 8.0
momentum-preserving transformation, 4.4

N

neutrino polarization, 5.0
Newton, 4.0, 13.3, A.0

O

O(2, 1), 2.4, 3.3, 5.1, 8.1, 9.3, 10.0–10.2, 11.0, 1.5.2, 2.4.1
O(2, 1)-like little group, 2.5, 4.2, 12.6
O(3), 2.6, 3.3, 4.0, 13.6, 2.4.1, 5.5.2
O(3)-like little group, 2.2, 2.6, 4.2, 12.4, 5.5.2
O(3, 1), 10.3, 13.6
O(3, 2), 8.0, 9.0, 9.4–9.6, 10.0, 10.1, 10.3, 10.4, 13.0, 13.1, 13.6, 1.5.3
O(3, 3), 8.4, 9.0, 9.5, 9.6, 1.5.3
optical activity, 12.5–12.7
optical beam, 11.6

optical filters, 12.7
oscillator well, 7.2

P
parity operation, 4.5, 5.5.2
parton model, 7.2
partons, 7.0, 7.2
Pauli matrices, 1.2, 2.0, 5.1
phase shifter, 12.0, 12.1, 12.3, 12.4, 12.7
phase space, 8.0, 8.1, 9.0, 9.5, 9.6, 10.0, 10.2
phase-shift matrix, 11.6
phase-space distribution function, 8.1, 9.5
photoelectric effect, 5.0, A.4
photon, 1.4, 2.6, 5.0, 5.5.3, 9.3
photon polarization, 5.0, 5.5.2
Poincaré group, 1.4, 1.5, 6.0, 6.2, 6.3, 10.1, 10.4
Poincaré sphere, 13.0, 13.1, 13.3, 13.5
Poincaré symmetry, 6.4, 10.4
Poincaré transformation, 6.3
Poisson brackets, 10.1
polarization axes, 12.1, 12.3
polarization - circular, 12.2, 12.3
polarization degree, 13.2, 13.4
polarization - elliptic, 12.2, 12.3
polarization optics, 12.0, 12.3, 12.6
polarization - projection operator, 12.1
polarization transformation, 12.4
polarized light, 12.0
polarized neutrinos, 5.4
proton form factor, 7.0, 7.4, 7.5
proton structure function, 7.3
pseudo-scalar, 5.5

Q
q-numbers, 1.6, 8.0
quantum bound states, 7.0
quantum coherence, 7.0
quantum mechanics, 2.0, 6.0–6.2, 6.7, 7.0–7.2, 8.0, 8.1, 9.0, 9.5, 9.6, 10.0, 10.1, 10.4, 2.4.1, A.0, A.4
quantum optics, 10.2
quark distribution, 7.2
quark model, 6.0, 7.0–7.2, 7.4
quark spin, 7.4
quark-parton puzzles, 7.2
quarks, 1.4, 4.0, 6.0, 6.3, 7.0–7.4
quarter-wave shifter, 12.7

R
radius of curvature, 3.5
recoherence, 13.7
rotation group, 1.0, 3.2, 5.1, 8.3, 12.7, 1.2.1, 5.5.3

S
Schrödinger equation, 7.3, 9.4
Schrödinger wave function, 8.0
shear transformation, 12.7
similarity transformation, 1.2, 4.3, 5.3, 11.2, 11.3, 11.6, 1.2.2
single-photon coherent state, 9.3
singular operation, 4.4
SL(2, *c*), 1.2, 4.5, 5.1, 5.3–5.5, 9.5, 11.0, 12.4, 13.0, 13.1, 1.2.1, 1.2.2
SL(2, *c*) subgroups, 5.4
SL(4, *r*), 8.0, 8.4, 8.6, 9.6
Sp(2), 4.5, 8.1, 8.2, 8.6, 9.0, 9.6, 10.0, 10.2, 11.0–11.2, 11.5, 12.4, 12.7, 1.5.2
Sp(4), 8.0, 8.3, 8.4, 9.0, 9.4–9.6, 10.0
space inversion, 5.5, 5.5.2
space-time, 1.2, 6.3
space-time coordinate, 5.4, 7.3
space-time extension, 7.0
space-time four-vector, 13.0, 13.1
space-time separation, 6.3
space-time structure, 2.0
space-time symmetry, 1.4, 2.0, 12.6, 12.7
space-time transformation, 6.3
space-time translations, 6.2
space-time wave function, 7.2
space-time width, 7.2
special relativity, 2.0, 4.0, 6.0, 6.1, 9.5, 12.0, A.0, A.4
spinors, 5.0–5.5, 13.0, 13.1, 5.5.1–5.5.3
squeeze transformations, 3.0, 3.1, 8.1, 9.6, 10.4, 12.2, 12.3
squeezed states of light, 3.0, 6.5, 8.6, 9.0, 9.5, 12.4

squeezed states of light - two-mode, 7.6
stability condition, 11.4
standing waves, 1.5
step-down operator, 9.4, 10.0, 1.5.1, 1.5.2
step-up operator, 9.4, 10.0, 1.5.1, 1.5.2
Stokes parameters, 13.0–13.2
SU(1, 1), 9.3
SU(2), 1.0, 4.2, 5.1, 12.4
subgroup of the Lorentz group, 2.0, 4.2

T

tangent cylinder, 2.5, 2.6
tangent plane, 2.5, 2.6, 3.2
tangential continuity, 3.3, 11.2, 11.7
Taoism, A.0, A.1, A.3
temperature, 7.6
thermal excitation, 7.6
three-body bound system, 7.3
time-like separation, 6.7
time-separation, 6.7, 7.4
time-separation variable, 6.6, 6.7, 7.1, 7.6, 9.3
translation group, 1.4
translation matrix, 11.4, 11.5
triangular matrix, 4.2, 11.2
two-dimensional phase space, 8.0, 10.0, 10.2
two-dimensional symplectic group, 8.1, 10.0
two-mode coherent photon state, 9.0
two-mode squeezed state, 9.3, 9.5
two-oscillator system, 9.5
two-photon coherent state, 9.2, 9.3, 1.5.3

W

Wigner decomposition, 4.4, 11.0, 11.4, 11.6
Wigner four-vector, 2.5, 2.6
Wigner frames, 2.5
Wigner function, 1.6, 6.7, 8.0–8.2, 9.5–9.7, 10.2
Wigner matrix, 4.2, 4.3, 4.6
Wigner phase-space, 9.4, 9.5
Wigner phase-space distribution, 6.7
Wigner phase-space distribution function, 8.0
Wigner transformation matrix, 2.6

Y

Yukawa's oscillator formalism, 7.4

Lightning Source UK Ltd.
Milton Keynes UK
UKHW050745040721
386574UK00003B/147

9 780750 336055